TREES, SHRUBS, AND WOODY VINES OF LOUISIANA

BY

CHARLES M. ALLEN, PhD
DAWN ALLEN NEWMAN, M.S.
AND
HARRY H. WINTERS, M.D.

Published by:
Allen's Native Ventures, LLC
5070 Hwy 399
Pitkin, LA 70656
337-328-2252
email = native@camtel.net

ISBN 0-9718625-0-8

Library of Congress Control Number 2002090807

Book is dedicated to All the future students of Dendrology; May you enjoy this as much as we enjoyed preparing it. And to the students who have me find, coin, and/or enjoy the following:

HONK IF YOU HATE RUBUS! OR SMILAX! OR CRATAEGUS!

IF YOU ARE NOT LIVING ON THE EDGE, YOU ARE TAKING UP TOO MUCH ROOM!

IF THE DATA DON'T WORK; WORK THE DATA!

JOGGING IS FOR PEOPLE WHO ARE NOT INTELLIGENT ENOUGH TO WATCH TV

EVERYTIME I THINK ABOUT EXERCISE, I LIE DOWN UNTIL THE THOUGHT GOES AWAY

DON'T ASK THE BARBER WHETHER YOU NEED A HAIRCUT

A MISTAKE IS EVIDENCE OF EFFORT

THE SECRET OF PATIENCE, DO SOMETHING ELSE IN THE MEANTIME

BEHAVE LIKE A DUCK; BE CALM ON THE SURFACE BUT PADDLE LIKE HELL UNDERNEATH

STRESS IS LIKE AN ICE CREAM CONE; YOU HAVE TO LICK IT

NEVER SERVE LOBSTER IN A MONTH WITH NO PAYCHECK

TOMORROW, I WILL WAKE UP AT THE CRACK OF NOON

EAT YOUR VEGETABLES, DON'T BECOME ONE

BRAIN: AN APPARATUS WITH WHICH WE THINK WE THINK

IF SWIMMING IS SUCH A GOOD EXERCISE, EXPLAIN WHALES

THERE ARE TWO KINDS OF STATISTICS; THE KIND YOU LOOK UP & THE KIND YOU MAKE UP

YOU HAVE TO GROW OLD, BUT YOU DON'T HAVE TO GROW UP

THERE ARE THREE KINDS OF PEOPLE; THOSE WHO CAN COUNT & THOSE WHO CAN'T

THERE ARE THREE KINDS OF LIES; LIES, DAMMED LIES, AND STATISTICS

PARADOX: A PEDIATRICIAN AND A VET

YOU CAN'T GET FIRED IF YOU DON'T HAVE A JOB

FOR PEACE AND QUIET, GET A PHONELESS CORD

NO PROBLEM IS SO BIG THAT IT CAN'T BE RUN AWAY FROM

ANY DAY ABOVE GROUND IS A GOOD ONE

COMPOST HAPPENS

GROW ANTENNAE NOT HORNS

IF WE KNEW WHAT WE WERE DOING, IT WOULDN'T BE RESEARCH

BASIC RESEARCH IS WHAT WE'RE DOING WHEN WE DON'T KNOW WHAT WE'RE DOING

GIVE ME AMBIGUITY OR GIVE ME SOMETHING ELSE

SUBDUED: NOT THE TOP DUDE

SHE LOOKS LIKE HELEN GREEN; AND SHE DON'T LOOK TOO GOOD IN BROWN EITHER

I CAN KEEP A SECRET; IT'S THE PEOPLE THAT I TELL THAT CAN'T

BEAUTY IS ONLY SIN DEEP

SHE'S SO DUMB; SHE OPENED SOMEONE ELSE'S MAIL AND THEN ANSWERED IT

OBESITY IS REALLY WIDESPREAD

HOW TIME FLIES WHEN YOU'RE DOING ALL THE TALKING

NEVER WRITE ON BOTH SIDES OF THE PAPER AT THE SAME TIME

GARDENERS HAVE A KEEN SENSE OF HUMUS

MANY BELIEVE THAT SLEEP IS FOR THE NIGHT; YOU SEEM TO BE OUT TO DISPROVE THEM

DON'T DIG YOUR GRAVE WITH A KNIFE AND FORK

IF YOU DO WHAT YOU'VE ALWAYS DONE; YOU WILL GET WHAT YOU'VE ALWAYS GOTTEN

SUBCANOPIES: NOT THE BEST CAN OF PEAS

NEVER GO TO A DOCTOR WHOSE HOUSE PLANTS HAVE DIED

CLEVERLY DISGUISED AS A RESPONSIBLE ADULT

NEVER MAKE FORECASTS; ESPECIALLY ABOUT THE FUTURE

WEED: A PLANT THAT HAS LEARNED ALL SURVIVAL SKILLS EXCEPT HOW TO GROW IN ROWS

I NEVER GET LOST; SOMEONE IS ALWAYS TELLING ME WHERE TO GO

A MIND AND A PARACHUTE ARE BOTH USEFUL, IF OPEN

LIST OF THINGS TO DO TODAY: 1. MAKE A LIST OF THINGS TO DO TOMORROW

NEVER PLAY LEAPFROG WITH A UNICORN

I AM NOT ILLEGITIMATE; I KNOW HOW TO READ!

DIFFERENCE BETWEEN ROAST BEEF AND PEA SOUP? ANYONE CAN ROAST BEEF!

I'VE GONE TO LOOK FOR MYSELF. IF I RETURN BEFORE I GET BACK, KEEP ME HERE.

MOST MINDS ARE LIKE CONCRETE; THOROUGHLY MIXED AND PERMANENTLY SET

I LOVE COOKING WITH WINE; SOMETIMES I EVEN PUT IT IN THE FOOD.

TAKE A TIP FROM NATURE: YOUR EARS AREN'T MADE TO SHUT BUT YOUR MOUTH IS.

THE FIRST THING THAT A GIRL WISHES FOR FROM THE GARDEN OF LOVE IS AT LEAST ONE CARAT.

A PESSIMIST SAYS THE LILY BELONGS TO THE ONION FAMILY; AN OPTIMIST DECLARES THAT THE ONION BELONGS TO THE LILY FAMILY.

EGOTISM: I STRAIN

A SMILE IS THE LIGHTING SYSTEM OF THE FACE AND THE HEATING SYSTEM OF THE HEART

YOU WILL FIND THE KEY TO SUCCESS UNDER THE ALARM CLOCK

THE BEST WAY TO BE UNDERSTOOD IS TO BE UNDERSTANDING

YOU ARE YOUNG AT ANY AGE IF YOU ARE PLANNING FOR TOMORROW

TO STAY YOUTHFUL, STAY USEFUL

MANY PEOPLE HATE ANY CHANGE WHICH DOESN'T JINGLE IN THEIR POCKET.

ANGER IS JUST ONE LETTER SHORT OF DANGER

THE RIGHT ANGLE FOR APPROACHING A DIFFICULT SITUATION IS THE "TRY ANGLE"

MOST PEOPLE HAVE TOO MUCH MONTH LEFT OVER AT THE END OF THE MONEY

YOU CAN'T CONTROL THE WEATHER BUT YOU CAN CONTROL THE ATMOSPHERE OF YOUR MIND

A DEAD FISH CAN FLOAT DOWNSTREAM BUT IT TAKES A LIVE FISH TO SWIM UPSTREAM

SUCCESS COMES IN CANS; FAILURE COMES IN CAN'TS

NO FAMILY SHOULD ATTEMPT AN AUTO TRIP IF THE KIDS OUTNUMBER THE CAR WINDOWS

A GOSSIP IS A PERSON WITH A KEEN SENSE OF RUMOR

AS LONG AS YOU LIVE, KEEP LEARNING HOW TO LIVE

HOW DO YOU CATCH A UNIQUE RABBIT? UNIQUE UP ON IT.

HOW DO YOU CATCH A TAME RABBIT? TAME WAY, UNIQUE UP ON IT.

WHERE DO YOU FIND A DOG WITH NO LEGS? RIGHT WHERE YOU LEFT HIM.

SIGN ON A SCIENTIST"S DOOR: GONE FISSION

HOKET ON FONIKS WERKED FUR ME

SUPPORT WILDLIFE, THROW A PARTY.

LEAVES BEGIN TO TURN THE NIGHT BEFORE A TEST

IF YOU SCATTER THORNS OR SPINES, DON'T GO BAREFOOT

IF THERE IS NO WIND, ROW

IN FRANCE, NEVER NAME A PIG NAPOLEON

PROFESSOR DOLL: WIND IT UP AND IT GETS AMBIGUOUS

SO MANY SPECIES SO LITTLE TIME

What tree are your branches connected to?

Find your birthday, find your tree and then read about yourself!

Jan 01 to Jan 11 - Fir Tree
Jan 12 to Jan 24 - Elm Tree
Jan 25 to Feb 03 - Cypress Tree
Feb 04 to Feb 08 - Poplar Tree
Feb 09 to Feb 18 - Cedar Tree
Feb 19 to Feb 28 - Pine Tree
Mar 01 to Mar 10 - Weeping Willow Tree
Mar 11 to Mar 20 - Lime Tree
Mar 21 - Oak Tree
Mar 22 to Mar 31 - Hazelnut Tree
Apr 01 to Apr 10 - Rowan Tree
Apr 11 to Apr 20 - Maple Tree
Apr 21 to Apr 30 - Walnut Tree
May 01 to May 14 - Poplar Tree
May 15 to May 24 - Chestnut Tree
May 25 to Jun 03 - Ash Tree
Jun 04 to Jun 13 - Hornbeam Tree
Jun 14 to Jun 23 - Fig Tree
Jun 24 - Birch Tree
Jun 25 to Jul 04 - Apple Tree
Jul 05 to Jul 14 - Fir Tree
Jul 15 to Jul 25 - Elm Tree
Jul 26 to Aug 04 - Cypress Tree
Aug 05 to Aug 13 - Poplar Tree
Aug 14 to Aug 23 - Cedar Tree
Aug 24 to Sep 02 - Pine Tree
Sep 03 to Sep 12 - Weeping Willow Tree
Sep 13 to Sep 22 - Lime Tree
Sep 23 - Olive Tree
Sep 24 to Oct 03 - Hazelnut Tree
Oct 04 to Oct 13 - Rowan Tree
Oct 14 to Oct 23 - Maple Tree
Oct 24 to Nov 11 - Walnut Tree
Nov 12 to Nov 21 - Chestnut Tree
Nov 22 to Dec 01 - Ash Tree
Dec 02 to Dec 11 - Hornbeam Tree
Dec 12 to Dec 21 - Fig Tree
Dec 22 - Beech Tree
Dec 23 to Dec 31 - Apple Tree

APPLE TREE (Love) - of slight build, lots of charm, appeal, and attraction, pleasant aura, flirtatious, adventurous, sensitive, always in love, wants to love and be loved, faithful and tender partner, very generous, scientific talents, lives for today, a carefree philosopher with imagination.

ASH TREE (Ambition) - uncommonly attractive, vivacious, impulsive, demanding, does not care for criticism, ambitious, intelligent, talented, likes to play with fate, can be egotistic, very reliable and trustworthy, faithful and prudent lover, sometimes brains rule over the heart, but takes partnership very seriously.

BEECH TREE (Creative) - has good taste, concerned about its looks, materialistic, good organization of life and career, economical, good leader, takes no unnecessary risks, reasonable, splendid lifetime companion, keen on keeping fit (diets, sports, etc.)

BIRCH TREE (Inspiration) - vivacious, attractive, elegant, friendly,pretentious, modest, does not like anything in excess, abhors the vulgar, loves life in nature and in calm, not very passionate, full of imagination, little ambition, creates a calm and content atmosphere.

CEDAR TREE (Confidence) - of rare beauty, knows how to adapt, likes luxury, of good health, not in the least shy, tends to look down on others,self-confident, determined, impatient, likes to impress others, many talents, industrious, healthy optimism, waiting for the one true love, able to make quick decisions.

CHESTNUT TREE (Honesty) - of unusual beauty, does not want to impress, well-developed sense of justice, vivacious, interested, a born diplomat, but irritates easily and sensitive in company, often due to a lack of self confidence, acts sometimes superior, feels not understood, loves only once, has difficulties in finding a partner.

CYPRESS TREE (Faithfulness) - strong, muscular, adaptable, takes what life has to give, content, optimistic, craves money and acknowledgment, hates loneliness, passionate lover which cannot be satisfied, faithful, quick-tempered, unruly, pedantic, and careless.

ELM TREE (Noble-Minded) - pleasant shape, tasteful clothes, loudest demands, tends not to forgive mistakes, cheerful, likes to lead but not to obey, honest and faithful partner, likes making decisions for others, noble-minded, generous, good sense of humor, practical.

FIG TREE (Sensibility) - very strong, a bit self-willed,independent, does not allow contradiction or arguments, loves life, its family, children and animals, a bit of a social butterfly, good sense of humor, likes idleness and laziness, of practical talent and intelligence.

FIR TREE (Mysterious) - extraordinary taste, dignity, sophisticated, loves anything beautiful, moody, stubborn, tends to egoism but cares for those close to them, rather modest, very ambitious, talented, industrious, uncontested lover, many friends, many foes, very reliable.

HAZELNUT TREE (Extraordinary) - charming, undemanding, very understanding, knows how to make an impression, active fighter for social cause, popular, moody, and capricious lover, honest, and tolerant partner, precise sense of judgment.

HORNBEAM TREE (Good Taste) - of cool beauty, cares for its looks and condition, good taste, is not egoistic, makes life as comfortable as possible, leads a reasonable and disciplined life, looks for kindness and acknowledgment in an emotional partner, dreams of unusual lovers, is seldom happy with its feelings, mistrusts most people, is never sure of its decisions, very conscientious.

LIME TREE (Doubt) - accepts what life dishes out in a composed way, hates fighting, stress, and labor, dislikes laziness and idleness, soft and relenting, makes sacrifices for friends, many talents but not tenacious enough to make them blossom, often wailing and complaining, very jealous but loyal.

MAPLE TREE (Independent) - no ordinary person, full of imagination and originality, shy and reserved, ambitious, proud, self-confident, hungers for new experiences, sometimes nervous, has many complexities, good memory, learns easily, complicated love life, wants to impress.

OAK TREE (Brave) - robust nature, courageous, strong, unrelenting, independent, sensible, does not like change, keeps its feet on the ground, person of action.

OLIVE TREE (Wisdom) - loves sun, warmth and kind feelings, reasonable, balanced, avoids aggression and violence, tolerant, cheerful, calm, well-developed sense of justice, sensitive, empathetic, free of jealousy, loves to read and the company of sophisticated people.

PINE TREE (Particular) - loves agreeable company, very robust, knows how to make life comfortable, very active, natural, good companion, but seldom friendly, falls easily in love but its passion burns out quickly, gives up easily, everything disappointments until it finds its ideal, trustworthy, practical.

POPLAR TREE (Uncertainty) - looks very decorative, not very self-confident, only courageous if necessary, needs goodwill and pleasant surroundings, very choosy, often lonely, great animosity, artistic nature, good organizer, tends to lean toward philosophy, reliable in any situation, takes partnership seriously.

ROWAN TREE or MOUNTAIN ASH (Sensitivity) - full of charm, cheerful, gifted without egoism, likes to draw attention, loves life, motion, unrest, and even complications, is both dependent and independent, good taste, artistic, passionate, emotional, good company, does not forgive.

WALNUT TREE (Passion) - unrelenting, strange and full of contrasts, often egotistic, aggressive, noble, broad horizon, unexpected reactions, spontaneous, unlimited ambition, no flexibility, difficult and uncommon partner, not always liked but often admired, ingenious strategist, very jealous and passionate, no compromise.

WEEPING WILLOW (Melancholy) - beautiful but full of melancholy, attractive, very empathetic, loves anything beautiful and tasteful, loves to travel, dreamer, restless, capricious, honest, can be influenced but is not easy to live with, demanding, good intuition, suffers in love but finds sometimes an anchoring partner

TABLE OF CONTENTS

ACKNOWLEDGEMENTS

The authors owe much to many people and to insure that we don't forget anyone, "Thanks to all who had a hand in this book". We appreciate the use of the line drawings from the sources listed on the copyright page.

We are also indebted to the many people who assisted in the preparation by typing, editing, or other important tasks. These include: Chris Doffitt, Lacy Ezell, Maraya Gates, Misty McCurry, GiGi Parker, Shane Rolen, Athena Starlard, Sara Thames, Dee Dee Walker, Sheyenne Wilborn, and Herbert Young. Thanks to the students who worked with Dr. Allen on grants and/or a thesis project including Shelly Ackel, Monica Barrios, David Brunet, Keith Cascio, Mary Cummings, David Geter, Cheryl Hampsten, Ryan Harper, Brent and Sherry Harrel, Mike Harrell, John Hayes, Thomas Hubbard, Nick Issac, Ning Kang, Duncan McKeithen, Phillip Paul, Chris Reid, Mike Renfrow, Cecil Slaughter, Tammy Soileau, Fen Su, Spencer Trichell, Ronnie Ulmer, Monica Wallace, Dan Webber, Michelle Weeks, Seth White, Jeremy White, Chris Williams, and J.J. Woodley. A special thanks to the dendrology instructors who trained Charles Allen including the late Dr. Clair Brown, the late Dr. William Reese, Dr. John Thieret, and Dr. R. Dale Thomas. Thanks to K. N. Gandhi for assistance in translating scientific names and to Dr. Barney Lipscomb for his assistance in finding and obtaining illustrations. A special thanks and remembrance go to the late Robert Murry for his notes on tree id. Thanks to the other children of Charles Allen: Tanya Allen Lueder and Andy Allen for putting up with all the plant searches. Thanks to our spouses: Susan Allen, Anthony Newman, and Mary Lou Winters for being patient as we worked on the book. Krislyn Newman and Baylie Newman, the grandchildren of Charles and children of Dawn, are thanked especially for calling pawpaw, ***Asimina triloba***.

A note of graditude is also expressed to the many Friends and Cohorts in the various Native Plant Organizations. We have shared many field trips and hashed over plants for many years. There are many but the following list includes the ones that we can remember today and we are sure this list will irritate those that we forget: Larry Allain, Carl Amason, Charles Ancona, Lynn and Rick Bailey, Gail Barton, Tom and Donna Barry, Kenny Benoit, Joyce Bennett, LaVerne Bodron, Phil Bourgeois, Alan Broussard, George Broussard, Joan Brown, David Campbell, Paul and Linda Chance, Marilyn and Fred Christian, Dean Clark, Lee and Deborah Clark, Olga and Walter Clifton, Jimmy Culpepper, Mary Grace Curry, Tom and Thayer Dodd, Ann Donnelly, Marion Drummond, Jackie Duncan, Joyce Dupree, Sandra Elsik, Larry Erbe, Beth and Terry Erwin, Kent Fagan, Bruce Fleing, Will Fleming, Bill Fontenot, Jim Foret, Avis Foster, Ray French, Julius and Susie Furr, Sandra Gibbs, Harland Guillory, Jim Harvey, Mona Hollier, Phil Hyatt, Margie Jenkins, Richard and Jesse Johnson, Celia Jones, Tim Kiphart, Diane LaBorde, Sterling and Loice Lacy, Garrie Landry, John Larkin and the Folsom Native Plant Society, Jerry Lenhard, David and Pat Lewis, Peter and Cass Loos, Ruth McDonald, Ben Martin, Therese Martin, John Mayronne, Scott Meredith, Jan Midgley, Joan Moncrief , David Moore, Alan Neuman, Katie Northrup, Scott Ogden, A.B. and Margaret Osborne, Kelby and Amy Ouchley, Steve Pagans, Marc and Darla Pastorek, Cale Pastorek, Helen Peebles, Larry Raymond, Don and Tina Reid, Jim Robbins, Ron Robbins, Corine Roberie, Jim Robinson, Charlotte Seidenberg, Sara Simmonds, Eric and Linda Sipco, Latimore and Nelywn Smith, Charles Stagg, Steve Strong, Don Thibodeaux, Al and Kathy Troy, Malcolm and Gail Vidrine, Karl Vincent, Rick Webb, and Ken and Beckie Wilson.

INTRODUCTION

What tree is that? Is this poison ivy? How many times have you wondered about the vegetation around you as you wandered down a trail or along a city street? Each plant species has its own id fingerprints. The key is to be able to sort out and recognize the characters. Most of you can recognize one or more groups of nature's miracles. There are a lot of people who can readily id birds even by their sounds, other can id snakes and other reptiles, others can readily id rocks, and the list goes on.

We decided a few years ago to attempt to create a book on the woody plants of Louisiana that would allow you to id them. This work is intended for everyone from the novice to the trained professional. We tried to split the difference between technical and too simple. We attempt to adequately describe each species, variety, or subspecies without using too many of the 25 cent words. In some cases, we had to go with the technical jargon or we just could not describe the plant. A glossary is provided with the hopes of you being able to look up all the unfamiliar words and then getting a grasp of the meaning.

To facilitate identification, we provide a listing of plants that have unusual characters: (Table 1) opposite or whorled leaf arrangement, (Table 2) compound leaves, (Table 3) palmate major veins, (Table 4) armed plants, (Table 5) strong odor, (Table 6) evergreen leaves and (Table 7) vines. We also include in a final table (Table 8) the genera without any of the unusual characteristics. If you are not sure of a characteristic, read the Basic Botany section or look up the word(s) in the glossary. Check out your plant and narrow the list of possible candidates. Most plants have alternate leaves and we provide a listing of those that have opposite or whorled leaves. The typical leaf is simple but a few plants produce compound leaves so we provide a list of plants with compound leaves. For the plants with simple leaves, most have pinnate major veins so we list those with palmate major veins. Most plants do not have "sticky" structures but a few do. We list all "armed" plants where armed could be with thorns, spines, prickles, or even spiny teeth on the leaves. We list all plants with an odor since most crushed leaves and/or stems do not have a strong odor but smell like a leaf. Some species have a pleasant or unpleasant (stinking) odor to the crushed stem or leaves. Most woody plants are trees or shrubs so we opted to list vine species. Take a close look at your plant and see if it has any of the characters listed. If it is on one or more of the lists, this will narrow your search. Conversely, you could also search our last list which is of plants that are not in any of the first seven lists. This last list is for the trees and shrubs, not vines, with alternate simple deciduous leaves and with pinnate major veins that are not armed and do not have a strong odor.

With the modern age of color photographs, some of you might wonder why we opted to go with line drawings. It is true that color photographs have a lot to offer and are very useful in identification of many common species and especially genera. And, color photographs are wonderful to look at. But for identification to species, line drawings can be used to emphasis the characters to use in separating species. And, perhaps the most important factor, it is much less expensive to produce a black and white book than one with color.

LOUISIANA AS A HABITAT FOR WOODY PLANTS

Louisiana, roughly 48,523 square miles, is politically divided into 64 parishes (Fig. 1). The state lies between 29 degrees N latitude (Gulf of Mexico) and 33 degrees N latitude (Arkansas). The eastern boundary is the Mississippi (approximately 91 degrees 15 minutes W longitude) and the Pearl (approximately 89 degrees 45 minutes W longitude) rivers which separate Louisiana from Mississippi. The western boundary is the Sabine River which separates Louisiana from Texas at about 94 degrees W longitude. Elevation within the state usually increases with increasing latitude. Mt. Driskill (535 feet) is the highest point in the state.

The climate of Louisiana is largely determined by its subtropical latitude and its proximity to the Gulf of Mexico (Dyke, 1941). The average January temperature ranges from 48 degrees F. in the north to 56 degrees F. in the extreme southeast corner. The average July temperature is 82 degrees F. The length of the growing season varies from an average of 350 days in the southeast corner to 230 days in the north. Average annual precipitation ranges from 46 inches in the northwest to 64 inches in the southeast. The majority of the precipitation is in the form of rain as the amounts of snow, hail, glaze, and sleet are negligible.

Louisiana lies entirely within the Coastal Plain physiographic province (Hunt, 1974). Based on the nature of the underlying sediments, the state can be divided into a Tertiary Upland section and a Quaternary Lowlands section (Holland, 1944) (Fig. 2). The Tertiary Uplands section is underlain by consolidated bedrock. The soils developed in this section are from Tertiary materials or from Tertiary and Pleistocene materials. The materials deposited during the Tertiary and Pleistocene are usually a mixture of marine and fluvial deposits. The uplands are usually much dissected and well drained.

The underlying sediments of the Quaternary Lowlands section are unconsolidated Pleistocene and Recent gravel, sand, silt, and clay. This section occurs in the southern part of the state and projects inland along the major streams.

The Lowlands are divided into a Pleistocene Terrace unit and a Recent unit. The Pleistocene unit includes four belt-like terraces which parallel the coast and form finger-like projections inland along the streams. These terraces were created when sea level rose during interglacial epochs of the Pleistocene and each interglacial epoch is represented by a different terrace. The oldest and most dissected terrace is located inland adjacent to the Tertiary Uplands. The youngest and least dissected terrace is located seaward adjacent to the Recent unit. The soils developed on the older terraces are from Pleistocene and Tertiary material. On the younger terraces, the soils are developed from Pleistocene material only or Pleistocene and limited Tertiary material.

The Recent unit consists of alluvial and deltaic plains. The alluvial plain is along the Red, Mississippi, and smaller rivers. The deltaic plain consists of the present Mississippi River delta and the coastal marsh which was formed of former deltas or delta material. The Recent unit is flat with little or no dissection, poorly drained, and has soils developed primarily from Recent materials. In the alluvial plain areas adjacent to the Pleistocene Terraces, the soil developed is often from a mixture of Pleistocene and Recent materials.

Based on characteristic, usually dominant, species, the state can be divided into five major vegetation regions: pine, floodplain, coastal marsh, prairie, and upland hardwoods (Fig. 3). The distribution and extent of these vegetation regions are correlated with the geology. Maps

Figure 1. Louisiana Parishes

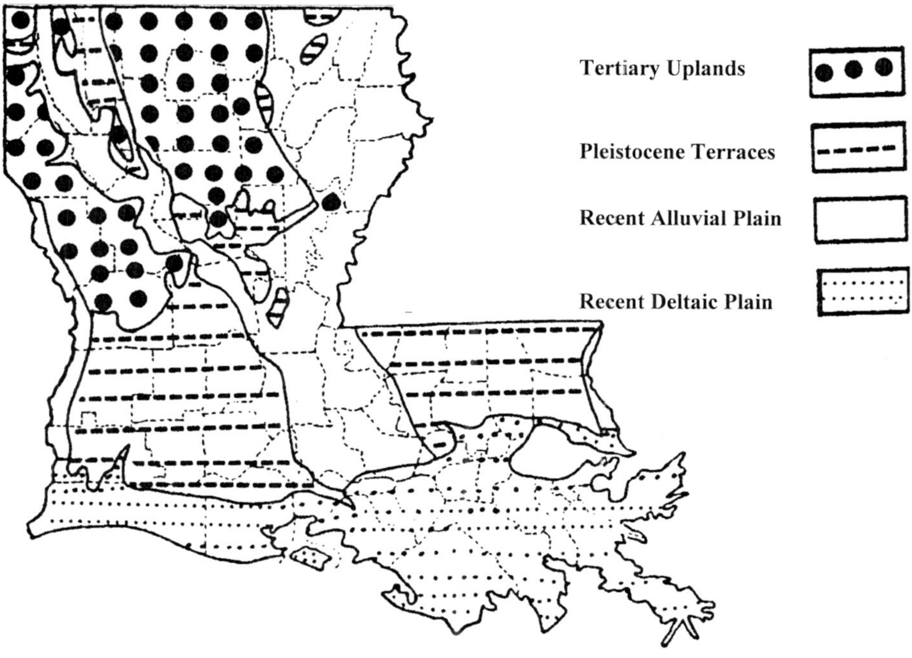

Figure 2. Louisiana Geological Regions

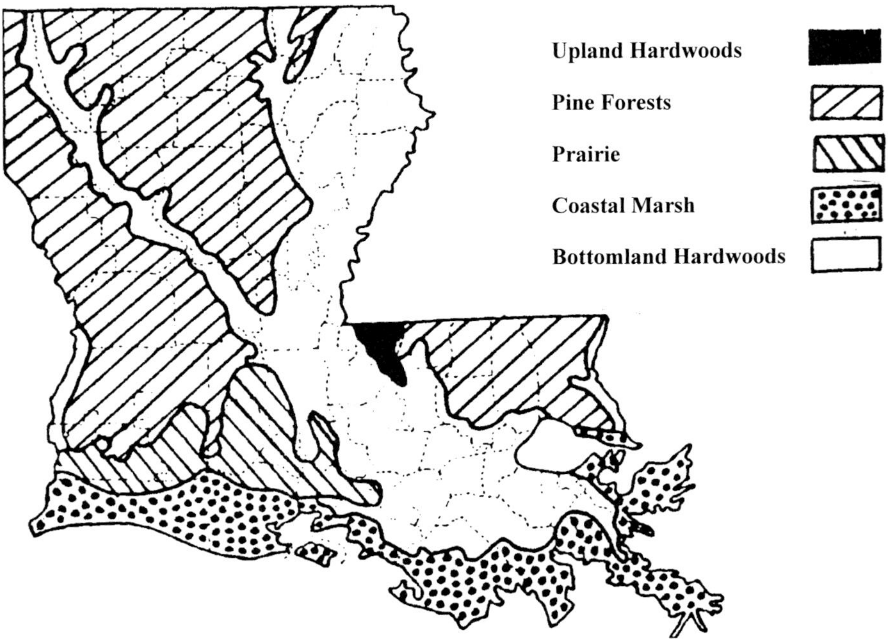

Figure 3. Louisiana Vegetation Regions

depicting the distribution of various categories of Louisiana's natural areas are available in Newton (1972).

The coastal marsh region lies adjacent to the Gulf of Mexico and extends inward irregularly. Most of the region is true marsh with wet soil conditions and is dominated by graminoid (grasses, sedges, and rushes), non woody plants. There are a few shrubs, for example ***Baccharis halimiifolia*** and ***Iva frutescens***, within the true marsh. A number of other shrubs, trees, and woody vines are found on the elevated areas within the marsh. These include coastal shrub thickets, the chenieres, and the man made spoil banks. The coastal shrub thickets are narrow strips of shrubs vegetation usually parallel to and very close to the Gulf of Mexico. Some shrubs commonly observed include ***Morella cerifera***, ***Tamarix*** spp., and ***Zanthoxylum clava-herculis***. Woody vine species include ***Cissus incisa, Smilax rotundifolia***, and ***Vitis mustangensis***. The chenieres are dominated by ***Celtis laevigata*** and ***Quercus virginiana*** and in recent year is being invaded by ***Triadica sebifera***. Common plants on the man made spoil banks include many ***Baccharis halimiifolia*** and recently many many ***Triadica sebifera***.

The prairie includes the large Cajun Prairie near the coast and the small inland prairies. Cajun Prairie once covered the nearly flat area just inland from the coastal marsh in southwestern Louisiana. A few railroad remnants are all that remain of this ecosystem. The Cajun Prairie, like the coastal marsh, is dominated by non woody graminoids. A few shrubs (***Ceanothus americanus, Baccharis halimiifolia, Morella cerifera***, and ***Styrax americana***) were scattered across the prairie. The prairie was not continuous but dissected by finger like forests along the streams called gallery forests. The gallery forest species are a reflection of the size of the stream, the elevation, and the distance from the Gulf of Mexico. In very large streams at low elevation, one would find ***Nyssa aquatica, Taxodium distichum*** and other wetland species but along smaller streams at higher elevations and especially inland, one would find ***Liquidambar styraciflua*** and ***Pinus taeda***, and other more upland species. There are many disturbed areas within the original prairie zone and these are being completely inundated by dense stands of ***Triadica sebifera***. There are also small inland prairies scattered through central and northern Louisiana such as Copenhagen in Caldwell Parish and Keiffer in Winn Parish. These are surrounded by a forest termed the Inland Prairie Forest. Some common trees and shrubs in these forests include ***Cornus drummondii***, several species of ***Crataegus, Frangula carolinensis, Fraxinus americana, Pinus echinata, Sideroxylon lanuginosum,*** and ***Viburnum rufidulum***.

The most extensive vegetation region in the state is the pine region and it can be subdivided into a hilly section to the north and flatwoods to the south. The hilly section is on the Tertiary Uplands and older Pleistocene Terraces and the flatwoods section is on those younger Pleistocene Terraces that have soils derived from Pleistocene and Tertiary materials. Many of the true pine forests are replanted and replanted and replanted loblolly pine ***(Pinus taeda***) plantations. The most common vegetation type within this area is obviously, pine (***Pinus*** spp.) forests but there are several vegetation types that are not dominated by pine. These include small patches of prairie as mentioned earlier or strips of upland forests along the slopes near streams or even strips or blocks of bottomland hardwood forests in wet areas long the streams or in depressions. There are also other vegetation zones within the pine region such as sandy woodlands, baygalls, and bogs. The sandy woodlands are on the highest, driest, and sandiest sites mostly in northwest Louisiana but a limited number occur in the central and west central area. There are a few pines but the areas are dominated by oaks (***Quercus***). The species commonly seen include ***Quercus incana, Quercus marilandica, Quercus margarettiae,*** and ***Quercus falcata.*** A common soil type for sandy woodlands is Betis loamy fine sand. A baygall is the vegetation developed along sandy bottom streams with acid ph soils in southeastern, southwestern, and west central Louisiana. The woody vegetation includes ***Nyssa sylvatica*** var. ***biflora*** plus a number of evergreen small trees and shrubs including ***Ilex coriacea, Lyonia lucida, Magnolia virginiana, Morella caroliniensis, Morella cerifera,*** and ***Persea palustris***. Some of the deciduous shrubs include ***Rhododendron*** spp., ***Toxicodendron vernix, Vaccinium fuscatum,*** and ***Viburnum nudum***. The woody vine, ***Smilax laurifolia***, is often very dominant in the baygall. The bogs are open areas and are usually associated with a baygall. Bogs are similar to prairie and marsh in being dominated by non woody graminoids but also include a number of carnivorous plants. The flat areas of southeastern and southwestern Louisiana produce the pine flatwoods and the savannahs. Both have an overstory of pines with a few scattered trees and shrubs. The flatwoods is usually drier, has a dense canopy of pines, and the floor has a sparse cover of graminoids. The savannah is wetter, often with bogs especially in southeast Louisiana, is more open and thus the floor has a dense cover of graminoids. The original range of ***Pinus elliottii*** was in these areas in southeast Louisiana. Some common hardwood trees in these areas include ***Nyssa sylvatica*** var. ***biflora***, ***Quercus nigra***, and ***Quercus laurifolia***.

The floodplain vegetation region is characterized by many wet or cultivated areas. This region is located in alluvial plain and part of the deltaic plain. The areas immediately adjacent to the larger streams often develop a vegetation type termed "batture". This is best developed along the Mississippi River and is commonly dominated by ***Populus deltoides***, ***Salix interior***, and ***Salix nigra.*** The most common vegetation type within the floodplain region is the bottomland hardwood forests. These are found on the wet to mesic sites throughout the state but are best developed in the Mississippi and Red River floodplains. There is a wide range of species associations that are controlled by soil type, elevation, geographical location, and human intervention. Some common tree species include ***Celtis laevigata, Fraxinus pennsylvanica, Platanus occidentalis, Quercus lyrata, Quercus nigra, Quercus phellos, Quercus similis, Ulmus americana,*** and ***Ulmus crassifolia***. Some shrub species include ***Amorpha frutescens, Crataegus viridis, Ilex decidua,*** and ***Styrax americana.*** Some vines in this vegetation type include ***Berchemia scandens, Brunnichia ovata,*** and ***Trachelospermum difforme***. In the very wet areas (depressions, old ox bow lakes etc.) the swamp vegetation develops. Two species (***Nyssa aquatica*** and ***Taxodium distichum***, in combinations of mixed to pure stands of each species) dominate the swamp vegetation. Other species often found include ***Acer rubrum*** var. ***drummondii*** and ***Cephalanthus occidentalis.***

The upland hardwood forest vegetation region occurs throughout the state but is only extensive in one area, in the southeast part of the state. This area is known as the Tunica Hills and is north of St. Francisville and east of the state prison at Angola. The other hardwood forests occur as small strips or patches paralleling streams, especially in the pine regions. The plant community nearer the stream and thus at lower elevation is often called the riparian vegetation and the plant community further upslope is termed hardwood slope forest. The type in the Tunica Hills is the hardwood slope forest and is sometimes called a southern mesophytic forest. Some of the hardwood slope forests grow in much dissected and well drained areas that are usually bluffs adjacent to the alluvial plain and the soils in some of these areas are apparently loess. The riparian forest usually has a well developed shrub layer that is absent or very small in the hardwood slope forest. Common trees in the riparian forest are ***Carpinus caroliniana, Fagus grandifolia, Ilex opaca, Magnolia grandiflora, Quercus michauxii, Pinus glabra*** (in southeastern Louisiana only), and ***Symplocos tinctoria***. Common shrubs of the riparian forest include ***Halesia diptera, Illicium floridanum*** (in southeastern Louisiana only), and ***Viburnum dentatum.*** The common trees of the hardwood slope forest are ***Cornus florida, Fagus grandifolia, Fraxinus americana, Magnolia grandifolia, Ostrya virginiana,*** and ***Quercus alba***.

The artificial or man made habitats include abandoned fields, old home sites, and trash piles. The species in these areas are often non native or introduced ones. Some only persisit at the sites while others spread into the surrounding areas. The species makeup is variable from site to site. Some species that are restricted to the disturbed area include ***Broussonetia papyrifera, Cunninghamia lanceolata, Elaeagnus pungens, Hedera helix, Lagerstroemia indica, Paulownia tomentosa, Phyllostachys aurea, Pyrus calleryana, Quercus acutissima,*** and ***Salix x sepulcralis.*** Some examples of species that are spreading from disturbed sites include ***Albizia julibrissin, Ligustrum sinense, Lonicera japonica, Melia azedarach, Poncirus trifoliata, Rosa bracteata, Rosa laevigata,*** and ***Triadica sebifera***

BASIC BOTANY

All true plants have roots, stems, and leaves with chlorophyll in the leaves. True plants include those without seeds like ferns, mosses, and liverworts and those with seeds. The plants with seeds include the Gymnosperms, cone plants like pine trees and Angiosperms, the flowering plants. Plants with seeds can be herbaceous or woody with most being separated on the obvious hard and woody character versus the soft and nonwoody character of the herbaceous. The technical distinction is that herbaceous plants die back to the ground level during the winter while woody plants maintain their above ground stems and/or leaves during the cold season. Some examples of plants that are not hard and woody but do maintain above ground stems or leaves during the winter are ***Sabal, Smilax***, and ***Yucca***. This book includes only woody plants; thus will include both Gymnosperms and Angiosperms. Angiosperms are divided into two major groups; the dicots and monocots. Dicots have broad leaves that are net veined (the veins branch and rebranch creating a net) while monocots have narrow leaves with veins arranged parallel to each other. The number of flower parts in dicots is four or five (or a multiple of four or five) while monocots produce flowers with three or six parts. The seed of dicots contain two cotyledons while monocots have a single cotyledon. The stem of monocots cannot produce true wood; there are some monocots that are considered woody, i.e. ***Smilax*** or ***Sabal***, because the above ground stems do survive the winter. Dicot stems can be true woody while others are herbaceous.

The roots of plants perform two major functions: absorption and anchorage but are generally not very useful for identification. The specialized or modified roots of a few species can be useful in identification; for example, the aerial roots of poison ivy.

The stem also performs two major functions: to transport water, nutrients, and food through the plant and to produce and support the leaves. Unlike the roots, the stem can be very useful in the identification of a species. Monocots can be separated from dicots based on the internal structure of the stem: monocots produce their vascular bundles in a random or nonpattern manner while dicots produce their vascular bundles in a circle around the center region known as the pith. The external portion of a stem can be very useful in identification. On older trees, the bark can be smooth or rough and vary greatly in color. Some trees like ***Celtis*** produce obvious corky proturbences while others may produce wings along the younger stems as in winged elm or sweet gum. The younger twigs can be often identified based on color (***Sassafras*** has bright green twigs) or shape (***Euonymus americanus*** has square twigs). A few species, i.e. ***Sideroxylon*** and ***Ilex decidua***, produce spur shoots where the internodes are very short and the leaves are all jammed together. Buds can also be useful in identification because of the variation in size, shape, number, texture, and location.

The leaf has only one major function, photosynthesis, but is the most important for identification of many woody plants. The point on a stem where the leaf is produced is the node. At each node, the leaf will have an associated lateral bud located just above the leaf. Lateral buds can grow into lateral stems or branches and produce leaves along the axis. The arrangement of the leaves along the stem or the number of leaves per node is a very important character for identifying plants. The most common arrangement is alternate (one leaf per node); the second most common type is opposite (two leaves per node); and the least common type is whorled (three or more leaves per node). In many species, the younger tips of the branches may appear whorled due to short internodes so it is best to check for arrangement on the older twigs. The typical leaf consists of three major parts; the blade (green flattened portion), the petiole (the stalk that connects the blade to the stem, and the stipules (paired structures located on the stem at the base of the petiole). Many plants do not have stipules so the ones with stipules can often be recognized based on this character. Stipules can be large and persistent as in ***Platanus occidentalis*** or much smaller and deciduous as in ***Prunus serotina***. Some stipules are small dots as in ***Ilex*** or a circular scar around the stem as in the ***Magnolia***. A few plants produce leaves without a petiole with this condition termed sessile while a leaf with very short petioles can be called subsessile. The arrangement of the veins in leaves can be netted (all dicots) or parallel (most monocots). The major veins refer to the number of veins at the base of the blade or leaflet. Some use the term venation for this but the term "major veins" is more descriptive and less likely to be confused with arrangement of veins. The very common number of major veins is one and this is termed "pinnate", Latin for feather. If you examine only the veins in this type of leaf, it does look like a feather. The one major vein is usually termed the "midrib". The other major vein choice is "palmate", more than one major vein, usually an odd number, especially three or five. Most plants with palmate major veins produce wide leaves. The term "palmate" comes from the similarity of the veins to the arrangement of the bones in the palm of the hand. Most plants produce simple leaves where the blade is not cut up into smaller segments. However, a few species produce compound leaves where the blade is cut up into smaller segments with each segment becoming a leaflet. The stalk of a leaflet is the petiolule. The two ways to distinguish between a simple and compound leaf is: (1) lateral buds are associated with the blade but not with the leaflet and (2) leaflets can all be flattened in the same plane. A good way to understand this is to take a large simple leaf and using scissors, cut the blade into leaflets. Almost all leaflets have pinnate major veins. Compound leaves can be palmate or pinnate with each type originating from a simple leaf with the corresponding major vein arrangement. In a palmate compound leaf, all the leaflets originate at the same point while in a pinnate compound leaf, the leaflets are scattered along the rachis, the axis that was the midrib. Pinnate compound leaves can have an odd number of leaflets if there is a terminal leaflet in addition to the lateral ones or an even number if there is no terminal leaflet. In some pinnate compound species, the leaflets are cut up into leaflets creating the two-pinnate compound situation and in a very few species the leaflets are further dissected into three or very rarely, even four pinnate compound leaves. The most common pinnate situation is one-pinnate, as described earlier, where the blade is cut up into leaflets only. The margins of blades or leaflets are also variable with three major categories: entire (smooth), toothed, and lobed. The size, shape, and number of teeth and lobes are very useful in identifying plants. The shade of green of the blade or leaflets can be useful but requires years of observing plants to acquire the knack for distinguishing the shades. Odor of the leaves, stems, or roots can be a very useful tool in identification. Odor can vary from a very pleasant (***Lindera bezoin***) to unpleasant (***Prunus serotina***). Plants that are armed with "stickers" on the stems or leaf can also be useful. Technically, the "sticker" can be a thorn-a modified stem, located at the node where a lateral bud should be (example ***Gleditsia***), or a spine-a modified leaf, located at the node where a leaf should be (example ***Robinia pseudoacacia***), or a prickle, an outgrowth from the epidermis of a leaf or a stem (examples ***Aralia spinosa, Rubus, Smilax***).

The habit of the plant is often a useful character. The habit refers to whether the plant is herbaceous (as described earlier) or woody. Woody plants can be divided into trees, shrubs, or vines. A tree is defined as a woody plant usually with one main stem and at maturity is taller than 20 feet and larger than 5" diameter. A shrub is a woody plant usually with several stems and at maturity is usually shorter than 20 feet and smaller than 5" diameter. Of course, trees are small prior to becoming a tree and a young tree is usually termed a sapling. Vines can be woody or herbaceous but climbs on other woody plants (or whatever is available-telephone pole, building, etc.) by tendrils, aerial roots, twining, or by sprawling. Trees and shrubs also may have three basic growth forms: (1) excurrent-the tree is cone shaped and very narrow at the tip and widens basally, example pine; (2) deliquescent-widely spreading to an extreme where it is as wide as tall, example live oak; and (3) columnar-the stem is not or only slightly tapered and the leaves are concentrated near the top, example devil's walkingstick.

The flower is a modified stem tip and the parts are modified leaves. The basic flower has four major parts: (1) sepals (calyx collectively); (2) petals (corolla collectively); (3) stamens (androecium collectively); and pistils (gynoecium collectively). The stamens and pistils are directly involved in reproduction while the calyx and corolla may attract insects or rarely birds for pollination. A complete flower has all four major parts while an incomplete flower is missing one or more parts. A perfect flower has both an androecium and gynoecium while an imperfect flower has only an androecium or only a gynoecium, but not both. Plants with imperfect flowers can be monoecious (produces both androecium flowers and gynoecium flowers on the same plant as in corn) or dioecious (produces androecium flowers on one plant and gynoecium flowers on a different plant) as in willow. The calyx and corolla collectively are called the perianth and a flower with no perianth is said to be naked. Flowers are very variable in woody plants and can be useful for identification but of course are only useful while the plant is in flower. The number of parts, shape, size, time of year produced, and color are some useful identifying flower variables.

A few plants produce their flowers singly as in ***Magnolia*** but most produce their flowers in an inflorescence. The inflorescence is the arrangement of the flowers on the plant and includes seven basic kinds; catkin, raceme, panicle, umbel, corymb, head, and cyme. The catkin has flowers without stalks and the first flower produced is at the base of the inflorescence. The catkin also droops and the flowers are imperfect; ***Salix*** produces both male and female flowers in catkins while ***Quercus*** and ***Carya*** produce their male flowers in catkins. The raceme is an inflorescence with stalked flowers and the first flower is at the base; black cherry and wisteria produce their flowers in racemes. The panicle is the largest inflorescence and is defined as an inflorescence with the oldest flowers at the base and the stalks are branched, often with several branches. The umbel, corymb, head, and often the cyme are flat topped inflorescences. In a cyme, the first flower produced is in the center while the first flower produced in the other three is on the outside. In the umbel and corymb, the flowers are stalked with all stalks starting at the same point in the umbel and not all at the same point in the corymb. In the head, the flowers do not have stalks. Another factor for identification is where the flowers are produced on the plant, in particular, on the top of the plant or in the axils of the leaves.

The fruit is the botanical term for the structure that develops from the pistil of the flower. The fruit wall is termed the pericarp and can be very important in classification and identification. The two big categories of fruits are simple and compound with simple being derived from a single pistil and compound from more than one pistil. The next division of fruits (both simple and compound) is into fleshy and dry fruits. In fleshy fruits, the pericarp is soft, succulent, fleshy, and usually edible raw while the dry fruits are hard, dry, and usually not edible raw. The fleshy fruits include: (1) berry which is fleshy throughout; (2) drupe (also called stone or pit fruit) that has a somewhat hard outer covering, the exocarp or skin, a fleshy middle portion, the mesocarp, and a very hard inner layer, the endocarp; and (3) pome that has an outer covering of the base of the flower and is fleshy inside of that. Among woody plants, ***Vitis*** is a good example of a berry, ***Prunus*** is a drupe, and ***Crataegus*** is a pome. Many consider all fleshy fruits to be berries and this might be a good artificial classification given that it is difficult to classify small fruits into berries, drupes, and pomes. The dry fruits are usually divided into the dehiscent and indehiscent groups; the dehiscent ones open up under natural conditions to release the seeds while the indehiscent ones do not. It should be noted that being stepped on by a elephant does not rate as a natural event. The dehiscent ones include the legume, follicle, and capsule. The legume and follicle both have one cavity while the legume opens along both sides to release the seeds while the follicle opens along one side only. The capsule is very popular among woody plants and has more than one cavity and usually splits along more than two lines to release the seeds. Members of the bean family (Fabaceae) have the legume fruit while ***Magnolia*** produces follicles. The indehiscent dry fruits include nuts, achenes, caryopses, and samaras. The nut is a dry indehiscent fruit where the pericarp or part of the pericarp is very hard. This is very popular among woody plants with the oak (***Quercus***) acorn and the hickory (***Carya***) nut being this type fruit. The achene and caryopsis both have a single seed with the seed attached to the pericarp in only one place in the achene and is completely attached in the caryopsis. Among all plants these two fruits are very important because the achene is found in the sunflower family (Asteraceae) and the caryopsis in the grass family (Poaceae). Due to the paucity of woody species in these two families, the fruit types are of limited importance in woody plants. Samaras are defined as achenes with wings and are the fruit type of maples (***Acer***) and elms (***Ulmus***).

As defined earlier, compound fruits are produced from more than one pistil and include aggregate fruits and multiple fruits. Aggregate fruits come from a single flower while multiple fruits are produced from multiple flowers. The individual fruit of a compound fruit can be classified as one of the simple fruits as discussed earlier. Many would simply refer to all compound fruits as just that, a compound fruit. Other than the type of fruit, other identifying characteristics include size, shape, color (hollies are often identified by the red or black colored fruits), and time of year for the fruit production.

Habitat or where the plant grows can be a very useful tool for identification. Some plants are restricted to wet habitats while others are found only in dry areas. Of course, as with all identification aids, there is overlap; persimmon can range from wet to dry. The range or distribution can also be used; some species are found only east of the Mississippi River and many are not found in the Mississippi River floodplain. Using range or distribution also has its drawbacks, especially with humans transplanting species.

Table 1. List of genera with opposite or whorled leaves. The * indicates that not all species in the genus and/or not all plants share this characteristic.

Acer	Cornus	Macfadyena
Aesculus	Decodon	Nerium
Avicennia	Decumaria	Osmanthus
Batis	Euonymus	Paulownia
Bignonia	Forestiera	Philadelphus
Borrichia	Forsythia	Phoradendron
Broussonetia*	Fraxinus	Ptelea*
Calamintha	Gelsemium	Punica*
Callicarpa	Hydrangea	Sambucus
Calycanthus	Hypericum	Scaevola
Campsis	Iva*	Staphylea
Catalpa	Jasminum	Symphoricarpos
Cephalanthus	Lagerstroemia*	Trachelospermum
Chionanthus	Lantana	Viburnum
Clematis	Ligustrum	Vinca
Clerodendrum	Lonicera	Vitex

Table 2. List of genera with compound leaves. The * indicates that not all species in the genus and/or not all plants share this characteristic.

Acacia	Erythrina	Pterocarya
Acer*	Fraxinus	Rhus
Aesculus	Gleditsia	Robinia
Ailanthus	Indigofera	Rosa
Albizia	Jasminum	Rubus
Amorpha	Juglans	Sambucus
Ampelopsis*	Koelreuteria	Sapindus
Aralia	Lespedeza	Sesbania
Bignonia	Macfadyena	Sophora
Campsis	Melia	Staphylea
Carya	Nandina	Toxicodendron
Cayratia	Parkinsonia	Vitex
Cissus	Parthenocissus	Wisteria
Cladrastis	Poncirus	Xanthorhiza
Clematis	Prosopis	Zanthoxylum
Desmanthus	Ptelea	

Table 3. List of genera with palmate major veins. The * indicates that not all species in the genus and/or not all plants share this characteristic.

Acer*	Ginkgo	Ribes
Ampelopsis*	Hedera	Ricinus
Aristilochia	Hibiscus	Sabal
Broussonetia	Liquidambar	Serenoa
Calycocarpum	Liriodendron	Smilax
Catalpa	Malvaviscus	Spiraea*
Celtis	Manihot	Tilia
Cercis	Morus	Vernicia
Clerodendrum*	Pachysandra	Vitis
Cocculus	Paulownia	Ziziphus
Ficus	Platanus	
Firmiana	Populus	

Table 4. List of armed (thorns, spines, or prickles or stems and/or leaves) genera. The * indicates that not all species in the genus and/or not all plants share this characteristic.

Acacia	Malus*	Rosa
Aralia	Opuntia	Rubus
Chaenomeles	Parkinsonia	Serenoa
Crataegus*	Pinus	Sideroxylon*
Elaeagnus*	Poncirus	Smilax
Erythrina	Prosopis	Yucca
Gleditsia	Punica	Zanthoxylum
Ilex*	Pyracantha	Ziziphus
Lycium	Ribes	
Maclura	Robinia	

Table 5. List of genera with strong odor to crushed leaves and/or stem. The * indicates that not all species in the genus and/or not all plants share this characteristic.

Acer*	Cryptomeria	Morella*
Ailanthus	Cunninghamia	Persea
Asimina	Halesia	Pinus
Batis	Illicium	Poncirus
Calamintha	Juglans	Prunus*
Callicarpa	Juniperus	Ptelea
Calycanthus	Lantana	Rhus
Carya*	Lindera	Sassafras
Catalpa*	Liquidambar	Vitex
Cinnamomum	Liriodendron	Zanthoxylum
Clerodendrum	Magnolia*	

Table 6. List of genera with evergreen leaves. The * indicates that not all species in the genus and/or not all plants share this characteristic.

Ardisia	Ilex*	Photinia*
Arundinaria	Illicium	Phyllostachys
Avicennia	Jasminum	Pinus
Baccharis*	Juniperus	Prunus*
Batis	Kalmia	Pyracantha*
Bignonia*	Lantana	Quercus*
Borrichia	Leucothoe	Ricinus
Cayratia?	Licania	Rosa*
Chionanthus*	Ligustrum	Sabal
Cinnamomum	Lonicera*	Salicornia
Cliftonia	Lyonia	Scaevola
Croton	Macfadyena	Sebastiania*
Cryptomeria	Magnolia*	Serenoa
Cunninghamia	Morella	Smilax*
Cyrilla*	Nandina	Symphoricarpos
Elaeagnus*	Nerium	Symplocos*
Eriobotrya	Opuntia	Tamarix
Euonymus*	Osmanthus	Vaccinium*
Gelsemium	Pachysandra	Vinca
Hedera	Persea	Yucca
Hypericum	Phoradendron	

Table 7. List of genera that are vines. The * indicates that not all species in the genus and/or not all plants share this characteristic.

Ampelopsis	Cissus	Rosa*
Aristilochia	Clematis	Rubus*
Berchemia	Cocculus	Schisandra
Bignonia	Decumaria	Smilax*
Borrichia	Gelsemium	Trachelospermum
Brunnichia	Hedera	Vinca
Calycocarpum	Lonicera*	Vitis
Campsis	Macfadyena*	Wisteria
Cayratia	Parthenocissus	
Celastrus	Pachysandra	

Table 8. List of genera that do not have any of the unusual characteristics. These plants have alternate, simple deciduous leaves with pinnate major veins. The leaf and/or stem does not have an odor, the plants are not armed and are shrubs or trees, not vines. The [1,2,3,4,5,6, or 7] indicates that some species or some plants do have one or more of the unusual characteristics and the number indicates the appropriate Table number.

Alnus	Hamamelis	Quercus[6]
Amelanchier	Ilex[4, 6]	Rhamnus
Baccharis[6]	Itea	Rhododendron
Betula	Iva[1]	Salix
Carpinus	Lagerstroemia[1]	Sideroxylon
Castanea	Licania	Solanum
Ceanothus	Magnolia[5, 6]	Spiraea[3]
Clethra	Malus[4]	Stewartia
Corylus	Nyssa	Styrax
Crataegus[4]	Ostrya	Symplocos[6]
Cyrilla[6]	Oxydendrum	Taxodium
Diospyros	Photinia[6]	Triadica
Dirca	Planera	Ulmus
Fagus	Polygonella	Vaccinium[6]
Frangula	Prunus[5, 6]	
Gaylussacia	Pyrus	

SWEET ACACIA - ***Acacia farnesiana*** - Fabaceae

The genus name, "*Acacia*", refers to the hard sharp spinescent stipules of some species. The specific epithet, "*farnesiana*", honors Cardinal Odoardo Farnese (1573-1626) of Rome. Other common names include Acacia Catclaw, Honey Ball, and many of Hispanic and/or Native American origin. This is a shrub to small tree that is armed with stipular spines. The leaves are alternate, deciduous, and two-pinnate even compound. The leaflets are asymmetrical with a small mucro apically and a circular pit on the petiole. The flowers are in compact heads, white to yellow, and sweet scented. The fruit is a legume that is cylindrical, straight or curved, and reddish brown to purple or black. The wood is reddish brown, hard, heavy, durable, and close-grained, with a specific gravity of 0.83. The wood is used for many purposes, including posts, agricultural instruments, and woodenware articles. It is considered a good winter forage plant and a desirable honey plant in semiarid areas, being resistant to drought and heat. The bark and fruit are used for tanning, dying, and ink making. Glue from the young pods is used to mend pottery. Medicinal uses reported include: (1) flowers-ointment for headaches and infusion for dyspepsia, (2) green fruits-astringent and used for dysentery and inflammation of skin and mucous membranes, (3) dried leaves-dressing for wounds, and (4) roots-decoction used to treat tuberculosis. It is infrequent to rare in raised areas (spoil banks etc.) along the coastal areas of Louisiana and extending inward (FACU). The current reports include 16 parishes, mostly along the coast.

MAPLES AND BOXELDER - ***Acer*** - Aceraceae

The genus name, "*Acer*", is an old Celtic word meaning "hard", in reference to the wood. The leaves are opposite, simple, deciduous, and with palmate major veins. Note: Boxelder (***Acer negundo***) has compound leaves with pinnate major veins in the leaflets. The fruit is a samara with wings and in pairs. The wood of all hard maples (Sugar Maple, Southern Sugar Maple, and Chalk Maple) is heavy, strong, stiff, has a high resistance to shock and is very hard. Shrinkage during seasoning is large. Hard maple takes a stain satisfactorily and polishes well. Soft maple (Red Maple, Drummond Red Maple, and Silver Maple) is 25% less hard than hard maple but suited to most of the same uses and has the same finishing properties. The three hard maples (Sugar Maple, Southern Sugar Maple, and Chalk Maple) have leaf sinuses that are U-shaped. All three contain a fair percentage of sugar and the sap was gathered in late winter to make maple syrup, sugar, and water. However, this practice may not have been as common in the southern areas as in the north.

SOUTHERN SUGAR MAPLE (***Acer barbatum***) is also known as Florida Sugar Maple and Florida Maple. The specific epithet, "*barbatum*", meaning "bearded", is from the beard in the flower. It is a tree that reaches over 100 ft. The bark is whitish, smooth when young, and becoming darker and shallow-fissured later. The leaves are 1.5-3" long with short, acute or obtuse, and entire lobes. The leaves are dark green and shiny above, glaucous and pubescent beneath, and turning red to yellow in autumn. The bases of the lobes are U-shaped. The samaras are green to reddish winged, flattened, joined at the seeded base and divergent-winged toward the apices, hairy at first but becoming glabrous later, hardly over 0.75" long. The seed is solitary, lustrous, and green. The wood is hard, strong, light reddish brown and individual trees sometimes found with a curly grain, which makes the bird's-eye maple of commerce. The wood of Southern Sugar Maple is used for furniture, flooring, and shoe lasts. This species is the most common of the sugar maple group (***Acer barbatum, Acer leucoderme, and Acer saccharum***) in Louisiana and is usually found on well-drained slopes or calcareous areas with American Beech. The current records include 20 parishes. (syn. *Acer floridanum*)

Acacia farnesiana (USDA-FS)

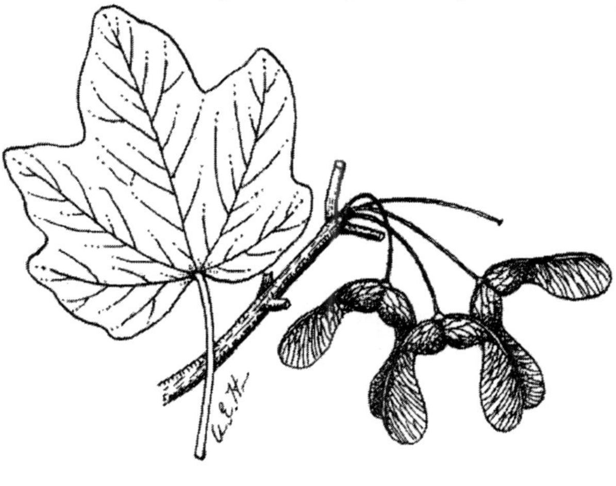

Acer barbatum (USDA-FS)

CHALK MAPLE (*Acer leucoderme*) is also known as Whitebark Maple. The specific epithet, "*leucoderme*", refers to the very white bark. Chalk Maple is a shrub or tree that reaches a height of 40 ft. The bark on young stems and large branches is smooth, light to dark gray or grayish brown. The trunk is very pale gray with narrow ridges, separated by deep fissures, and with closely appressed scales. Chalk Maple has small leaves and light colored bark like Southern Sugar Maple. However, the leaves are green beneath with pilose pubescence that feels like velvet, and the leaf lobes are blunt and not narrowed at the base. The bases of the lobes are U-shaped. The leaves appear with the flowers and are up to 6" wide. The small yellow flowers are found in April in nearly sessile corymbs. The samaras mature in September and are reddish brown, drooping, at first with pale hairs, and then later glabrous. Chalk Maple is sometimes planted as a shade tree in the South and has been cultivated since 1900. This species is apparently only in western Louisiana in similar habitats as Southern Sugar Maple with reports from seven parishes mostly west of the Mississippi River.

BOXELDER (*Acer negundo*) is also known as Boxelder Maple, Ash Leaved Maple, Maple Ash, Water Ash, Sugar Ash, Sugar Maple, Red River Maple, Black Maple, Manitoba Maple, and Three Leaved Maple. The specific epithet, "*negundo*," is obscure (used by John Ray in 1688). The common name indicates the resemblance of the foliage to that of the elders (***Sambucus***) and the whitish wood to that of Boxwood (***Buxus sempervirens***). Boxelder is a tree that reaches a height of 70 ft. The bark on young branches is bright shiny green, often purplish in winter, and bitter to taste. The bark on older trunks is thin, grayish to light brown, and deeply fissured. The leaves are compound (3, 5 or less commonly 7 leaflets). The lower leaflets are lobed, coarsely toothed, 2-4" long, pointed, smooth, shiny green above and paler beneath, and prominently veined. The leaves resemble Poison Ivy. The flowers are very small, without petals, yellow green, of two kinds on different trees, and in drooping clusters. The flowers appear just before unfolding of the leaves from March to May. The fruit is a pair of almost parallel winged achenes 0.75-1" long, reaching full size in early summer but not maturing until autumn (August to October). The fruits often remain on the tree after the leaves are shed. The wood is creamy white, soft, and light (27 lb per cu ft). The wood is of little value except for fuel, boxes, crates, woodenware, pulp, ties, and furniture. The seeds are eaten by evening grosbeak, purple finch, many other bird species and squirrels. Boxelder furnishes deer-browse. Beavers eat the bark. It was first cultivated in 1688. The pollen of this wind-pollinated species may cause hay fever. It is occasionally tapped for its sugary sap, which is inferior to that of Sugar Maple. It is widely distributed in Louisiana and is most abundant on the floodplain soils (FACW) but occurs also on better-drained terrace soils. It probably occurs in all parishes with records from 58 parishes currently documented.

DRUMMOND RED MAPLE (*Acer rubrum* var. *drummondii*) is often referred to as Swamp Red Maple. This variety of Red Maple is a large tree that reaches up to 60 ft. The bark is light gray and smooth when young, becoming darker, furrowed and flaky on old trunks. The twigs are dark red, shiny, and odorless. The leaves are 3 to lobed, broader than long, bright green above and dense wood-like hairiness below especially along the veins. The bases of the lobes are V-shaped. Flowers appear in the early spring, followed by showy fruit. The samaras are bright red, double and 1.5-2.5" long. Color varies from red to yellowish green. The tree is short-lived and the wood is weak. It is restricted to wet areas (OBL) throughout the state and is common in shallow swamps with Bald Cypress. It occurs in all parishes with records from 61 parishes presently documented. (syn. *Acer drummondii*)

Acer leucoderme (USDA-FS)

Acer negundo (USDA-FS)

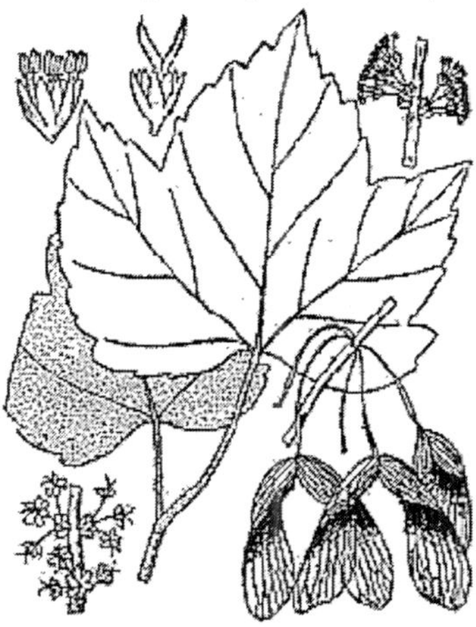

Acer rubrum var. ***drummondii*** (Britton and Brown 1943)

RED MAPLE (***Acer rubrum*** var. ***rubrum***) is also known as Scarlet Maple, Shoe Peg Maple, Soft Maple, and Hard Maple. The specific epithet," *rubrum*", means red for the flowers, fruit, and leaves. It is a small to medium sized tree that reaches 100 ft. The twigs are dark red, shiny, and odorless. The bark is light gray and smooth but getting darker and furrowed with age. The leaves are irregularly toothed, 2-6" long, light green above and lighter below and 3 to 5 lobed. The bases of the lobes are V-shaped. The leaves are brightly colored orange to red in the fall. It is the first tree to break winter's grasp on the forest. The bright red or yellowish green flowers are found from January to March before the leaves. The 0.75" long red or yellowish green double samaras are also found before the leaves in March to April and are on slender drooping stems. The fruit eventually turns brown and falls to the forest floor where it usually immediately germinates. The wood is heavy (38 lb per cu ft), close-grained, not strong, light reddish-brown, with thick lighter-colored sapwood. The wood is used for furniture, turnery, veneer, crates, boxes, flooring, slack cooperage, interior finish, novelties, fuel, gunstocks, woodenware, and wood pulp. Birds and squirrels eat the seeds. Cottontail and white-tailed deer eat the foliage. It is the state tree of Rhode Island. The flower is a valuable source of nectar and pollen in very early spring stimulation of bees. It is widely distributed in Louisiana in well-drained upland areas (FAC). It is absent from the coastal marsh and the wetter bottomland sites; the current records include 49 parishes.

TRIDENT RED MAPLE (***Acer rubrum*** var. ***trilobum***) is also known as Carolina Maple. The varietal name "***trilobum***" refers to the three lobed leaves. The bark is remarkably smooth even on old trunks, reminding one of American Beech. The leaves are thicker, almost leathery, a darker green and shining above, and coated below with a heavy bloom and with persistent cottony tufts in the axils of the main nerves which are very prominent beneath. The bases of the lobes are V-shaped. In autumn the foliage commonly turns yellow not crimson. The lobes of the leaves are usually three, rarely five, and rather broad, the two lateral ones somewhat divergent, with wide-angled sinuses between them. The teeth are not nearly so close or numerous as in the other Red Maples, except near the base, and are not especially sharp or fine. The flowers are often yellowish green. Trident Red Maple and Drummond Red Maple share the same habitats (OBL). There are only ten parishes with this variety but it is more widespread as many do not recognize this variety. (syn. *Acer rubrum* var. *tridens*)

SILVER MAPLE (***Acer saccharinum***) is also known as Creek Maple, Swamp Maple, White Maple, Soft Maple, River Maple, Water Maple, and Silver Leaved Maple. The specific epithet, "*saccharinum*", refers to the sweet sap. It is a large, rapid-growing tree that reaches 120 ft. The bark is gray and tends to flake and leave brown spots. The broken twigs have an unpleasant odor. The leaves are 3-6" across, very long petioled, deeply 5 lobed, with deep indentations that are V-shaped, lobes acuminate, remotely dentate, green above, silvery white beneath, and pubescent when young. The flowers appear before the leaves in February or March. They are greenish yellow or red, of two kinds, and in nearly sessile lateral corymbs. The fruits are on slender drooping pedicels, pubescent when young, and with large divergent wings. They appear in April or May. The wood is soft, close-grained, even-textured, pale brown with thick sapwood, and weighs 33 lb per cu ft. The wood is used for furniture, turnery, woodenware, fuel, veneer, crates, boxes, flooring, slack cooperage, interior finish, and novelties. The tendency of this fast growing tree to develop cavities is useful for nesting and shelter for raccoon, opossum, squirrels, owls, woodpeckers, and other large birds. The fruit is eaten by a number of species of birds, and by squirrels. The sap is sometimes used in the manufacture of syrup and sugar. It has been cultivated since 1725. It is usually found along streams in the floodplain (FACW) but has been widely planted in Louisiana. The current distribution is in 26 parishes.

Acer rubrum var. ***rubrum*** (USDA-FS)

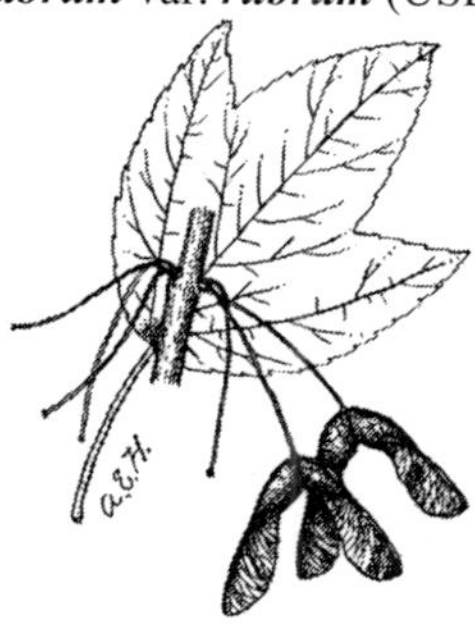

Acer rubrum var. ***trilobum*** (USDA-FS)

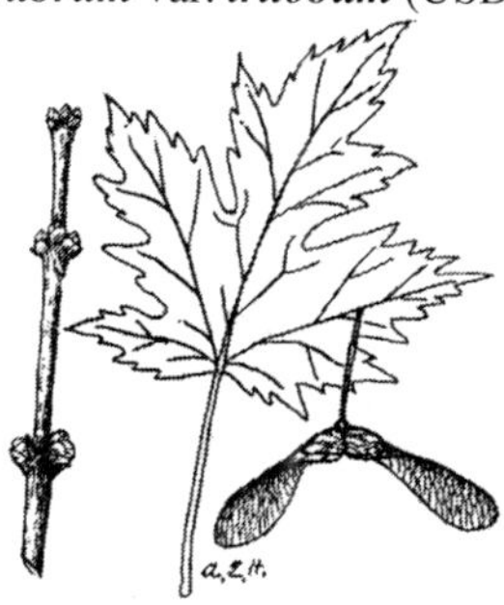

Acer saccharinum (USDA-FS)

SUGAR MAPLE (*Acer saccharum*) is also known as Hard Maple, Rock Maple, Sweet Maple, Black Maple, Sugar Tree, and White Maple. The specific epithet, "*saccharum*", refers to the sugar content of the sap. It is the state tree of New York and Vermont. Sugar Maple is a large tree and reaches a height of 100 ft. The bark on young trees is thin and grayish-brown. On old trunks, it becomes much darker and deeply furrowed. The leaves, with 3 to 5 lobes, are thin, bright green above, downy below, and are 1.5-3" long and wide. The leaves turn yellow and scarlet before falling in fall. The upper surface of the leaf is usually glossier than Drummond Red Maple and the underside is usually not as whitish. The tiny green flowers come before the leaves, or with their unfolding, and are borne in dainty, drooping clusters. The fruit is green, borne in pairs, with the usual maple wings. It turns light reddish-brown in late spring when it matures. The fruit is similar to Drummond Red Maple but not as colorful or showy. The wood is light brown tinged with red, heavy (44 lb per cu ft), and hard. The wood is used for flooring, furniture, and veneer. Individual trees can produce the bird's eye and curly grains used in furniture. The seeds, buds, and flowers provide food for many kinds of birds and other animals. Squirrels eat the seeds, frequently storing them in caches after removing the hull and wing. Birds use the leaves and seed stalks commonly in nest building. It has been in cultivation since 1753. It is infrequent in Louisiana in well-drained upland areas (FACU-). The current record includes 24 parishes

RED BUCKEYE - *Aesculus pavia* - Hippocastanaceae

Red Buckeye is also known as Scarlet Buckeye, Wooly Buckeye, Firecracker Plant, and Fishpoison Bush. The genus, "*Aesculus*", is from the Latin name for an oak with edible acorns. The specific epithet, "*pavia*", honors Peter Paaw (Petris Pavius), a Dutch botanist who died in 1616. It is a shrub but does rarely make a tree up to 30 ft tall and 10" diameter. The bark is gray to brown, smooth at first with age breaking into short fissures. The leaves are deciduous, opposite, and palmately compound. The five (rarely 3 or 7) oval to obovate leaflets are dark green above, with velvety hairs beneath, are conspicuously veined, have irregular toothed margins, and are 3-6" long. The petioles are often red and 3-7" long. The leaves turn yellow and fall from the plant early in the season; the plants may be naked or partially naked before the end of August. The very showy upright terminal cluster of dark red flowers, each about 6-10" long, is found from March to April. The large shrub begins flowering while still as small as three ft. The fruits are 1 to 3 seeded capsules with a pale brown, smooth, leathery cover. The globe like capsule is up to 2.5" across. The seeds are large and brown and with a distinct eye. The wood is soft and light. It is seldom large enough for use. The flowers are used by hummingbirds. Squirrels occasionally eat the fruit in the "soft" stage apparently without fatal consequences. Deer make very little use of the foliage as browse. The powdered bark is used for toothaches and ulcers. The root is used for washing clothes. The crushed fruits are used as fish poison. Young shoots, young leaves, and fruit are poisonous to humans and livestock. The Red Buckeye is the state tree of Ohio. Some people consider it good luck to carry the buckeye seed in your pocket. It is widely distributed in the state mostly on well-drained soils and is absent from the coastal marsh and the Mississippi and Red River and other large floodplains (FAC). It is currently reported from 46 parishes.

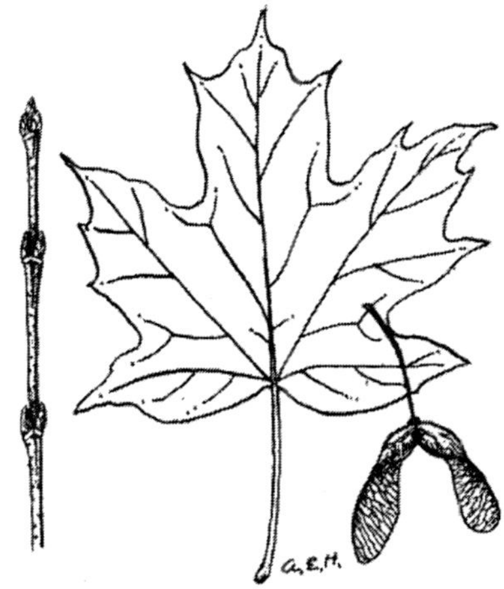

Acer saccharum (USDA-FS)

Aesculus pavia (USDA-NRCS)

TREE OF HEAVEN - *Ailanthus altissima* - Simaroubaceae

The genus name, "*Ailanthus*," is from a Chinese name, *Ailanto*, meaning "Tree-of-Heaven," in reference to its height. The specific epithet, "*altissima*", means "very tall." Other common names include Copal, Tree of the Gods, Chinese Sumac, Heavenward Tree, False Varnish Tree, and Devil's Walkingstick. This species is a tall tree to 100 ft with pale grayish brown bark with shallow fissures. The leaves are alternate, deciduous, and odd pinnately compound. The leaflets are oblong to lanceolate, sometimes asymmetrical, and entire except for a few teeth basally. The crushed leaflets have an unpleasant odor. The flowers are yellowish green and produced in large panicles. The fruits are samaras that are elliptic often curved and with the seed located centrally. William Hamilton introduced the tree into the United States from China in 1784. The wood is pale yellowish brown, medium hard, not durable, weak, coarse, and open-grained. The wood is used for cabinets but does make fairly good fuel. It was once used in erosion control work in the dune areas of the Black Sea, and was planted for timber in New Zealand. The seeds are eaten by a number of birds, including the pine grosbeak and crossbill, and occasionally browsed by white-tailed deer. The tree is sometimes planted in China as host to a species of silkworm, *Attacus cynthia*, which produces a coarse inferior silk. In China, the powdered bark is reported to be effective against tapeworms and a popular remedy for dysentery. The root bark is used to treat asthma, heart ailments, and epilepsy. The chewed bark is reported to produce nausea, vertigo, weakness, shivering, and cold sweats. In Louisiana, this plant persists and perhaps spreads from abandoned home sites. The records for this species are widespread throughout the state. It is currently reported from 16 parishes.

MIMOSA - *Albizia julibrissin* - Fabaceae

Mimosa is a native of Asia and is sometimes called the Silk Tree, Pink Siris, and Pink Mimosa. The genus, "*Albizia*", honors F. Degli Albizzi, an Italian nobleman and naturalist, who introduced it. The specific epithet, "*julibrissin*", is a form of the Persian name for this species. It is a small flat-topped tree with spreading branches that is often wider than tall. The tree may reach 40 ft but is often shorter. The bark is smooth and blotched with gray. The fern-like leaves are alternate, deciduous, 8-15" long, and bi-pinnately even compound. The leaflets are dull green, asymmetrical, and about 3/8" long. The pink flowers are in head-like clusters and are found from May to August. The fruit is a 5-8" long and 1" wide legume. The seeds are of limited value to birds, squirrels, and other wildlife. Mimosa furnishes low quality deer-browse. Hummingbirds use the flowers. The wood is used for cabinetmaking in Asia. Mimosa was first introduced into cultivation in 1745. The tree is not very aggressive but does escape spread to a limited extent. It is widely distributed in Louisiana and has probably escaped or is persisting in all 64 parishes.

ALDER - *Alnus serrulata* - Betulaceae

Alder is also known as Hazel Alder, Smooth Alder, Tag Alder, and Common Alder. The genus name, "*Alnus*", is the classical name of the alder, and the specific epithet, "*serrulata*", refers to the finely toothed leaves. It is a deciduous shrub or small tree up to 15 ft with a crooked trunk and young twigs bearing dense, brown pubescence. The bark is brown and marked by dark short lenticels. The leaves are alternate, simple, deciduous, and with pinnate major veins. The blades are elliptic to obovate, upper surfaces green and lower surfaces brownish. The margins are slightly wavy and unevenly toothed. The veins are conspicuously depressed on the upper surfaces of leaves and protruding on the lower. The male flowers are tiny, green to reddish, and borne in conspicuous, pendulous catkins that are produced in the fall but do not open until late winter. The female flowers are borne in tiny, deep red, cone-like catkins. The fruits are nuts borne in a structure reminiscent of a tiny pinecone. The wood is light brown, soft, brittle, and

Ailanthus altissima (USDA-FS)

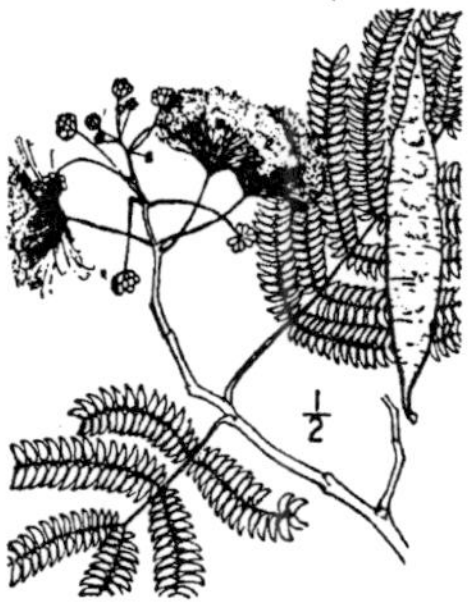

Albizia julibrissin (Britton and Brown 1943)

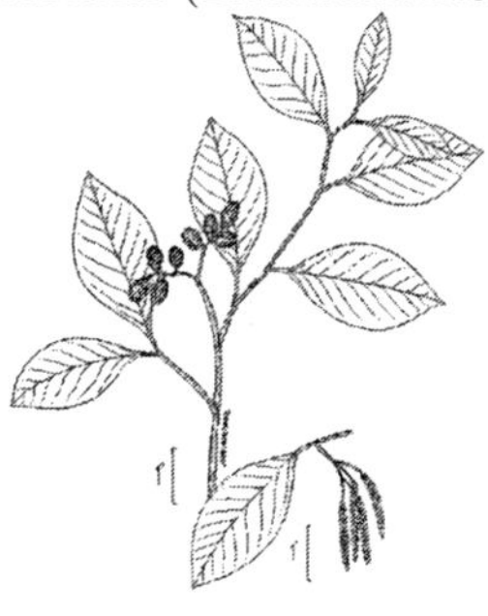

Alnus serrulata (USDA-NRCS)

weighs 29 lb per cu ft. The inner bark can be ground into crude flour in an emergency. Also, the bark yields tannic acid and is an astringent. It was formerly used in the treatment of intermittent fever, relieve pain of childbirth, blood tonic, emetic, purgative, coughs, fevers, to stimulate kidneys, to bathe hives or piles, for eye trouble, indigestion, jaundice, heart trouble, toothaches, to lower blood pressure, and to clear milky urine. Deer eat the twigs but it is not a favorite food. The seeds are eaten by several bird species including the goldfinch. Beaver eat the bark. Alder is found forming thickets in swamps, wet woods, baygalls, and other wet areas (FACW+) with an acid pH in the pine forest regions of the state. It is notably absent from the Mississippi and Red River and other large floodplains. It is a species that seems to prefer disturbed and other open areas and is usually along the edges of baygalls not in the central portion (personal observation by the late Robert Murry that was verified with research (Allen et al 1990). It is currently reported from 23 parishes.

SERVICE BERRY - *Amelanchier arborea* - Rosaceae

The genus name, *"Amelanchier"*, is derived from the French name of a European species. The specific epithet, *"arborea"*, refers to the treelike character of this species. Other common names include Shadblow Service Berry, Boxwood, Billberry, June Plum, Indian Cherry, Swamp Shadbush, Indian Pear, Juice Pear, Sugar Pear, Plum Pear, and Berry Pear. This is a small tree or shrub seldom taller than 25 ft. The bark is gray to black and smooth but becoming shallowly fissured with scaly longitudinal ridges with age. The leaves are alternate, simple, deciduous, and with pinnate major veins. The blades are oval or obovate, slightly cordate at base, and with finely serrate margins. The flowers are very fragrant and in racemes with five green sepals, five white petals, and 15-20 stamens. The fruit is a reddish purple pome. It is occasionally cultivated in gardens for the showy white flowers. The berries may be eaten uncooked or made into pies. This plant is a valuable wildlife plant, its fruit bearing eaten by at least 35 species of birds, and its foliage browsed by cottontail and white-tailed deer. The wood is brown, close grained, hard, strong, tough, and elastic, weighs 49 lb per cu ft and is sometimes used for making handles. This species inhabits well-drained soils and is found throughout the pine regions of the state on well-drained soils (FACU). It is currently reported from 25 parishes.

LEAD PLANT, FALSE INDIGO - *Amorpha* - Fabaceae

The genus, *Amorpha* is from the Greek, "amorphous" (deformed), meaning the flowers have only one petal. The genus includes shrubs to small trees. The leaves are alternate, deciduous, and odd pinnately compound. The flowers consist of the purplish blue standard only and are in racemes. The fruit is a glandular dotted legume.

DESERT FALSE INDIGO (***Amorpha fruticosa***) is also called Lead Plant, Indigo Bush, Bastard Indigo, False Indigo, and Dull Leaf Indigo. The specific epithet, "fruticosa", means shrubby. It is a shrub to 18 ft. The twigs are brown to gray and sparingly pubescent. The leaves are usually ascending and 2.5-10" long. The 9-27 leaflets are oval or elliptic to oblong, 0.5-1.5" long, apex obtuse to rounded and often mucronate, base rounded or cuneate, upper surface dull green and glabrous, lower surface pubescent to glabrous with glandular dots, and margins entire. It has been in cultivation since 1724. Quail and a few other birds eat the seeds. Edges of streams, ponds and other moist areas throughout the state, especially in the large floodplains (FACW), usually on clay soils. The current records include 46 parishes. (syn. *Amorpha bushii, Amorpha croceolanata)*

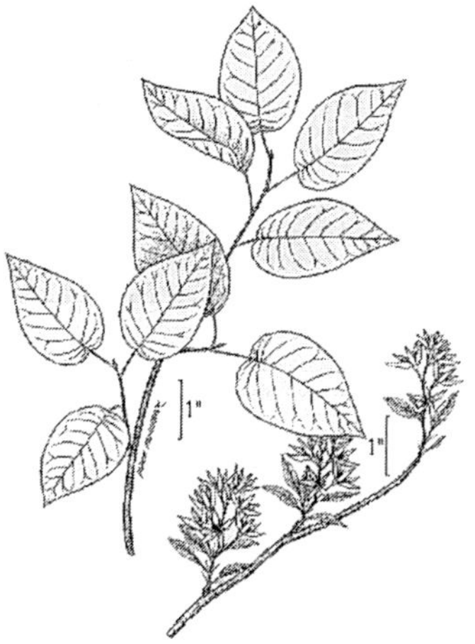

Amelanchier arborea (USDA-NRCS)

Amorpha fruticosa (USDA-NRCS)

SMOOTH LEAD PLANT (***Amorpha laevigata***) is similar to ***Amorpha fruticosa*** and is questionable in Louisiana. The specific epithet, "*laevigata*", refers to the smooth glabrous leaves. The major characteristic is the smooth glabrous plant. Isely (1990) rejects this species.

SHINING FALSE INDIGO (***Amorpha nitens***) is also called Georgia Amorpha and Shining Indigo Bush. It is similar to ***Amorpha fruticosa*** and differs in having lustrous leaflets that turn black on drying. Due to the limited number of specimens, the habitat is difficult to ascribe but it is listed as FAC for a wetland ranking. The current distribution in Louisiana includes Acadia, Plaquemines, and St. Tammany parishes but it could be more wide spread.

PANICLED FALSE INDIGO (***Amorpha paniculata***) is also called Panicled Amorpha and Panicled Indigo Bush. It is very similar to ***Amorpha fruticosa*** but differs in being a more robust plant, having larger leaflets with the veins strongly reticulate and in the racemes being terminal in distinct clusters like a panicle. Edges of streams and hardwood forests (FACW) in parishes west of the Mississippi River. It has been recorded in central and northern Louisiana from eight parishes.

PEPPER VINE - *Ampelopsis* - Vitaceae

This genus includes woody vines with tendrils and alternate deciduous leaves. The pith is white and continuous through nodes. The genus, "*Ampelopsis*", is from the Greek "ampelos" (vine) and "opsis" indicating resemblance. The flowers are in cymes rather than in panicles like the grapes. The fruit is a berry with differing opinions as to edibility; some report the fruits as poisonous and others as edible.

PEPPER VINE (***Ampelopsis arborea***) is also known as Ampelopsis, Pinnate Leaved Ampelopsis, Cow Vine, and Wild Sarsaparilla. The specific epithet, "*arborea*", refers to its climbing habit in trees. It is an attractive sprawling or climbing vine with double or triple odd pinnate compound leaves. The leaflets are in groups of 3 to 5, with large teeth and are dissected to the leafstalk into separate lobes. The leaves at the ends of the branches are red; especially so in early spring. Some plants display a two-colored leaf in early spring: reddish and cream or tan. The flowers are very small, greenish white or yellowish and in cymes on long stalks from the main stem of branchlets and are found from June to July. The berries are in compact groups, green at first, then white to pink, and finally, shiny black when mature, usually with small raised dots on the outside. The fruits are produced from July to October. It is found in a wide variety of habitats from upland fencerows on moist sites to winter-flooded bottomlands throughout Louisiana (FAC+). It occurs in all 64 parishes with current records from 63 parishes.

HEART LEAF PEPPER VINE (***Ampelopsis cordata***) is also known as False Grape, Simple Leaf Ampelopsis and Heart Leaf Ampelopsis. The specific epithet, "*cordata*", refers to the cordate leaf. This plant is sprawling or climbing but tendrils may be few or even lacking. The heart-shaped or flat-based leaves are simple with palmate major veins. The flowers appear from May to July. The fruits are dry and in thick groups on long, green shoots from the main stem, which is tan, thick and woody. As the fruits ripe, they turn white, pink, or red and finally, a deep metallic blue. All of these colors are often present at the same time making this a colorful and attractive plant. The fruits appear from August to September. Quail and other birds eat the fruit. This species is also widely distributed in many habitats but is usually in more disturbed habitats than Peppervine. It is sometimes locally abundant in open areas along rivers, sandbars, and fencerows (FAC+). It is currently reported from 49 parishes.

Amorpha nitens (Vines 1960)

Amorpha paniculata (Vines 1960)

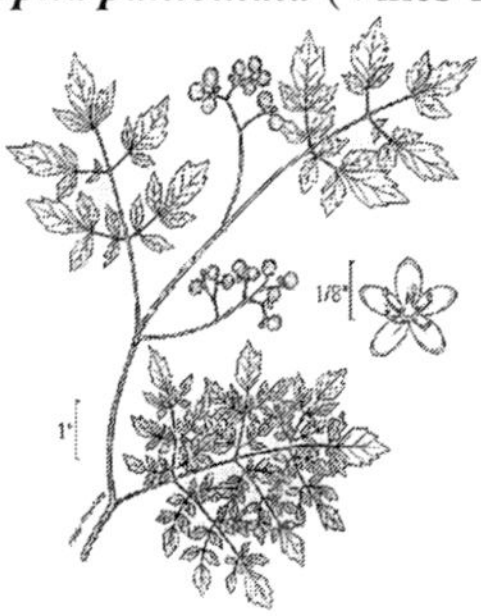

Ampelopsis arborea (USDA-NRCS)

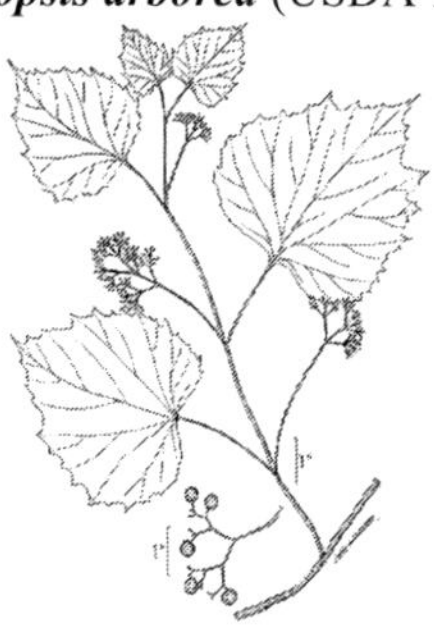

Ampelopsis cordata (USDA-NRCS)

DEVIL'S WALKINGSTICK - ***Aralia spinosa*** - Araliaceae

Devil's Walking Stick is also known as Shot Bush, Hercules Club, Angelica Tree, Mississippi Hoe Handle, Spikenard Tree, Prickly Ash, Prickly Elder, Tear Blanket, Toothache Tree, Pigeon Tree, and Pick Tree. Origin of the genus name, "*Aralia*", is from the French-Canadian Aralie, the name appended to the original specimens sent to Tournefort by the Quebec physician, Sarrasin. The specific epithet," *spinosa*", refers to the prickles on the stem. It is a tall armed shrub or small tree that reaches up to 40 ft tall. The bark is dark brown with ridges and irregular orange to brown prickles with the inner bark yellow leaf scar abundant and conspicuous. Most trees are palm-like with leaves concentrated near the top of the plant. The leaves are alternate, deciduous, large to 4 ft long and wide, two or three pinnate, and with prickles. The oval leaflets are toothed, pointed, dark green above, paler beneath, and turn yellow in autumn. The base of the petiole clasps the stem. After the leaf falls, a conspicuous scar is created. The flowers are yellow and white, small, in dense umbellate clusters, arranged in a panicle above the leaves and are found from July to August. The fruit is a black, egg-shaped, 1/6" diameter drupe that is found from September to October. The tree presents a striking appearance in autumn, when the fruit ripens and the inflorescence branches turn a rich, wine-red. The wood is soft, light, weak, close-grained, and brown streaked with yellow and with thin, lighter colored sapwood. It is used for small shop articles, like button boxes, photograph frames, pen racks, stools and arms for rocking chairs. Often a large hollow pith column occupies the center of the trunk. The seeds are eaten by many birds, fox, skunk, squirrels, and rodents and the leaves are browsed by white-tailed deer. The bark roots and berries are used for medicine, as a stimulant, or diaphoretic. The entire plant was used in medicines of earlier times. Early settlers used the aromatic and spicy roots for home remedies and it was thought to be a cure for toothaches. The Choctaw Indians used it for boils and sore eyes. Occasionally used as a grotesque ornamental in the Victorian era, it has been cultivated since 1688. The large taproot makes transplanting large specimens difficult. The fruits are edible by man and many animals. It is widely distributed in Louisiana on well drained to moist sites (FAC) and is absent from the marsh areas. It is currently reported from 45 parishes.

HEN'S EYES - ***Ardisia crenata*** - Myrsinaceae

Hen's Eyes is also known as Christmas Berry, Coral Berry, and Coral Ardisia. The genus name, "*Ardisia*", is Greek from Schefflera ardis (a point) referring to the pointed antlers. The specific epithet "*crenata*" is for the crenate margins of the leaves. It is a small shrub to two ft tall. The leaves are simple, alternate, evergreen, and with pinnate major veins. The margins are crenate and the upper surface is dark green while the lower surface is a paler green. The flowers are white and the fruit is a red drupe. It is a native of China and is cultivated and escaping and/or persisting primarily in south Louisiana. It is currently reported from six parishes. (Not Illustrated)

DUTCHMAN'S PIPE VINE - ***Aristolochia tomentosa*** - Aristolochiaceae

Dutchman's Pipe Vine is also known as Woolly Dutchman's Pipe Vine and Pipe Vine. The genus name, "*Aristolochia*", is from its supposed medicinal properties. The specific epithet, "*tomentosa*", refers to the tomentose leaves and stems. It is a perennial high climbing vine with thickened rootstocks. The stems are twining, gray to brown or black, densely hairy, and up to 75 ft. The leaves are alternate, simple, deciduous and with palmate major veins. The blades are broadly ovate, pointed at the tip, heart-shaped at the base, without teeth, densely soft-hairy, up to 8" long, and up to 6" wide. The flowers are usually solitary or paired in axils of the leaves, hairy, yellow and purple, and very symmetrical. They are found from April to June. The

Aralia spinosa (USDA-FS)

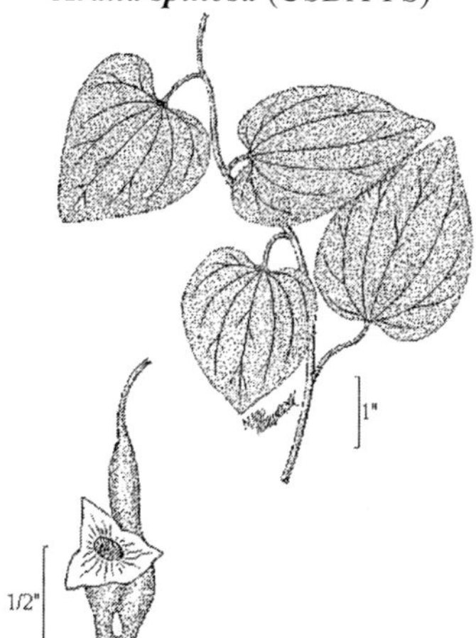

Aristolochia tomentosa (USDA-NRCS)

flowers are somewhat reminiscent in shape to a miniature version of Sherlock Holmes's pipe (hence the common name). Flies trapped at the bottom of the floral tube pollinate the flowers. The fruit is a capsule that is oblong-cylindric, up to 4" long, with numerous flat, triangular seeds arranged in layers. The fruits are found from September to October. The aromatic rootstock can be used for seasoning. It is often cultivated for ornament and is the host for the caterpillar of the pipevine swallowtail butterfly. It is found in shaded woodlands, along streams, bottomlands, slopes, and into moist upland sites (FAC). It is rare to infrequent and has been reported from only nine parishes.

BAMBOO - *Arundinaria gigantea* - Poaceae

BAMBOO (***Arundinaria gigantea***) is native and is also known as Giant Cane, Switch Cane, and Southern Cane. It is a monocot (not true wood). The genus "*Arundinaria*" comes from the Latin word "arundo", which means "reed". The specific epithet "*gigantea*", is for its unusually tall or large habit. There are two subspecies in Louisiana, ***gigantea*** and ***tecta***. Bamboo is a rhizomatous perennial forming extensive stands. It grows up to 30 ft in erect culms in dense thickets or canebrakes of tough, woody stems. The leaves are elliptical to lance-shaped, glabrous to pubescent, alternate, and parallel veined. The margins are scabrous and ciliate. The base of the leaf clasps the hollow stem. The flowers (panicle, raceme, or solitary spikelet) are found sporadically from January to March. The fruit is a caryopsis and found from April to May. All the cane plants of the same species in a given area flower, set seed and die together. Within a four to five year period, all the individuals of the same species throughout the world do the same. In the period between flowerings, this species multiplies by sending up new shoots from underground rhizomes, creating the characteristic thickets of cane. The young shoots can be prepared like bamboo shoots. The large seeds can be used as a cereal or ground into flour. Indians and early settlers ate the seeds. The stems are used for fishing poles, mats, and whistles. Livestock eat the shoots and fruit. The ssp. ***gigantea*** is the typical one and is widely distributed in the state in moist to somewhat well drained soil with reports from 58 parishes. The ssp. ***tecta*** is often called Switch Cane and is restricted to wet seepage areas of extreme southeastern Louisiana (St. Tammany, Tangipahoa, and Washington parishes). The best character to separate the two species is the presence of a ring of air spaces in the rhizome of ssp. ***tecta*** that is not found in ssp. ***gigantea***. Both are rated as FACW but ssp. ***tecta*** seems to occupy wetter sites with the air spaces in the rhizome indicating so.

PAWPAWS - *Asimina* - Annonaceae

The genus includes shrubs to small trees. The genus name, "*Asimina*", is from the early French name Asiminier, which in turn was derived from the Indian Arsimin. The plants have rusty pubescent twigs with chambered pith and buds that are soft, velvety, chocolate brown, and non-scaly. The leaves are simple, alternate, deciduous, and with pinnate major veins. The blades are obovate, acuminate, base cuneate, short petiolate, and with entire margins. The crushed leaf has a strong odor that has been described as smelling like green peppers or an oil well. The flowers usually appear before leaves from February to April. The flowers are solitary with three green sepals and six purple or purplish green petals. The fruit is a greenish-yellow to dark brown berry with several flattened seeds. The caterpillar of the zebra swallowtail lives only on the leaves of pawpaw. Many birds and other wildlife eat the mature fruit.

Arundinaria gigantea ssp. ***gigantea*** (Hitchcock 1950)

Arundinaria gigantea ssp. ***tecta*** (Hitchcock 1950)

DWARF PAWPAW (***Asimina parviflora***) is also called Small Fruited Pawpaw, Small Custard Apple, Custard Banana, and Small Flowered Pawpaw. The specific epithet, "*parviflora*", refers to the small flowers. It is a small shrub or small deciduous tree to 12 ft. The leaves are up to 6" long but generally smaller. The new leaves on flowering branches are often held somewhat erect and only marginally resembling mature leaves. The fruit is about the size of peanuts in the shell (3" or less). They are found in clusters of up to five and appear from June to September. The bark was at one time used in domestic medicine as a bitter. It is found mostly in dry sandy upland areas (FACU) and is notably absent from the Red and Mississippi River floodplains. The current reports include 22 parishes.

PAWPAW (***Asimina triloba***) is also known as Custard Apple, Fetid Shrub, Indiana Banana, Wild Banana, and Arkansas Banana. The specific epithet, "*triloba*", refers to the petals which are in sets of three. It is a small tree that does not reach over 40 ft high. The bark is rather thin, smooth, gray brown in younger stems becoming somewhat fissured on older trunks. Large, ash-colored spots mark the smoother trunks. The leaves are large generally 8-10" long. The fruit is from 2 to 5" long, about 1.5" in diameter and found from July to October and soft when ripe. The flesh is edible, nutritious, and considered delicious by some. It may be baked into pies, made into dessert, made into a jelly, or eaten raw with cream as breakfast food. The wood is light (25 lb per cu ft), soft, weak, coarse-grained, spongy and is seldom used except for poles. The rough bark is sometimes used as a rope substitute and woven into fiber cloth. Deer are reported to browse the foliage and beaver the bark. It often grow in thickets because of the development of root suckers hence the term "pawpaw patch". It was first recorded by the lower Mississippi Valley Expedition of DeSoto in 1541. It is reported that Lewis and Clark survived on this fruit on their return trip from the west. Research has isolated a powerful chemical found in the tree that is one million times as potent as the widely used cancer drug, Adriamycin. The seeds of the Pawpaw contain an alkaloid, asimine, which is reported to have emetic properties and bark contains the alkaloid, analobine. Its habitat ranges from well drained beech upland forests to the moist edges of ponds and streams (FAC). The current records include 44 parishes.

BLACK MANGROVE - *Avicennia germinans* - Verbenaceae

The genus name, "*Avicennia*", honors Abdallah Ibn Sina, Arabian philosopher and scientist. The specific epithet, "*germinans*", refers to the viviparous nature of the seeds of germinating while still attached to the plant. Other common names include White Mangrove or Mangle Blanco. Black Mangrove is a shrub to small tree with dark brown bark on older stems and green on younger stems. In more tropical areas, this is the plant that forms the branching roots or pneumatophores. The leaves are opposite, simple, evergreen, and with pinnate major veins. The petioles are short and the blade is elliptic, oblong, or obovate and entire. The upper surface is glabrous and shiny while the lower surface is densely pale pubescent. The flowers are in spikes and are fragrant. There are 4-5 sepals, 4-5 creamy white petals with a yellow throat, and four stamens. The fruit is a capsule with two valves. The wood is dark brown, close grained, and hard. The bark is reported to be used for tanning. This species is restricted to the sandy tidal flats and lagoons of coastal Louisiana (OBL). It is currently reported from Jefferson, Lafourche, Plaquemines, St. Bernard, and Terrebonne parishes.

BACCHARIS - *Baccharis* - Asteraceae

The genus name, "*Baccharis*", is possibly for Bacchus the Greek God of Wine. This genus includes shrubs with alternate, simple, deciduous to evergreen leaves with pinnate major veins. The plants are dioecious with male flowers on one plant and the female flowers on a different

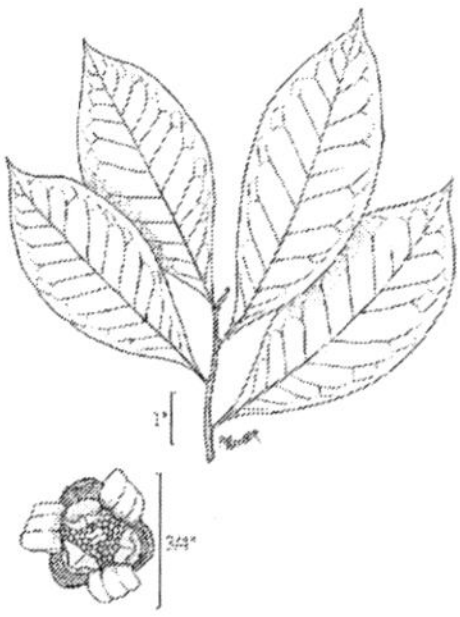

Asimina parviflora (USDA-NRCS)

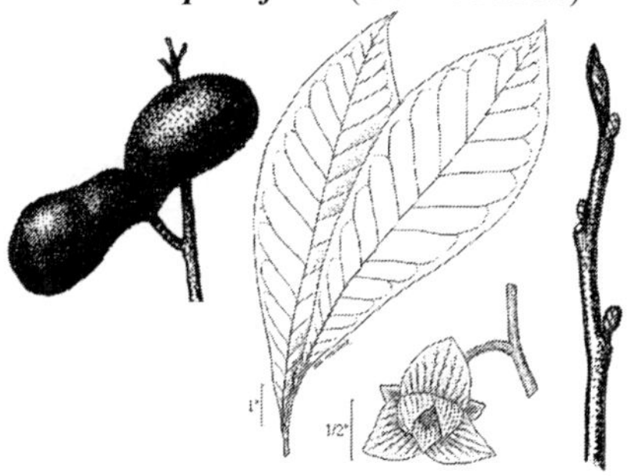

Asimina triloba (USDA-FS and USDA-NRCS)

Avicennia germinans (USDA-FS)

plant. Baccharis usually flowers from September to November. Both male and female flowers are in heads with disc flowers only that have a yellowish corolla. The outstanding flower characteristic perhaps is the large and distinct capillary bristle pappus. The male flowers produce an odor of rotten honey. The fruit is an achene that is ten ribbed.

EASTERN BACCHARIS (***Baccharis halimifolia***) is also known as Salt Bush, Silverling, Eastern Groundsel, Cottonseed Tree, Consumption Weed, Sea Myrtle, Bush Groundsel, Groundsel Tree, and Ploughman's Spikewood. The specific epithet, *"halmifolia"*, refers to the Halimus-like leaves, with which there is a marked similarity to those of the orach plant, at one time known under the name *Halimus*. It is a shrub that reaches 10 ft with toothed, wedge-shaped, leaves. The twigs are green and angled. It is usually found on roadsides, ditches, near streams, open woodlands, and wet areas (FAC) and is one of the most common and aggressive shrubs. It is thought that the original range of this species was only along the coast but was spread inland with the movement of oil drilling equipment. The seeds are dispersed by the wind and this species is very commonly the first pioneer species in early secondary succession throughout the state but especially in the non-pine regions. It is found in all with current records from 62 parishes.

SALTWATER FALSE WILLOW (***Baccharis angustifolia***) is also known as Narrowleaf Silverling, Narrowleaf Baccharis, False Willow and Salt Bush. The specific epithet, *"angustifolia"*, refers to the narrow leaves. This shrub reaches a height of 12 ft. The leaves are linear to needle-like, 1/4" wide (narrower than ***Baccharis halmifolia***), dark shiny green, and entire (sometimes slightly toothed). The twigs are resinous, as are the leaves sometimes. This species is restricted to the coastal marsh area (FACW+) and is infrequent to rare and reported from only six coastal parishes.

SALT WORT - *Batis maritima* - Bataceae

Salt Wort is also known as Maritime Salt Wort, Sea Sapphire, Beach Wort, Lechuga de Mar, and Barrilla. The genus name," *Batis*", is from a Greek word meaning prickly roach, and refers to the fruit. The specific epithet," *maritima*", refers to its habitat close to the sea. It is a low prostrate shrub of sandy beaches, mud flats, and salt marshes. The crushed stems and leaves have a strong odor. The stems are usually creeping, but occasionally sending up shoots as much as 6 ft in height. The leaves are opposite, simple, evergreen, entire, and succulent. The flowers are small, yellow and borne on a fleshy cone-like structure (actually a spike) in the leaf axis and appear from April throughout the summer. The fruits are green, fleshy, rounded, cylindrical, and rough surfaced. The succulent leaves crunch when trodden upon. They have a salty flavor and have been used in salads, in domestic medicine for treating ulcers, and as a diuretic. The ashes of the burned plant have been used in the West Indies in the manufacture of soap and glass. This species is infrequent along the coast (OBL) and is reported from seven coastal parishes.

RATTAN VINE - *Berchemia scandens* - Rhamnaceae

Rattan Vine is also known as Blackjack Vine, Supple Jack, and Alabama Supple Jack. The genus name, *"Berchemia"*, honors Berthout van Berchem, eighteenth century Dutch botanist. The specific epithet, "*scandens*", refers to the climbing or leaning habit. It is a woody, high climbing vine that climbs by twining. It often wraps tightly around trees and can create indentations in the bark producing odd-shaped trunks. The vine can girdle and in time actually kill a tree. The bark is smooth and when young is green with gray to brown-colored streaks and changes to black or red as it gets older. The leaves are alternate, simple, deciduous, and with pinnate major veins. The blades are dark green with the veins impressed above, and lateral veins distinctly parallel.

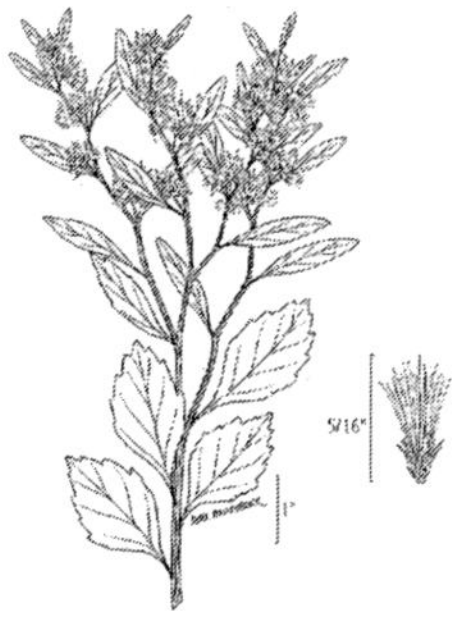

Baccharis halimifolia (USDA-NRCS)

Baccharis angustifolia (USDA-NRCS)

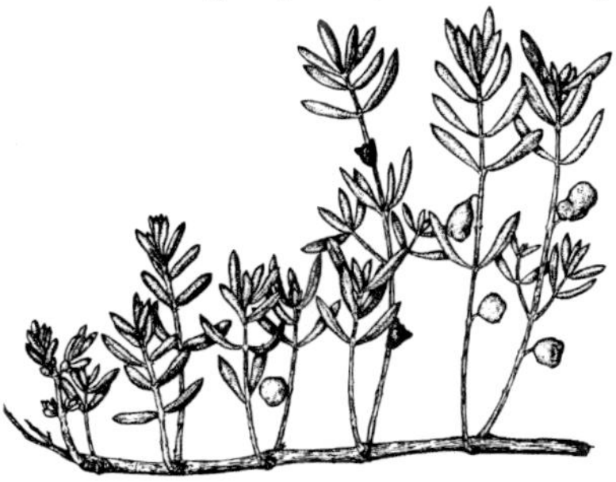

Batis maritima (Vines 1960)

Berchemia scandens (USDA-NRCS)

The flowers are in green to yellow panicles and are produced from March to June. The fruit is a bluish black, slightly flattened, oblong drupe which is produced from July to October. The vines are used for wickerwork. The Indians used bark or leaf teas as a "blood purifier" and to restore vigor and sexual vitality. The fruit is used by 17 species of birds including turkey, bobwhite, quail, and duck. Deer are reported to browse this vine. This vine is found throughout the state in a variety of habitats (FACW). It is currently reported from 61 parishes, but probably occurs in all parishes.

RIVER BIRCH - *Betula nigra* - Betulaceae

River Birch is also known as Red Birch, Black Birch, and Water Birch. The genus name "*Betula*" is the ancient classical name of the birch, and the specific epithet, "*nigra*", means "black", but no reason can be found for the name. River Birch is a small to medium-sized tree that reaches 90 ft. It is native to low, sandy wet areas. It has thin reddish bark on young trees soon becoming white or salmon-pink and papery; ultimately gray to gray-brown and coarsely scaly. At any season, River Birch may be easily identified by its bark, which hangs in loose, papery flakes, giving the trunk an odd, ragged appearance. When these tags are pulled off, the bark underneath is satiny smooth and beautifully tinted cream, tan and pink rose. One can write on the papery sheets of bark. The leaves are alternate, simple, deciduous, 2-3" long, oval, with pinnate major veins, and with doubly toothed edges. The upper surface is dark green and the lower much lighter. The leaves are elm-like but with flat "V" shaped bases. The flowers of River Birch come in the winter in drooping catkins. The female flowers are much shorter. The fruit is a samara that is produced in a cone-like structure. The wood is light (36 lb per cu ft), hard, strong, close-grained, and is used chiefly in turnery and the manufacture of woodenware and sometimes for wagon hubs. Because of its rather scattered distribution, it is not very important commercially. Birds sometimes eat the seeds and the foliage is browsed by white-tailed deer. Native Americans are reported to have used river birch to treat dysentery, colds, and milky urine. It has been cultivated since 1736. River birch is widely distributed in Louisiana primarily on the sandy soils of the natural levees of streams (FACW); it is notably absent from the Mississippi floodplain and the marsh. Since its popularity as an ornamental has increased, river birch can be found in planted situations virtually throughout the state. It is currently reported from 31 parishes.

CROSS VINE - *Bignonia capreolata* - Bignoniaceae

Cross Vine is also known as Trumpet Flower, Tendriled Trumpet Creeper, Smoke Vine, and Quarter Vine. It is a woody, high climbing vine via tendrils with discs on the end. "*Bignonia*" honors Abbe Jean Paul Bignon (1662-1743), court librarian to Louis XV. The specific epithet, "*capreolata*", means bearing tendrils (from the ends of the petioles). The bark is green when young and as it gets older, it becomes finely grooved and light brown. A cross section of the stem shows a cross formed by the pith. The leaves are opposite, tardily deciduous, 4-7" long, and compound (2 leaflets and tendril). The trumpet shaped flowers are large, up to 2" long, red-orange on the outside and yellow inside, usually with 5 yellowish lobes, individually stalked, and smell like Bit-O-Honey or beef buillion. The flowers are in axillary clusters and produced from February to April. The fruits are flat, linear, 4-7" long capsules and produced from June to October. The fruits split open to release thin papery seeds with large flat wings. The roots are a source of sarsaparilla. It is used as an alterative and a cleaning agent. The sections of stems are smoked like cigars in some localities hence the name Smoke Vine. This vine is widely distributed in the state on moderately well drained soils (FAC) but is absent from the marsh. It is currently reported from 57 parishes. (syn. *Anisostichus capreolata)*

Betula nigra (USDA-FS)

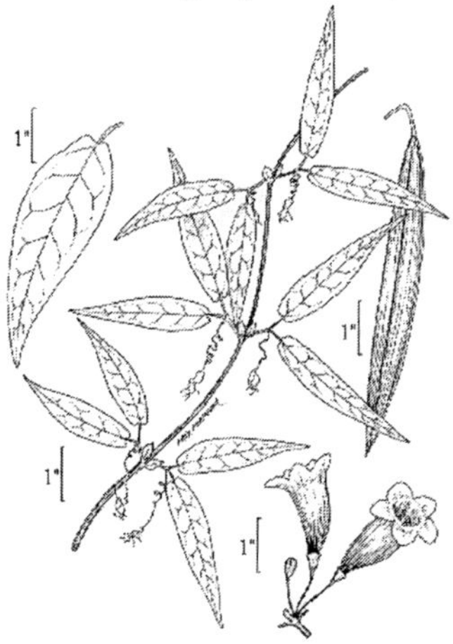

Bignonia capreolata (USDA-NRCS)

WOOLY SEA OXEYE - *Borrichia frutescens* - Asteraceae

Wooly Sea Oxeye is also known as Sea Oxeye and Sea Daisy. The genus name, "*Borrichia*", honors Ole Borch (Latinized Olaus Borrichius), a Danish botanist (1626-1690). The specific epithet, "*frutescens*", refers to the shrubby habit. It is a low, erect, little-branched, gray-hairy, evergreen shrub that reaches 4 ft. It has narrow leaf scar pairs encircling the branchlets. The leaves are simple, opposite, thick, fleshy, elliptic to oblanceolate, entire, and have a few remote teeth below the midleaf. The inflorescences are large conspicuous yellow flower heads that are found from April to September. The heads contain both ray and disc flowers. The fruit is an achene and dry, angled, and grayish. This species is restricted to salt marshes (OBL) and sandy habitats of the coastal parishes. It is currently reported from 11 parishes.

COMMON PAPER MULBERRY - *Broussonetia papyrifera* - Moraceae

The genus name, *"Broussonetia"*, honors Pierre Marie Auguste Broussonet, 1761-1807, a French Biologist at Montpellier. The specific epithet, *"papyrifera"*, refers to the use of the inner bark in papermaking. Paper mulberry is a small tree or shrub. The bark is smooth, tight, reticulate, and green to yellow. The leaves are alternate, opposite or whorled, deciduous, simple, and with palmate (3) major veins. The stipules are large and ovate but soon deciduous. The petioles are long and the blade is ovate with coarsely dentate and often lobed margins. The blades are cordate basally with an acuminate apex and rough above but conspicuously veined and velvety pubescent below. The plants are dioecious with male and female plants. The male flowers are in catkins and the female in globose heads. The fruit is a multiple of reddish orange drupes. The wood is coarse grained, soft, light, and easily worked. The inner bark is also used for making cloth in the tree's native home of Japan and China. The famous tapa cloth of the South Pacific Islands is also made from the bark by macerating it and pounding with a wooden mallet. The fruit is also eaten by a number of species of birds. This species persists and escapes from cultivation and is widely distributed across the state. It is currently reported from 37 parishes.

LADIES EARDROP VINE - *Brunnichia ovata* - Polygonaceae

Ladies' Eardrop Vine is also known as American Buckwheat Vine and Red Vine. The genus, "*Brunnichia*", honors M.T. Brunnich, a Norwegian naturalist of the eighteenth century and the specific epithet "*ovata*", is for the ovate blades. This is a vine that climbs by tendrils. The bark is greenish to reddish brown and finely grooved. The leaves are alternate, simple, deciduous, and with pinnate major veins. The shape is ovate with a truncate base and acuminate tip. The tubular, cream-colored, fragrant, five-pointed flowers occur in thick, rounded groups or in strings of hanging masses. They are produced from June to August. The fruits are elongated with a narrowed section at the base and 5 pointed lobes at the tip. The fruits are surrounded, by old flower parts, and look like an earring or eardrop. The flowers are very attractive to insects. This vine is common in bottomland hardwoods (FACW) throughout the state with current records from 62 parishes. (syn. *Brunnichia cirrhosa)*

GEORGIA CALAMINT - *Calamintha georgiana* - Lamiaceae

The genus," *Calamintha*", is from "cala" for reed and "mintha" for mint. The specific epithet, "*georgiana*", is for the state of Georgia. It is a small shrub with pubescent twigs. The leaves are opposite, deciduous, simple, and with pinnate major veins. The petiole is 1-3" long and the blade is ovate to ovate-elliptic, with crenate to serrate margins, a cuneate base, and an acute to obtuse apex. The crushed leaves have the typical mint smell. The flowers are terminal and axillary and zygomorphic with a two lipped green calyx and a two lipped pink to lavender

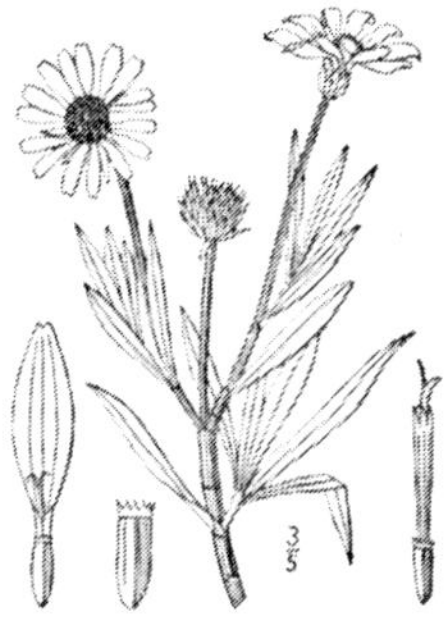

Borrichia frutescens (Britton and Brown 1943)

Broussonetia papyrifera (USDA-FS)

Brunnichia ovata (Britton and Brown 1943)

Calamintha georgiana (Godfrey 1988)

corolla. The four stamens are exserted. The fruit is a schizocarp with four nutlets or mericarps that are dark brown and smooth. It is found infrequent in sandy soils of southeastern Louisiana (St. Tammany and Washington parishes). (syn. *Clinopodium georgianum, Satureja georgiana)*

FRENCH MULBERRY - *Callicarpa americana* - Verbenaceae

French Mulberry is also known as American Beauty Berry, Spanish Mulberry, Bermuda Mulberry, Sour Bush, and Sow Berry. The genus name, "*Callicarpa*", is from the Greek word "kallos", meaning "beauty", and "kapos", meaning "fruit", both of which refer to the beautiful purple berries which appear in the fall on the plant. The specific epithet, "*americana*", is for America. It is a large shrub that reaches up to 9 ft but is mostly smaller. The leaves are simple, opposite, deciduous, and with pinnate major veins. The blades are 3-6" long, yellow green, hairy on the underside, soft textured, aromatic, and with rounded teeth on margins. The small flowers are rose to pink or pale blue, borne in axillary cymes on new growth and appear in May and June. The fruits are purple to violet (rarely white) and encircle the nodes of the stem from August to November. The leaves are used as an edema. The fruits are eaten by 10 species of birds, especially the bobwhite quail, squirrel, white-tailed deer, raccoon, opossum, gray fox, and sometimes man. It is a pioneer species in plant succession, appearing early in cleared, open, sunny sites near woodlands. It is very abundant in many parts of Louisiana in well-drained forests (FACU-). It is reported from 58 parishes but is especially common and widespread in the pine regions.

SWEET SHRUB - *Calycanthus floridus* - Calycanthaceae

Sweet Shrub is also known as Hairy Allspice, Carolina Allspice, Sweet Scented Shrub, Bubby Bush, and Bubby Blossoms. The genus name, "*Calycanthus*", is from the Greek kalyx (calyx) and anthos (flower), the calyx being large and conspicuous. The specific epithet, "*floridus*", refers to the plants conspicuous flowering habit. It is a shrub to 9 ft with reddish brown bark. The leaves are opposite, simple, deciduous, and with pinnate major veins. The blades are entire, elliptic to oval, dark green and veiny above, somewhat whitish below, and aromatic when bruised. The flowers are sweetly fragrant, reddish to purple-brown, showy, and with many, ribbon-like tepals. They are found from April to June. The fruit is a fleshy, pendent, rough-textured capsule bearing many one-seeded nutlets. It is often cultivated for its very fragrant flowers. This species is widely cultivated in the state with the only? non-cultivated report from an upland area (FACU+) in West Feliciana Parish. (syn. *Calycanthus fertilis)*

CUP SEED - *Calycocarpum lyonii* - Menispermaceae

The genus, "*Calycocarpum*", is from Greek: "calyx" for cup and "carpos" for fruit. The specific epithet, "*lyonii*", honors John Lyon, a British Explorer. This is a vine that climbs by twining. The stems are often bluish green. The leaves are alternate, simple, deciduous, and with palmate major veins. The petioles are to 10" long and the blade is 4-5 lobed with the margins of the lobes entire or coarsely dentate. The blades are broadly ovate to almost orbicular with a cordate base and the apex is long acuminate. The plants are dioecious with the flowers in axillary racemes or panicles. The flowers are without petals and the sepals are three greenish structures. The fruit is a green drupe that turns deeper green or black on drying. The single seed is cup shaped. This vine is found along streams (FACW) in upland forests and is infrequent in Louisiana with records from only 12 parishes.

Callicarpa americana (USDA-NRCS)

Calycanthus floridus (Britton and Brown 1943)

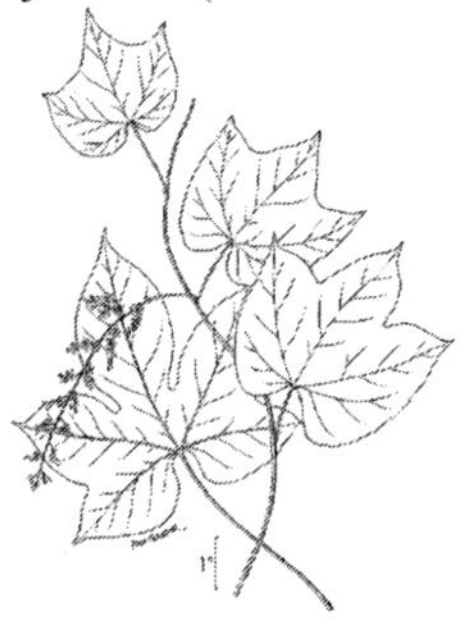

Calycocarpum lyonii (USDA-NRCS)

TRUMPET CREEPER - ***Campsis radicans*** - Bignoniaceae

Trumpet Creeper is also known as Trumpet Vine, Trumpet Ash, Trumpet Flower, Devil's Shoestring, Fox Glove Vine, and Cow Itch Vine. Trumpet creeper is the state flower of Kentucky. The genus, "*Campsis*", is from the Greek "kampe" (something bent) referring to the curved stamens or the curved capsules. The specific epithet, "*radicans*", refers to the aerial rooting stems. This common, deciduous, high-climbing, and widespread vine climbs trees or travels on the ground in moist woods and thickets and also lives on poles and fencerows. It makes a colorful ornamental, but getting rid of its troublesome due to sprouting roots. It is a vine that climbs by aerial roots. The leaves are opposite and odd-pinnate compound. The 7 to 15 egg-shaped leaflets are 8-15" long and serrate. The large trumpet-shaped flowers are orange red, 3-4" long, and have 5 flared or slightly backward bent lobes and appear from May to September. They mature into a persistent, large pod, 2-6" long, tapered on both ends, which upon maturity split along 2 ridges to release the many thin winged seeds, and appear from July to December. Trumpet Creeper furnishes deer browse mainly in the spring and summer. Hummingbirds visit the flowers. The roots cause sweating and are used for healing wounds. A lot of people believe that Trumpet Creeper has the same harmful attributes of Poison Ivy and this understanding has given it the name of Cow Itch Vine. Trumpet Creeper is widely distributed in Louisiana and is reported from all 64 parishes. It is one of the most common vines in drier areas (FAC) in bottomland hardwood forests.

IRON WOOD - ***Carpinus caroliniana*** - Betulaceae

Iron Wood is also known as Hornbeam, Water Beech, Smooth Barked Iron Wood, Blue Beech, Musclewood, American Hornbeam, Charme de Caroline, Bois de Fer, and Lean Tree. The genus name," *Carpinus*", is the classical name for hornbeam, and the specific epithet, "*carolinana*", refers to North or South Carolina. It is a small, slow growing tree that reaches up to 50 ft and with flattened gray, twisted trunk. Ironwood grows naturally in moist rich loams along streams and bottomlands and in hardwood slope forests; it is slightly wetter and down slope from Hop Hornbeam. It is recognized by its smooth mottled bluish gray "muscled" bark. It is called Ironwood because of its hardness. The common name Hornbeam derives from European Hornbeam for word "horn" for toughness and "beam" for tree and refers to the very hard wood. The name "Beech" was applied to this member of the Birch family because of its bark. The leaves are alternate, simple, deciduous, with pinnate major veins, ovate, 2-4" long, dull blue-green above, pale yellow-green below and doubly serrate. The leaves resemble those of Beech but have more numerous, small teeth and are darker and elm-like with the veins close together. The buds are square in a cross-section, which distinguishes it from Hop Hornbeam. The leaves of Blue Beech also do not have the "y" shaped lateral veins. The flowers are green catkins that appear from April to June. Separate male and female flowers appear on the same trees. The male catkins are 1-2" long and evident all winter long; the female catkins are about 3/4" long with small, leaf-like, three-lobed scales, appearing in spring on ends of twigs. The fruits are nutlets in spike-like clusters that are 3-6" long; each nutlet is attached to a 3 lobed, leafy bract. The fruits are produced from July to October. The strong, close-grained, hard, tough, light brown with whitish sapwood, heavy wood weighs 45 lb per cu ft. The wood is used for posts, golf clubs, tool handles, fuel, mallets, cogs, levers, and wedges. The wood has very little commercial value but deer browse on stem tips and succulent growth. The fruit and nutlets provide food for small mammals and nine species of birds. Native Americans are reported to have used ironwood for flux, navel yellowness, cloudy urine, consumption, diarrhea, and constipation. It is also reported to be used as an astringent, a tonic, a wash, and to facilitate childbirth. In autumn, the dew-heavy yellow leaves float through the air in a very interesting manner. It has been cultivated since 1812. Ironwood is widely distributed in Louisiana

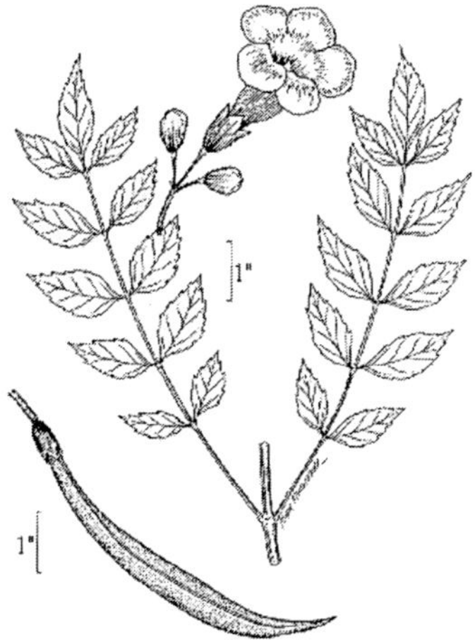

Campsis radicans (USDA-NRCS)

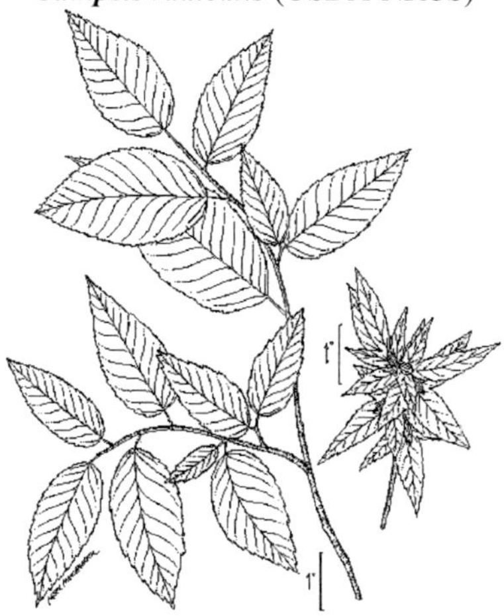

Carpinus caroliniana (USDA-NRCS)

primarily along the floodplains of streams and extending upslope (FAC); it is usually at lower elevation and slightly wetter sites than Hop Hornbeam. Ironwood is notably absent from the marsh and most of the Mississippi and Red River floodplain. It is currently reported from 47 parishes.

HICKORY, PECAN - *Carya* - Juglandaceae

The genus, "*Carya*", is the ancient word for walnut. This genus includes trees with alternate, deciduous, and once odd pinnately compound leaves. The leaflet margins are serrate. The pith is continuous compared to chambered in walnuts (***Juglans***). The lateral and/or the branch veins in the leaflets of ***Carya*** run to the margins while in ***Juglans***, the lateral and/or branch veins curve and mesh near the margins. The plants are monoecious with male and female flowers on the same plant. The flowers are small and green with the staminate flowers in catkins and the female flowers solitary or in short spikes. There are ten species of this genus in the state. Four of the species (***Carya aquatica, Carya cordiformis, Carya illinoinensis***, and ***Carya myristiciformis***) have buds that are valvate while the other six species have buds that are imbricate. Hickory wood is the favorite for the curing of meat. It is burned green for this purpose. Woodpeckers and songbirds seek cracked nuts of hickories. Hickories are rated high in humus building capacity.

MOCKERNUT HICKORY (***Carya alba***) is also known as White Hickory, White Heart, Hognut, Fragrant Hickory, Hardbark Hickory, Big Bud, Red Hickory, and Bull Nut. The specific epithet, *"alba"*, means white and probably refers to the wood. It may attain a height of 130 ft although in the uplands it may be much smaller. Mockernut Hickory is long-lived and can live up to 500 years. On young trees, the light gray, mottled bark is almost smooth. On old trees, the bark is grayish-brown, close, slightly ridged by shallow, irregular fissures and covered with thin, close, soft scales. The crushed twigs produce an odor. The nine (5-9) oval (sometimes obovate), leaflets are pubescent and strong scented. The fruit is large, slightly oblong, with a very thick husk. The nut is smooth, usually four-ridged, with a thick shell and partitions and a small, sweet kernel. The wood is heavy, hard, strong, tough, flexible, close-grained and all white, except the very small heart, and weighs 51 lb per cu ft. It is used in the manufacture of wagons, handles, and farm implements and for fuel. Mockernut Hickory has the most strength in bending than any other hickory. The foliage is occasionally browsed by deer. It was first cultivated in 1766. In well drained soils in the pine region of the state. It is currently reported from 37 parishes. (syn. *Carya tomentosa)*

BITTER PECAN (***Carya aquatica)*** is also known as Water Hickory, Wild Pecan, Bitter Water Hickory, Bitter Hickory, Swamp Hickory, and Water Pignut. The specific epithet, "*aquatic*", refers to wet or aquatic habitat. It is a large tree reaching 90 ft or taller. The light brown bark separates freely into long, loose, thick, plate-like scales. The buds are dark red brown and valvate. The 7-11 (or 13) leaflets are dark green and shiny above, more or less hairy beneath, and thin in texture. The buds are brown to black and valvate. The lateral leaflets are distinctly asymmetrical while the terminal leaflet is oblong lanceolate to ovate lanceolate. The nut, often borne in clusters of 3 or 4, is rather oblong and conspicuously four-angled, with a thin hull that splits tardily. The nut itself is nearly as broad as long, four-angled and ridged, with a thin shell and very bitter kernel. The wood is heavy (43 lb per cu ft), strong, close-grained, dark reddish brown and rather brittle. It is not used as much as other hickories and only for fuel, posts, and props. Mammals and
birds are reported to eat the fruit reluctantly. Non fruiting specimens of Bitter Pecan can be separated from Sweet Pecan by (1) habitat; Bitter Pecan will be on wetter sites (OBL) compared

Carya alba (USDA-FS)

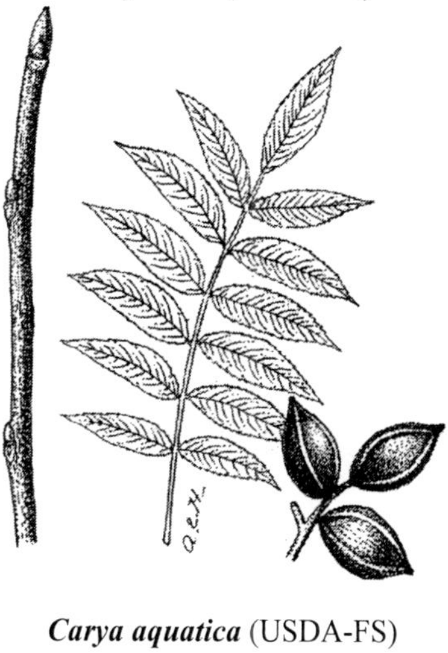

Carya aquatica (USDA-FS)

to drier sites (FAC+) for Sweet Pecan and (2) leaflets; the number is usually more for Sweet Pecan (13-15) and fewer for Bitter Pecan (9-11), the leaflets are sessile in Bitter Pecan and have a short petiolule in Sweet Pecan; the leaflets tend to be wider in Sweet Pecan and narrower in Bitter Pecan; the leaflet margins are always serrate in Sweet Pecan but can vary from entire to serrate in Bitter Pecan and the midrib in Bitter Pecan is pubescent and is usually glabrous in Sweet Pecan. Bitter Pecan is reported from almost all parishes and probably occurs in all. It is currently reported from 54 parishes.

BITTERNUT HICKORY (***Carya cordiformis***) is also known as Bitternut, Swamp Hickory, Bitter Walnut, Bitter Pecan, Pig Hickory, White Hickory, Pignut Hickory, and Red Hickory. The specific epithet, "*cordiformis*", means "heart shaped" and is for the terminal end of the nuts. It is a medium to large tree to 100 ft tall. The buds are valvate and yellow. The 7-11 leaflets (lower surface, especially along the veins), petiolules, and rachis are pubescent. The lateral leaflets are nearly symmetrical and the terminal leaflet is obovate to ovate-lanceolate. The covering over the nut is yellowish brown and the nut is globose to ovate and apiculate terminally from a depressed or obcordate end. The nuts are so bitter, so much so that they are hardly eaten by wildlife. It was first cultivated in 1689. It is the most rapid grower of the hickories. Bitternut Hickory is widely distributed throughout Louisiana on mesic to well drained sites (FAC). It is currently reported from 35 parishes.

PIGNUT HICKORY (***Carya glabra*** var. ***glabra***) is also known as Broom Hickory, Sweet Pignut, Coast Pignut Hickory, Smoothbark Hickory, Swamp Hickory, Switch Hickory, Black Hickory, Red Hickory, White Hickory, Brown Hickory, and Switchbud Hickory. The specific epithet, "glabra", refers to the smooth character of the leaves and petioles. The tree reaches a height of 100 ft with tight grayish bark. The five, rarely seven, leaflets are glabrous. The fruit is pear-shaped (obovate) with husks that split halfway to the base. The wood is brown, heavy, hard, strong, tough, elastic, and weighs 51 lb per cu ft. It is used for fuel, tool handles, wagons, and agricultural implements. Settlers made brooms from the tough, flexible wood by splitting it into narrow strips. The wood is occasionally used for smoking meats. Sapsuckers frequently girdle the bole. It has been in cultivation since 1750. Pignut hickory is widely distributed in Louisiana mostly on well-drained sites (FACU) and is absent from the marsh and large floodplains. It is currently reported from 36 parishes. (syn. *Carya microcarpa)*

SWAMP HICKORY (***Carya glabra*** var. ***hirsuta***) is also known as Pignut Hickory. The specific epithet, "*glabra*", refers to the smooth character of the leaves and petioles while "*hirsuta*" refers to the hairs on the leaves. Swamp Hickory reaches a height of 80 ft with grayish brown rather smooth bark. The seven, rarely five, leaflets are smooth above and sparsely pubescent below. The fruit is pear shaped (obovate) with husks that split halfway to the base. The wood is light reddish brown, hard, strong, and used for same purposes as that of other hickories. Swamp Hickory has a similar distribution pattern as Pignut Hickory but usually on wetter sites although the wetland ranking is the same (FACU). It is currently reported from 23 parishes. (syn. *Carya leiodermis)*

SWEET PECAN (***Carya illinoinensis***) is also known as Nogal Morado, and Nuez Encarcelada. It is a large tree reaching 130 ft tall. Sweet Pecan is one of the few native to the present day United States cultivated crop plants. DeSoto first observed it in 1541 in the present state of Arkansas after crossing the Mississippi River and emerging from the swamps onto higher ground. It was reported to be a major food for the Native Americans that was quickly adopted by Europeans including Thomas Jefferson who planted some at Mount Vernon; some of

Carya cordiformis (USDA-FS)

Carya glabra var. ***glabra*** (USDA-FS)

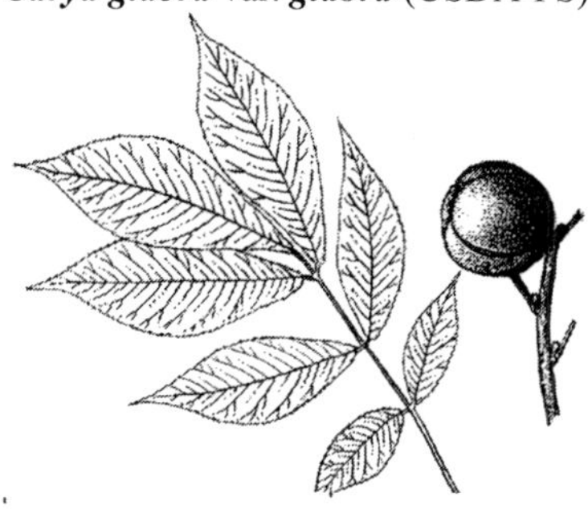

Carya glabra var. ***hirsuta*** (USDA-FS)

Carya illinoinensis (USDA-FS)

those are still alive today and are reputed to be the oldest living trees there. Pecan nuts were traded by trappers in the eastern United States, where they were known as Mississippi nuts or Illinois nuts. It is the state tree of Texas. There are many cultivars of this species (some reports of more than 500) with the size and thinness of the shell being the major selection qualities. It is widely planted with the first selection being reported in 1846. It can easily live 300 years or more. The bark varies, but is usually gray and rough, hard and tight, but sometimes broken into scales. The buds are brown to yellowish brown and valvate. The lateral leaflets are distinctly asymmetrical while the terminal leaflet is oblong lanceolate to ovate lanceolate. The nuts are terete in cross section and are sweet; for the separation of Sweet Pecan from Bitter Pecan see the discussion under Bitter Pecan. The nuts are used in the making of candles and cakes; the famous French pralines are well known in Louisiana. The nuts mature in September and October. It is used by many wildlife species including turkey, most of the game and furbearing animals and browsed to some extent by deer. It is important food for crow, cardinal, blue jay, woodpeckers, grackles, Carolina wren, titmice, chickadee, and nuthatches. The sap can be used for syrup and water. The wood is brittle, tough, whitish, heavy, hard and not strong, and weighs 47 lb per cu ft. It is used for the same purposes as the wood of other hickories. The bark and leaves are sometimes used medicinally as an astringent. Sweet Pecan and Bitter Pecan are two of the last trees to green up (produce new leaves) in the spring and there is a saying that winter is definitely over when the pecan trees leaf out. Due to its widespread cultivation, it is difficult to determine the original habitat but is on drier sites (FAC+) than Bitter Pecan. Sweet Pecan is reported from almost all parishes and grows in all. It is currently reported from 55 parishes.

SHELLBARK HICKORY (***Carya laciniosa***) is also known as King Nut, Bottom Shellbark, Western Shellbark, Big Shellbark, and Thick Shellbark. The specific epithet, "*laciniosa*", refers to the deep furrowing and splitting of the bark. It can reach a height of 140 ft with long loose plated bark. The twigs are light orange, usually hairy with large, rough, winter end buds. The 7 (5-9) leaflets and petioles are covered with curly hairs. The wood is dark brown, close-grained, hard, strong, tough, heavy, very flexible, and weighs 50 lb per cu ft. It is used as that of other hickories, particularly for tool handles, baskets, and fuel. The large nuts are edible and sweet, but are considered to lack the flavor of the Shagbark Hickory. It is rather similar in appearance to Shagbark Hickory, but has larger leaves and nuts, and the bark is somewhat less shaggy in appearance. However, the two species are known to hybridize. It has been in cultivation since 1800. This species is reported from only Madison and West Feliciana parishes and is reported from bottomland situations (FACW-).

NUTMEG HICKORY (***Carya myristiciformis***) is also called Bitter Waternut, Bitter Water Hickory, Swamp Hickory and Nogal. The specific epithet, "*myristiciformis*", comes from the fruit that is shaped like nutmegs which are in the Myristicacae Family. It is a medium to large tree to 100 ft tall. The buds are valvate and yellow. The 5-9 leaflets (lower surface), petioles, and rachis are covered with flattened yellow scales. The lateral leaflets are nearly symmetrical and the terminal leaflet is ovate lanceolate to obovate. During the growing season the foliage has a yellow cast that turns golden brown in autumn. The covering over the nut is bright yellow and the nut is ellipsoid and rounded and apiculate at both ends. It is easily recognized by the football shaped nuts which are usually solitary with yellowish brown hairs on the very thin husks. The wood is light brown, close-grained, tough, and strong with lighter sapwood. The wood is considered inferior to that of other hickories and is of little commercial importance. Nutmeg Hickory is widely distributed in central and northern Louisiana especially on calcareous soils (FACW-). It is currently reported from 13 parishes.

Carya laciniosa (USDA-FS)

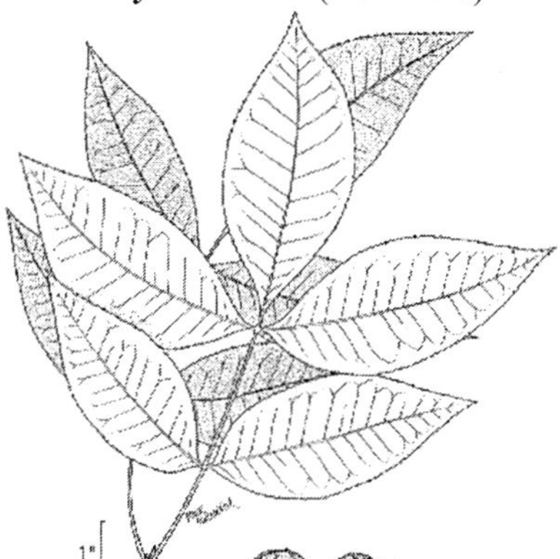

Carya myristiciformis (USDA-NRCS)

RED HICKORY (***Carya ovalis***) is also known as Oval Pignut. The specific epithet "*ovalis*", refers to the ovoid shape of the fruit. The tree can reach of height of 100 ft and the bark resembles Shagbark Hickory but the plates are much smaller. The leaves resemble Pignut Hickory but differ in usually having seven leaflets. The fruit is ovoid to obovate with husks that split all
the way to the base as compared to halfway in Pignut Hickory. The wood is very heavy, very hard, tough, flexible, with light or dark brown heartwood and thick paler sapwood. Red Hickory is reported only from ten parishes primarily from well-drained (FACU) upland areas.

SHAGBARK HICKORY (***Carya ovata***) is also known as Scaly Bark Hickory, Red Heart Hickory, Shellbark, Carolina Hickory, White Hickory, Upland Hickory, Sweet Walnut, and White Walnut. The specific epithet, "*ovata*", refers to the ovate-shaped leaflets. The tree is long-lived, slow growing, and can reach a height of 140 ft. It is easily identified by its bark, which breaks into strips up to 12" long and 8" wide, with the ends bowing away from the tree. On younger trees, the bark is tight and light gray. The compound leaves usually have 5 leaflets. The leaves have tufts of hairs at the tips of the serrations of the leaves. The nut is four-angled with sutures at the corners of the angles. The wood is light brown, close-grained, heavy, hard, strong, tough, flexible, and weighs 52 lb. per cu ft. The sapwood is lighter. The wood is used for fuel, tool handles, baskets, wagons, and other general uses. This nut, next to the Pecan, is the best of any American tree, and is the common Hickory Nut of commerce. It has been in cultivation since 1911. General Andrew Jackson was given the nickname "Old Hickory" (because the soldiers he commanded said that he was tough as Hickory) and six towering Shagbark Hickories surround his grave. Shagbark Hickory is widely distributed on well drained soils (FACU) throughout the state but is absent from the marsh and the bottomland forests. It is currently reported from 33 parishes.

SAND HICKORY (***Carya pallida***) is also known as Pale Hickory and Brown Hickory. The specific epithet, "*pallida*", refers to the pale lower surface of the leaves. Sand Hickory grows to a height of 30 to 40 ft (rarely 100 ft). The bark is tight, dark gray to black with deep furrows. The 7 (5-9) leaflets have dense clumps of hairs on the rachis and midrib plus silver plate-like scales on the lower leaf surface. The nuts are tan to reddish brown, slightly flattened, with thick shells and sweet kernels. The wood is brown, tough, moderately strong and little used except for posts and fuel. This species is rare in southeastern Louisiana and reported from sandy soils in St. Helena, Tangipahoa, and Washington parishes.

BLACK HICKORY (***Carya texana***) is also known as Arkansas Black Hickory, Buckley Hickory, Ozark Hickory, Texas Hickory, and Pignut Hickory. The specific epithet, "*texana*", is for the state of Texas. It is a medium sized tree reaching heights of 80 ft. The bark may be dark gray, irregularly fissured; separating into thin scales or may be nearly black and deeply divided into rough ridges. The buds are imbricate and the twigs are slender. The 7 to 13 leaflets are rusty pubescent below especially along the midribs. The twigs and buds are hairy also. The fruits are pear-shaped, about 1" long, with the hull rather thick, splitting to, or nearly to, the base. The nut is somewhat flattened and slightly four-angled; shell about 1/8" thick. The kernel is edible. Birds and mammals relish the fruits. The wood is brown, hard, tough, brittle, and the sapwood is paler. There is little use for the wood except for fuel. Black Hickory is widely distributed throughout the state on well drained often, sandy soils and has not been reported from wetlands. It is currently reported from 33 parishes.

Carya ovalis (USDA-FS)

Carya ovata (USDA-FS)

Carya pallida (USDA-FS)

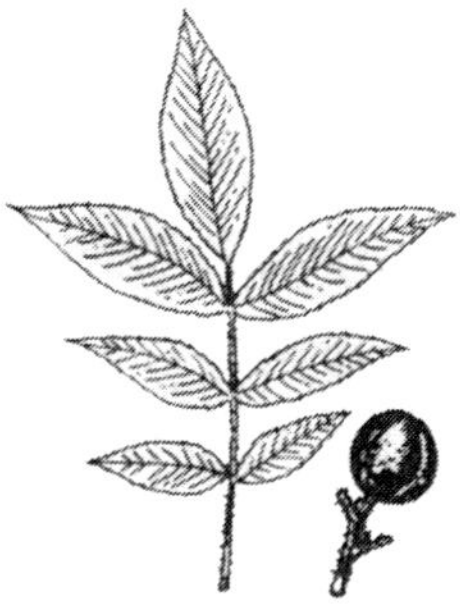

Carya texana (Main and Mills 1981)

CHESTNUT, CHINQUAPIN - *Castanea* - Fagaceae

The genus, "*Castanea*", name is for a town in Thessaly and includes woody plants that are trees or shrubs. The leaves are alternate, deciduous, simple, with pinnate major veins, margins serrate, teeth pointed forward, and with small bristles on the marginal teeth. The plants are monoecious with the staminate flowers in slender long white catkins that produce a distinctive odor. The pistillate flowers are in short catkins with naked flowers; some of the shorter staminate catkins also bear pistillate flowers near the base. The fruit is a nut that is surrounded by a prickly involucre usually called a bur.

AMERICAN CHESTNUT (***Castanea dentata***) is a tree that reaches a height of 50 ft. The bark is dark brown with broad flat ridges. The leaves are coarsely serrate with pointed, sharp teeth, and smooth and green on both sides. The flowers are long, slender whitish catkins that open in midsummer. The fruit is a prickly bur with 2 to 3 nuts per involucre. American Chestnut was once a source of income to people living within its range-bark for tanning, nuts for the table, and as lumber is one of the most durable hardwoods. However, *Endothia parasitica* (chestnut blight) has almost caused the extinction of American Chestnuts. Sprouts may continue from some old stumps, and these flower and produce fruits. As soon as these shoots attain a moderate size, the bark shattering blight usually girdles them near their bases. The fruits are eaten by squirrels and other rodents. The foliage and twigs are browsed by whitetail deer. There are only two reports of American Chestnut from Louisiana; Morehouse and Webster parishes. In widely scattered locations across Louisiana, there are planted and persisting chestnut trees that are probably the Chinese Chestnut (***Castanea mollissima***).

CHINQUAPIN (***Castanea pumila***) is a shrub or small tree that reaches a height of 50 ft. The chestnut blight also affects the Chinquapin killing many trees but the plant sprouts from the base so most Chinquapins in Louisiana today are small stump sprouts. The bark is smoothish, reddish brown, with shallow furrows, and ridges with platelike scales. The leaves are typically elliptic to lance elliptic and with a pubescent undersurface and mostly shorter than 15 cm. The bur is small, mostly smaller than 1.5" diameter with one nut per bur. The nut is black at maturity, round, and mostly smaller than ½" in diameter. The wood is dark brown, coarse-grained, light, hard, strong, rather brittle, durable, and weighs 31-37 lb per cu ft. The tree is generally too small for commercial use but is occasionally used for posts, railroad crossties, and fuel. During the depression of the early 1930's, Chinquapin nuts were considered a rare delicacy and were sold and bartered on school lots across the South. The going price was five cents per handful, with the buyer allowed to use his or her hand, whatever the size. There are two varieties in Louisiana, var. ***ozarkensis*** and var. ***pumila***. The variety ***pumila*** has pubescent twigs, a petiole mostly shorter than 8 mm, lighter colored and smooth bark, and produces burs with spines shorter than 10 mm while the var. ***ozarkensis*** has glabrous twigs, a petiole mostly longer than 8 mm, darker and fissured bark, and produces burs with spines mostly longer than 10 mm. The variety ***ozarkensis*** is called Ozark Chinquapin with the specific epithet, "*ozarkensis*", for the Ozark Mountains of Arkansas, where it is most abundant. The typical chinquapin (variety ***pumila***) is widely distributed in the state usually on well-drained and often sandy soils; it is absent from the large floodplains and the marsh. Ozark chinquapin (var. ***ozarkensis***) is rare and reported only from Claiborne and Union parishes in North Louisiana.

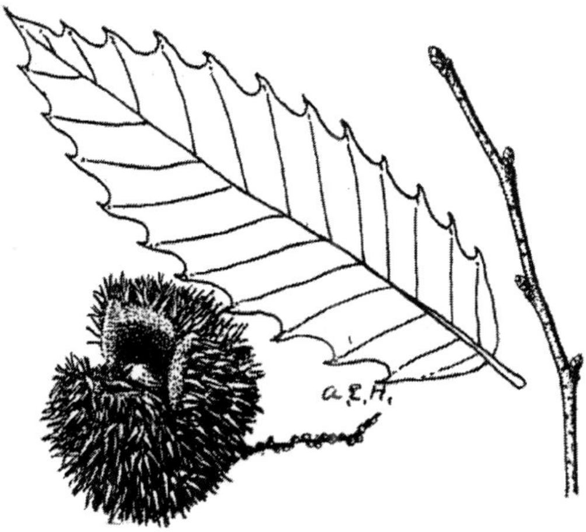

Castanea dentata (USDA-FS)

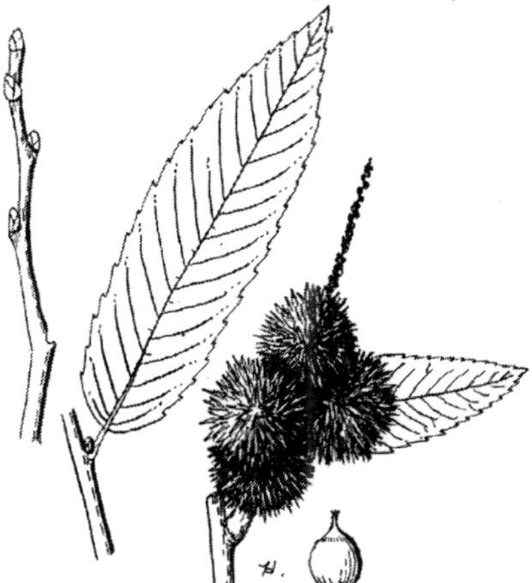

Castanea pumila (USDA-FS)

CATALPA - ***Catalpa*** - *Bignoniaceae*

The genus name, "*Catalpa*", is the American Indian name. The genus includes small to large fast growing trees. The bark is usually brownish and the young twigs have prominent white lenticels. The leaves are whorled or opposite, simple, deciduous, and with palmate major veins. The blades are large 4-12" long, ovate, cordate at the base, with a petiole as long as the blade, light green, pubescent on the lower surface. The flowers appear from May to July in large panicles and are very showy. The corollas are zygomorphic, white, and marked inside with two yellow lines and purple spots. The fruit is an elongated cylindric capsule and the seeds are flat, large and winged.

SOUTHERN CATALPA (***Catalpa bignonioides***) is also known as Candle Tree, Cigar Tree, Smoking Bean, Catalpa, Caterpillar Tree, Bean Tree, Catawba, Indian Bean, and Indian Cigar. The specific epithet, "*bignonioides*", refers to its flowers resembling the Bignonia vine. It is a tree that usually reaches a height of 60 ft. The bark is gray-brown and separates into thin flakes. The leaves have a somewhat foul odor when crushed. The flowers of this species are smaller (3 cm or less wide) and the lower lobe is entire. The flowers are also reported to have more purple spots than those of the northern catalpa. The panicles in this species are smaller with fewer flowers and the fruits have a thinner wall. Groves of Southern Catalpa trees have been planted just to attract Catalpa worms (larvae of Sphinx moths) which are popular as fish bait. The seed contains a bitter glycoside, catalpin, which has reportedly been used for asthmatic, cardiac, and antispasmodic purposes. Large doses of the seed are said to produce weak pulse, nausea, and vomiting. The plant has also been used medicinally in the treatment of snakebites and as an antiseptic for cuts and bruises. Some people develop a skin allergy when handling the flowers. The wood is grayish brown to lavender-tinged, sapwood lighter, coarse and straight-grained, no characteristic odor or taste, very durable, weighs about 28 lb per cu ft, shrinks little, weak in endwise compression, soft, moderately high in shock resistance, and weak in bending. The wood is widely used because of its resistance to decay for fence posts, poles, and railroad ties. However, it is not suitable for fireplaces, because it produces minor explosions when burning. Southern Catalpa probably occurred in or has been planted in all parishes in the state but its natural habitat seems to have been along streams on the natural levees (FAC-). It is currently reported from 52 parishes.

NORTHERN CATALPA (***Catalpa speciosa***) is also known as Catalpa, Hardy Catalpa, Larger Indian Bean, Bois Plant, Indian Cigar, Smoking Bean, Cigar Tree, Shawnee Wood, Catawba Tree, and Western Catalpa. The specific epithet, "*speciosa*", is given for the large showy flowers. It can be a large tree to 100 ft and the bark is usually dark gray to brown and fissured. The leaves have a much less, often absent, foul odor when crushed. The flowers of this species are larger (4 cm or more wide) and the lower lobe is notched. The flowers are also reported to have fewer purple spots and the panicles are larger with more flowers. The fruits have a thicker wall. The heartwood is light brown, soft, weak, and coarse-grained, with narrow, whitish sapwood. It is moderately weak in bending, weak in endwise compression, and moderately high in shock resistance. It weighs 26 lb per cu ft and is very durable in contact with the soil and thus used for fence posts. Other uses include poles and interior finish. It has the same larva of Sphinx moths that can be used as fish bait. The distribution of this species is difficult to determine but is currently reported from 13 parishes widely scattered across the state (FACU).

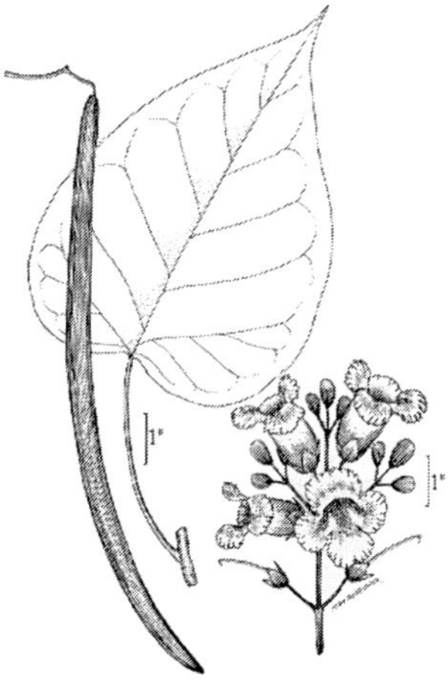

Catalpa bignonioides (USDA-NRCS)

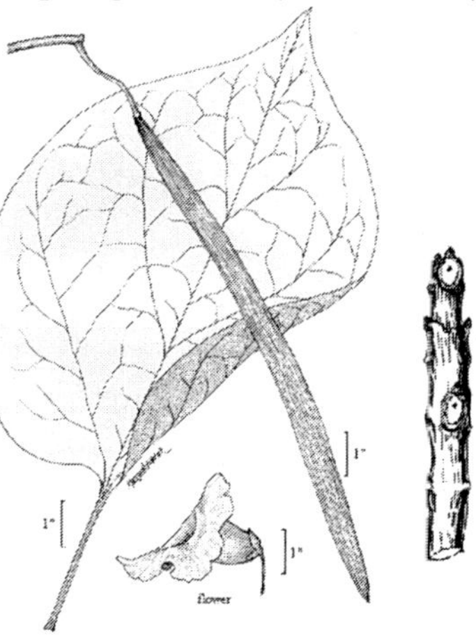

Catalpa speciosa (USDA-FS and USDA-NRCS)

BUSHKILLER - ***Cayratia japonica*** - *Vitaceae*

Bushkiller is also called Sorrel Vine. The genus, "*Cayratia*", is a apparently a Vietnamese common name. The specific epithet "*japonica*" is for Japan, one of the countries where it is native. It is technically herbaceous but is included due to its close resemblance to Virginia Creeper. Bushkiller gets its name from its ability to grow rapidly and shade out and kill the trees and shrubs on which it grows; some reports indicate breaking of trees and shrubs by the excess weight of this vine. It is a native of southeast Asia and was first reported in North America from St. Tammany Parish (Shinners 1964). It is an herbaceous vine with tendrils opposite the leaves; the tendrils are usually branched. The leaves are alternate, evergreen?, and palmate compound with five leaflets. The leaflets are toothed, glabrous, and with an acuminate to acute tip. The inflorescence is axillary and flat-topped. The flowers are of white (salmon) color. The fruit is a black berry with 2 or 4 triangular shaped seeds. The current Louisiana distribution includes 10 parishes in the southern part of the state. (Not Illustrated)

JERSEY TEA - ***Ceanothus*** - Rhamnaceae

The genus name, "*Ceanothus*", is from an obscure name given in Theophrastus, probably not for this genus. This genus includes shrubs with alternate simple, deciduous leaves with pinnate major veins but with a pair of large lateral veins that arise from near the base and extend toward the apex. The margins are serrate with glandular tipped teeth and the upper surface is green but the lower surface has a whitish cast. The white flowers are produced in racemes and the fruit is a capsule. The leaves can be dried and made into a tea by steeping in hot water; the tea is reported to contain caffeine. The best time to harvest leaves is reported to be when the flowers are in full bloom usually in May-June. This tea was reported to have been very popular when tea from China had problems being imported before and during the Revolutionary War. Some medicinal use has been made of this plant with the root being reported to be astringent and also imparts a red color to water. The plants were reported to be used to treat syphilis and also reported to cause a fall in blood pressure and to decrease the time required for blood to coagulate. A recent report indicates this plant inhibits tumor growth. The plants are used as an ornamental and the flowers are a good nectar source for butterflies.

NEW JERSEY TEA (***Ceanothus americanus***) is also known as Wild Snowball, Sprangles, Walpole Tea, Wild Pepper, Mountain Sweet, and Red Root. The specific epithet, "*americanus*", refers to its growth in North America. This is the most widespread and best known of the species. It has pubescent leaves that are ovate to broadly elliptic that are acute to acuminate at the tips and pubescent above. The inflorescences are axillary with peduncles longer than 2". This species is the widespread and common one that is widely distributed throughout the state on well-drained soils. It is absent from the floodplains and the marsh. It is currently reported from 33 parishes.

JERSEY TEA (***Ceanothus herbaceus***) is also known as Inland Ceanothus and Red Root. The specific epithet, "*herbaceus*", refers to the herbaceous nature of the branchlets. The leaves are narrowly elliptic to elliptic lanceolate with an obtuse, rarely acute tip. The upper surface is glabrous or nearly so. The inflorescences are terminal on the ends of branchlets with peduncles shorter than 2". This species is restricted to calcareous areas in northwest Louisiana with reports from Caddo, Natchitoches, and Winn parishes.

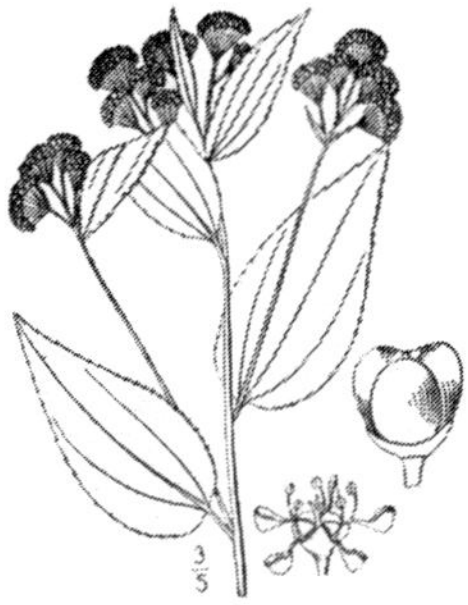

Ceanothus americanus (Britton and Brown 1943)

Ceanothus herbaceus (Vines 1960)

Celastrus scandens (Britton and Brown 1943)

BITTERSWEET - *Celastrus scandens* - Celastraceae

Bittersweet is also known as American Bittersweet, False Bittersweet, Shrubby Bittersweet, Climbing Bittersweet, Staff Vine, Fever Twig, Gnome's Gold, Waxwork, Roxbury Waxwork, Jacob's Ladder, and Climbing Orangeroot. The genus name, "*Celastrus*", is from the ancient Greek name kelastrus, and the specific epithet, "*scandens*", refers to its climbing habit. Bittersweet is a tangled or high-climbing vine that reaches 90 ft. It sometimes injures trees by constriction. The stems are round, hairless, thornless, and brown. The buds are small, pointed, and sharply set nearly at right angles to the stems. The leaves are simple, deciduous, alternate, and with pinnate major veins. The blades are light green, egg to wedge-shape, somewhat long-pointed, and finely wavy-toothed. The flowers are small and green. The fruits are clustered orange capsules that open to reveal scarlet or red orange aril covered seeds. The orange colored fruit and especially the scarlet seeds are used in flower decorations and make this an important ornamental especially in the northern United States. The fruits are eaten by several birds and mammals. The twigs and bark are eaten by rabbit, squirrel and deer. It is considered poisonous to horses. Bittersweet has been in cultivation since 1736. The root bark is used for liver affections and secondary syphilis. It is said to be emetic, diaphoretic, and alterative. This is a rare vine in Louisiana (six parishes) and is only known from well-drained areas or as an escape from cultivation.

HACK BERRY, SUGAR BERRY - *Celtis* - Ulmaceae

The genus, "*Celtis*", is a Latin word that was used by Pliny for a sweet fruited African lotus. The genus includes trees with alternate simple deciduous leaves and with palmate major veins. This plant has the smallest leaf with palmate major veins. The blade is ovate to lanceolate and the margins vary from entire to toothed. The flowers are small and greenish. The fruit is an orange red to black drupe. The bark is grayish and often with numerous corky protuberances. The wood is yellowish with thick, lighter colored sapwood, and is close-grained, soft, weak, and weighs 49-50 lb per cu ft. It is used for furniture, flooring, boxes, baskets, burial boxes, crating, fuel, cooperage, and posts. Indians and 10 species of birds ate the fruits. In addition, the fruits attract mealy bugs. After the mealy bugs are overstuffed, they secrete a "ghost rain". The pits can be used as a spice.

HACK BERRY ***(Celtis laevigata)*** is also known as Sugar Hack Berry, Bois Inconna (unknown wood), Sugar Berry, and Palo Blanco. The specific epithet,"*laevigata*," means "smooth". It is a tall tree that reaches up to 100 ft. The leaves are lanceolate, and mostly longer than 2". The apex is long acuminate; the surface is light green and rough. The drupes are mostly smaller than 8 mm in diameter and beakless. It is reported that the bark was sued to treat sore throat and venereal disease. This species is very much related and difficult to separate from ***Celtis occidentalis***, which is a somewhat more northern species. It is widely distributed in Louisiana and probably occurs in all with current records from 60 parishes. It is most common along streams especially in the larger floodplains (FACW).

NET LEAF HACK BERRY (***Celtis reticulata***) is also known as Palo Blanco. The specific epithet, "*reticulata*", refers to the reticulate nature of the veins of the leaves. It is a small tree with small leaves (mostly shorter than 2") that are broadly ovate, stiff, and with distinctly reticulate veins on the lower surface. The margins tend to be entire basally with teeth distally. The veins are pubescent, as is the petiole. The drupes are mostly larger than 8 mm in diameter and with a beak. It is reported that this plant was used to treat indigestion. This species is restricted to dry sandy habitats of western and northwestern Louisiana with reports from ten parishes. (syn. *Celtis laevigata* var. *reticulata)*

Celtis laevigata (USDA-NRCS)

Celtis reticulata (USDA-FS)

SMALL HACK BERRY (***Celtis tenuifolia***) is also called Dwarf Hack Berry, Sugar Berry, Nettle Berry, and Nettle Tree. The specific epithet, "*tenuifolia*", refers to the thin leaves. Other common names include Sugar Berry, Nettle Berry, and Nettle Tree. The leaves are ovate with an obtuse apex and mostly longer than 2". The leaves are dark green above and lighter green below. The flowers are usually single, and the fruit is an orange to brown drupe 5-8 mm long. The fruits are smaller than 8 mm in diameter and beakless. It is used by 25 species of birds. This species is restricted to western and northern Louisiana with reports from 16 parishes. (syn. *Celtis pumila)*

BUTTON BUSH - *Cephalanthus occidentalis* - Rubiaceae

Button Bush is also known as Little Snowbell, Pinball, Box, Button Wood, Pond Dogwood, Uvero, Globe Flower, Crouper Brush, Spanish Pincushion, River Brush, Swamp Wood, Button Willow, and Crane Willow. The genus, "*Cephalanthus*", means "head-flower." The specific epithet, "*occidentalis*", refers to the Western Hemisphere. Buttonbush is a shrub to small tree that reaches up to 18 ft. The thin, smooth bark is gray to brown that becomes furrowed with age. The leaves are opposite or whorled (3's), deciduous, short petioled, and with pinnate major veins. The blades are oblong, entire, glabrous above and hairy or glabrous underneath. The flowers are creamy-white, crowded in globular heads measuring 1-1.5" diameter, and are produced from June to September. The button-like fruits are clusters of nutlets that are produced from September to November. The bark is used to treat bronchial, skin, and venereal diseases, laxative, astringent, tonic, and fever remedy. It is also said to be poisonous, causing paralysis, convulsions, and vomiting. The nutlets are eaten by 25 species of birds (mostly water birds) and the flowers produce nectar for bees and butterflies. Bark is regularly eaten by beaver and it is a low quality deer-browse. The poisonous foliage is unpalatable to livestock. It is frequently cultivated as an ornamental shrub and provides good bee food. Buttonbush is widely distributed throughout the state in wet sites (OBL), often along the edge of and in ponds, lakes and other bodies of water. It occurs in all 64 with current reports from 63 parishes.

REDBUD - *Cercis canadensis* - Fabaceae

Redbud is also known as Eastern Redbud and Judas Tree. The genus name, "*Cercis*", is the ancient Greek name of the closely related Judas Tree of Europe and Asia which means "a weaver's shuttle", descriptive of the woody fruits. According to tradition, Judas hung himself from a branch of the tree. The specific epithet, "*canadensis*", literally means "of Canada". It is a shrub or small tree that reaches 50 ft. The bark is reddish brown to gray and smooth. As the bark gets older, fissures that separate into plates are produced. The bark on young branches is smooth, brownish, and marked with evident light lenticels. The leaves are alternate, simple, deciduous, cordate to ovate, and 3-5" long. The leaves have palmate major veins and the entire petiole is swollen at the tip. The leaves are glossy green turning in autumn to a bright, clear yellow. The conspicuous, rose-purple rarely white, pea shaped flowers are in clusters along the twigs and small branches and are produced from March to April before the leaves. Flowers can be added to salads and the young fruits can be eaten raw or cooked. The fruit is a flat, oblong, many-seeded legume, 2-4" long, and are produced from May to October. It is reddish during the summer and often hangs on the tree most of the winter. The wood is moderately heavy (30-40 lb per cu ft), hard, not strong, close-grained, and rich dark brown tinged with red and thin, lighter colored sapwood. The bark is used as an astringent and to treat dysentery. The flowers are pickled for salads, fried in Mexico, and used by bees for nectar. Birds eat the seeds and white-tailed deer browse the foliage. Redbud also has some value as a source of honey. George Washington and Thomas Jefferson called it Redbud. Redbud is fast growing but seems to be short lived, usually due to leaf spot, fire blight, or borers. The tree has been in cultivation since

Celtis tenuifolia (Vines 1960)

Cephalanthus occidentalis (USDA-FS)

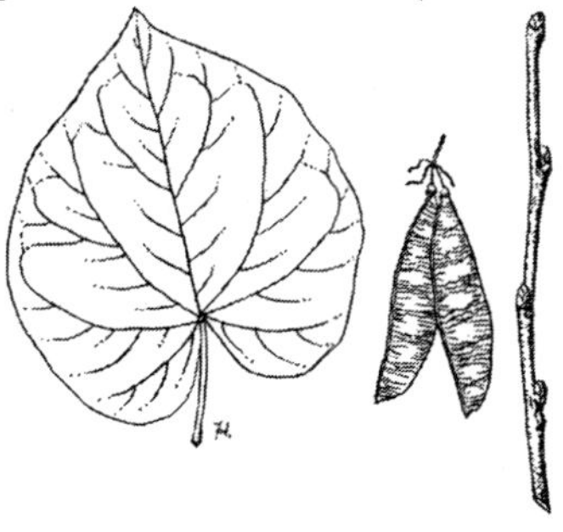

Cercis canadensis (USDA-FS)

1641. The original habitat in Louisiana was probably the well-drained upland areas (FACU) and the Inland Prairie Forests. The present day distribution includes many abandoned house sites and other human influenced habitats with reports from 50 parishes.

FLOWERING QUINCE - *Chaenomeles speciosa* - Rosaceae

Flowering Quince is also called Japanese Quince. The genus, "*Chaenomeles*", comes from chaen = open or split and meles = apple or any fruit thus split fruit. The specific epithet "*speciosa*" refers to the beautiful, showy, good-looking flowers. This shrub is introduced from the Orient. It reaches a height of 10 ft and has branch tips that are modified into thorns with some lateral thorns occasionally. The leaves are alternate, simple, deciduous, and with pinnate major veins. The stipules are large. The blades are ovate to oblong, sharply serrate, dark green and lustrous above, and paler below. The flowers are produced before the leaves, solitary or in clusters of 2-4, and are white, pink, or red. The fruit is a pome that turns yellow green. The quince fruit is not very edible but can be used to make jelly. Planted throughout Louisiana and persisting around old home sites throughout the state and reported from 18 parishes. (syn. *Chaenomeles lagenaria*)

FRINGE TREE - *Chionanthus virginicus* - Oleaceae

The genus name, "*Chionanthus*", is a combination of two Greek words meaning "snow flower." The specific epithet "*virginicus*" refers to the state of Virginia. Other common names include Grancy Gray Beard, Flowering Ash, White Ash, Old Man's Beard, American Fringe, Grandfather

Gray Beard, Snowflower Tree, Sunflower Tree, Poison Ash, White Fringe, Shavings, and Gray Beard Tree. It is a shrub to small tree with opposite, deciduous to tardily deciduous, simple leaves, and with pinnate major veins. The blades are glabrous above, glabrous to densely pubescent below, and entire. The petioles are purple and the flowers are in clusters on pendant fringelike inflorescences with green leaflike bracts. The corolla is white with four narrow, elongate petals. The fruit is a blue black ovoid drupe that is 0.4-0.8" long. The wood is light brown, sapwood lighter, hard, heavy, close-grained, and weighs about 39 lb per cu ft. The bark has medicinal uses as a diuretic and fever remedy. The boiled bark is reported to treat skin inflammations and to bathe wounds. The tree is cultivated to some extent for the fragile panicles of flowers in spring and for the dark green foliage. It has been cultivated since 1736. The leaves are persistent in winter in the Gulf Coast area, but farther north fall after turning bright yellow. Fringe Tree is found on better drained sites (FACU) throughout the state but is absent from large floodplains and the marsh. It is reported from 34 parishes with some reports possibly from planted specimens.

CAMPHOR TREE - *Cinnamomum camphora* - Lauraceae

The genus name, "*Cinnamomum*", means "resembling cinnamon odor" or "light brown with red and yellow". The specific epithet, "*camphora*", refers to the camphorous aromatic resin which the tree contains. It is a small to medium sized tree that reaches a height of 40 ft and is a native of Japan, China, and Malaya. The bark on young branches is smooth, green to reddish brown or gray and on old trunks is light to dark gray, broken into deep fissures and flattened with confluent, wider ridges. The leaves are alternate, simple, evergreen, and with pinnate major veins but with larger lower pair of lateral veins that run parallel to the blade edge. The blade is green above and paler green below, the veins are distinctly yellow on the lower surface, and two large callus-like growths are often located in the axils of the two larger lateral veins. The margins are entire with a distinct yellowish rim. The crushed leaves are very aromatic. The flowers are small, greenish white, and inconspicuous, and the fruits are small blackish drupes.

Chaenomeles speciosa (Bailey 1949)

Chionanthus virginicus (USDA-FS)

Cinnamomum camphora (Main and Mills 1981)

The wood yields the camphor of commerce. The camphor is obtained by passing a current of steam through the chips, and the volatilized camphor is then condensed into crystals and oil. Later the crystals and oil are separated by filtration under pressure. About 30 lb of chips are required for 1 lb of camphor.Camphor is used for many purposes in the Orient, but in the United States its main use is as a mild counterirritant in muscular strains and in rheumatic conditions. It is sometimes used in conjunction with phenol or menthol for local anesthetic effects to relieve skin itching or acute rhinitis. Persisting and escaping mainly in south Louisiana on well drained sites (FACU-) with reports from 27 parishes.

IVY TREEBINE - *Cissus trifoliata* - Vitaceae

The genus name, *"Cissus"*, is the Greek name for "ivy". The specific epithet, *"trifoliata"*, refers to the three lobed or three leafleted leaves. Ivy Treebine is also called Sorrel Vine. It is a fleshy to woody vine from tuberous roots that sprawls over shrubs and small trees. The stems are four or six angled with orange red lenticels, becoming warty with age, and with white pith. The leaves are alternate, evergreen, and compound with three leaflets or sometimes simple and deeply 3-lobed. The leaflets are ovate to obovate with coarsely dentate margins. The crushed leafets have a strong disagreeable odor. The flowers are small, green, in a 3-parted cyme. The inedible berry is 0.4-0.6" long and black. It is widely distributed throughout the state but more commonly in sandy soil (FAC) along the coast with reports from 18 parishes. (syn. *Cissus incisa*)

YELLOW WOOD - *Cladrastis kentuckea* - Fabaceae

Yellow Wood is also known as Kentucky Yellow Wood, American Yellow Wood, Yellow Tree, Fustic Tree, Yellow Locust, Yellow Ash, Gopher Wood, and Virgilia. The genus name, "*Cladrastis*", is from the Greek and means "brittle-branched." The specific epithet, "*kentuckea*", refers to Kentucky, one of the states where the species is common. It is a small tree reaching a height of 60 ft. The bark is thin, smooth, and beech-like. The inner bark has delicate light streaks as the outer bark becomes fissured. The leaves are one-odd-pinnately compound, alternate, 8-12" long, deciduous, and with a leaf base covering the lateral bud. The 5-11 leaflets are dark green and turn bright yellow in autumn. The flowers are pea-shaped, hang in long stems, and are pure white delicately marked with yellow. The fruit is a small legume. The wood is with distinct but not prominent growth rings, tasteless, odorless, and yellow to brown or yellow streaked with brown, sapwood nearly white, surface smooth, close-grained, hard, strong, very hard and difficult to work, and weighs about 39 lb per cu ft. It is used for fuel, cabinetwork, and gunstocks. It also yields a clear yellow dye. Yellow Wood was first introduced into cultivation in 1812. It is known from only one location (Copenhagen) in Caldwell Parish. (syn. *Cladrastis lutea)*

LEATHER FLOWER, VIRGIN'S BOWER - *Clematis* - Ranunculaceae

The genus name, *Clematis*, is a Greek name for a climbing plant that was used by Dioscorides. The Louisiana species are all herbaceous to slightly woody vines. The leaves are opposite, pinnately compound, and deciduous. The flowers are showy with the petals absent and the four sepals petaloid and with many stamens. The fruit is an achene, often in clusters and usually plumose.

SATIN CURLS (***Clematis catesbyana***) is also known as Leather Flower and Catesby Clematis. The specific epithet, "*catesbyana*", honors Mark Catesby (1697-1749), author of natural histories of parts of North America. It is very similar to ***Clematis virginiana*** but has five

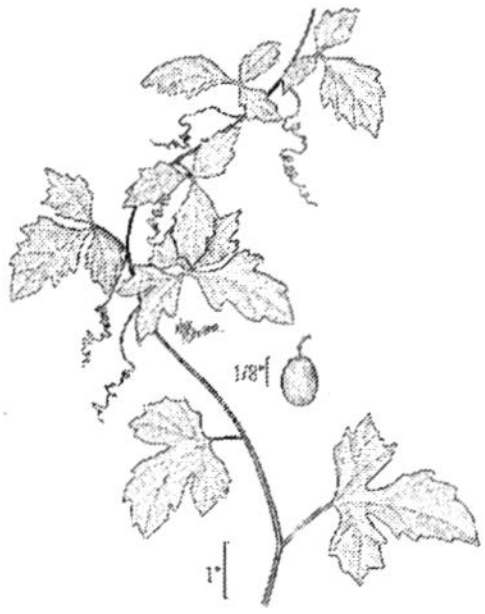

Cissus trifoliata (USDA-NRCS)

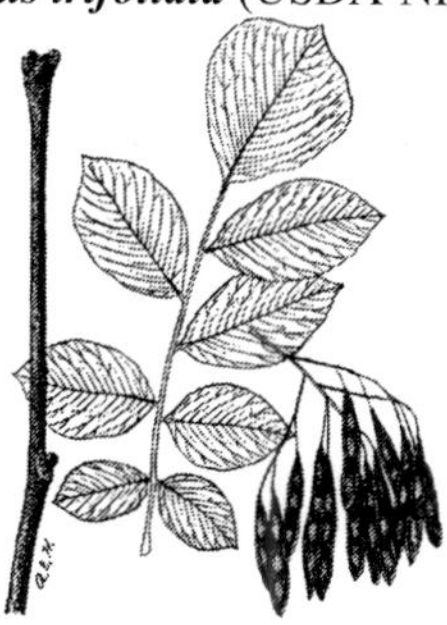

Cladrastis kentuckea (USDA-FS)

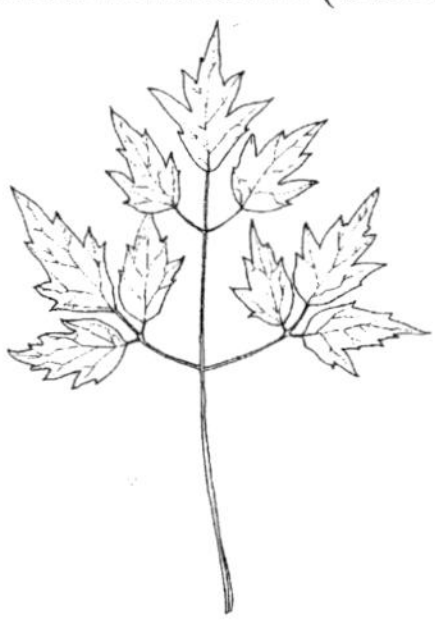

Clematis catesbyana (Nelson 1996)

leaflets and a darker achene that is reddish brown to dark purplish black. Rare in Louisiana with only reports from the loess bluffs of St. Helena and West Feliciana parishes.

SWAMP LEATHER FLOWER (***Clematis crispa***) is also known as Curl Flower, March Clematis, and Blue Jasmine. The specific epithet, "*crispa*", is for the crinkled margin of the sepals. The leaves have 3-5 linear to ovate leaflets that are cuneate to truncate at the base and acute to obtuse apically. The margins are entire and revolute, sometimes lobed. The flowers are solitary and bluish to white. Common in floodplains and edges of bodies of water (FACW+) throughout the state. It probably occurs in all with current reports from 58 parishes.

WHITE LEAF LEATHER FLOWER (***Clematis glaucophylla***) is also known as Glaucous Leather Flower. The specific epithet, "*glaucophylla*", refers to the pale undersurface of the leaves. The leaves have 2-5 ovate, entire leaflets that are usually cordate at the base. The lower surface is grayish white when young, hence the name glaucophylla, but becoming green with age. The flowers are solitary and rose purple on the outside and white inside. Very rare in Louisiana with the only report from Copenhagen (FAC-) in Caldwell Parish.

NET LEAF LEATHER FLOWER (***Clematis reticulata***) is also known as Net Leaf Clematis. The specific epithet, "*reticulata*", refers to the strongly net-veined leaves. The leaves have 3-9 ovate to elliptic, leathery, leaflets that are usually cuneate at the base and obtuse apically. The veins are distinctly reticulate and the margins are revolute, frequently lobed. The flowers are solitary and bluish. Infrequent in west and northwest Louisiana on well drained, often sandy soils with reports from seven parishes.

SWEET AUTUMN VIRGINS'S BOWER (***Clematis terniflora***) is also known as Sweet Autumn Clematis. The specific epithet, "*terniflora*", may refer to the secondary panicle-branches, which often have three flowers. The leaves have 3-5 ovate to elliptic leaflets that are entire to lobed. The base of the leaflets is subcordate or rounded and the apex acuminate to acute. The flowers are white and in large clusters. The achenes are brown and obovoid. Introduced from Japan and Korea and escaping and spreading (FAC-). The current records include 32 parishes. (syn. *Clematis dioscoreifolia, Clematis maximowicziana, Clematis paniculata*)

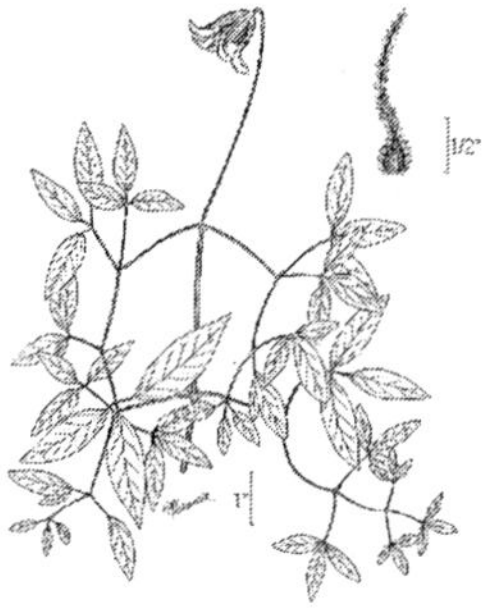

Clematis crispa (USDA-NRCS)

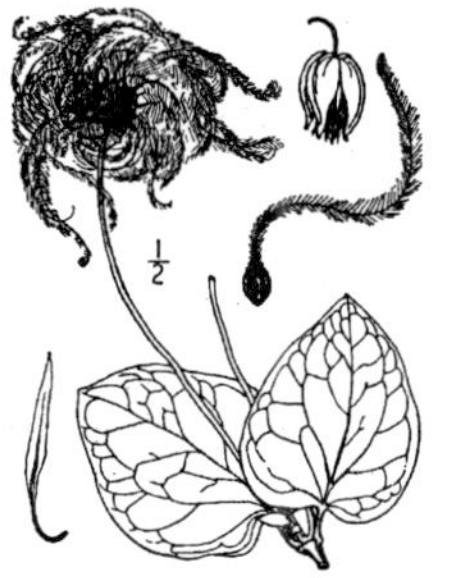

Clematis glaucophylla (Britton and Brown 1943)

Clematis reticulata (Vines 1960 as ***Clematis paniculata***)

Clematis terniflora (Vines 1960)

VIRGIN'S BOWER (***Clematis virginiana***) is also known as Devil's Darning Needle, Herbe and Gueux, and Gander Vine. The specific epithet, "*virginiana*", is for the state of Virginia. The leaves have three, rarely five, ovate leaflets that are serrate to lobed, rarely entire. The base of the leaflets is truncate to cordate and both surfaces are light green. The flowers are white and in large clusters. The achenes are light to dark brown or greenish brown. Various medicinal uses are reported including a liniment once used for skin eruptions and itching and a leaf tea used for insomnia, nervous headaches and uterine disease. There are also warnings that the plant can be toxic and highly irritating to the skin and ingestion may cause vomiting, diarrhea, and convulsions. The plant was first introduced into cultivation in 1720. This species is widely distributed in Louisiana and ranges from moist to well drained habitats (FAC+). It is absent from the marsh and is currently reported from 51 parishes.

GLORYBOWER - *Clerodendrum* - Verbenaceae

The genus name, "*Clerodendrum*", is from the Greek "kleros", "chance", and "dendron", "a tree", and alludes to a native legend that the trees possessed medicinal properties. The genus includes shrubs that are native to tropical or subtropical areas. The leaves are opposite or whorled, simple, and deciduous. The flowers are salverform, large and showy and in terminal heads or corymbs.

ROSE GLORYBOWER (***Clerodendrum bungii***) is also known as Brocamelia and Camelia Americana. The specific epithet, "*bungii*", honors Alexander Andrejewitsch von Bunge, a Russian botanist and explorer from 1803-1890. He was a professor of botany at Dorpat and author of "Flora of Russia and Central Asia" in 1851. The stems are dark and have distinct lenticels. The leaves are opposite, broad, deltoid-ovate, and with palmate major veins and the margins are dentate. The leaves are dark green above and paler below and the crushed leaf has an unpleasant odor. The flowers are rose to red or purple. It is native to China and northern India and is cultivated in Louisiana and persisting or perhaps escaping mostly in south Louisiana. The current records include 15 parishes. (Not Illustrated)

TURK'S TURBAN (***Clerodendrum indicum***) is also known as Skyrocket, Turk's Head, Bead Flower, and Tube Flower. The specific epithet, "*indicum*", is for India, where it is native. The stems are sulcate and hollow. The leaves are whorled, oblong or elliptic to oblanceolate, with pinnate major veins and entire margins. The blades are pale green on both surfaces. The flowers are white to yellow. The roots and leaves are used to treat asthma in Burma and Java. It is native to India and Burma and is cultivated in Louisiana and persisting or perhaps escaping mostly in south Louisiana. It is currently reported from 11 parishes.

SWEET PEPPER BUSH - *Clethra alnifolia* - Clethraceae

Sweet Pepper Bush is also known as White Alder, Coast Pepper Bush, Summersweet Clethra, White Bush, and Spice Bush. The genus name, "*Clethra*", is the Greek name for Alder, and the specific epithet, "*alnifolia*", means "alder-leaved". It is a shrub or small tree, up to 13 ft, with slender, mostly erect stems. The stems are woody, densely hairy at first, and becoming smooth and purple-brown at maturity. The leaves are alternate, simple, deciduous, and with pinnate major veins. Some plants may have leaves that are densely tomentose beneath. The blades are smooth on the upper surface, smooth or hairy on the lower surface, thin, obovate, strongly narrowed at the base and sharply toothed towards the tip. The flowers are small, white or pinkish, in erect and usually branching racemes, with a spicy odor, and appear in mid-summer. The fruit is a small, smooth, nearly spherical hairy capsule. The racemes, bearing empty capsules, persist this plant during the fall and winter months. White Alder has been in

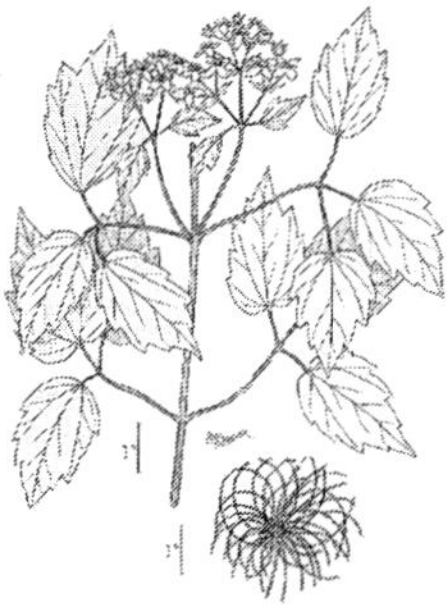

Clematis virginiana (USDA-NRCS)

Clerodendrum indicum (Radford, Ahles, and Bell 1968)

Clethra alnifolia (USDA-NRCS)

cultivation since 1906. It is found in baygalls, bogs, around ponds and along streams (FACW) in southeastern Louisiana. It is reported from uncultivated sites only from St. Tammany, Tangipahoa, and Washington parishes. (syn. *Clethra tomentosa*)

BUCKWHEAT BUSH - *Cliftonia monophylla* - Cyrillaceae

Buckwheat Bush is also called Black Titi, Iron Wood, and Titi. The genus name, "*Cliftonia*", honors Francis Clifton (d. 1736), an English physician. The specific epithet, "*monophylla*", means one leaf. The fruit resembles that of buckwheat, and hence the common name of the plant. It is a medium-sized tree with dark, sometimes blackish bark. The leaves are alternate, simple, evergreen, and with pinnate major veins. The blades are entire, normally sessile, rarely with short petioles, and with the veins on the lower surface not conspicuous. The mature leaves are dark green and the young leaves on fruiting branches are pale and bluish green. The flowers are white or rose, and fragrant. They are borne in conspicuous, upright clusters in February or March. The fruit is a two to four winged capsule that is golden amber in color and borne in conspicuous clusters from August to September. It is attractive to bees and butterflies. It is a source of excellent honey. Deer and livestock browse the foliage. The pithy twigs are used for pipe stems. It was first cultivated in 1806. The wood is reddish brown, lighter-colored sapwood, close-grained, fairly hard, weak, and brittle. It is used for fuel locally. It inhabits wet areas along baygalls and bogs (OBL). It is rare in southeastern Louisiana with collections only from St. Tammany and Washington parishes.

CAROLINA SNAIL SEED - *Cocculus carolinus* - Menispermaceae

Carolina Snail Seed is also known as Carolina Coral Bead, Moon Seed, Red Berry Moon Seed, Red Moon Seed, Snail Seed, Coral Seed, Coral Vine, Carolina Moon Seed, Wild Sarsaparilla, Margil, and Hierba del Ojo. The genus, "*Cocculus*", refers to the curled, snail-like seed. The specific epithet, "*carolinus*", is for Carolina. It is a dioecious twining, slender, semi-woody vine that has a green to brown stem. The leaves are alternate, simple, deciduous, and with palmtae major veins. The blades are 1-4" long, ovate to triangular, cordate at base, acute or obtuse apically, and entire or three shallowly lobed. The petioles are hairy. The bright green color makes the leaves stand out in winter months. The flowers are green and the male plants have large branched racemes while the female plants have smaller unbranched racemes. The fruit are bright red drupes with spiraled seeds (like a snail) and is flattened on one side. The flowers appear from April to May with the fruit maturing by August to October. The cluster or red fruit give these vine good ornamental possibilities. The fruit is consumed by a number of species of birds. The root is used as a diuretic and for arthritis and blood disorders. This vine is widely distributed in Louisiana (FAC) but is most common on clay soils of bottomlands and also in the Inland Prairie Forests. It is currently reported from 57 but probably occurs in all 64 parishes.

DOGWOOD – *Cornus* - Cornaceae

The genus name, "*Cornus*", is from cornu (a horn), in reference to the hard wood. The common name, Dogwood, comes from the fact that a decoction of the bark of *Cornus sanguinea* was used in England to wash mangy dogs. The genus includes trees or shrubs with opposite, simple, deciduous leaves. The lateral veins are arcuate that is they curve forward along the margins. The blades have entire margins and an acuminate apex. The blades of dogwoods can be carefully torn in half and the two torn halves will stay attached by material, the primary xylem, from the veins. The flowers are produced in a cyme or head with a four-parted flower. The fruit is a drupe with a hard pit in which there are two seeds. The color of the drupes for the Louisiana species is like the American flag with one species red, one white, and one blue.

Cliftonia monophylla (Vines 1960)

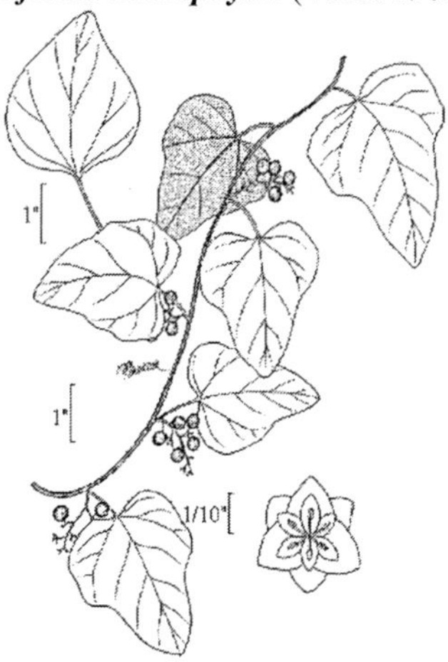

Cocculus carolinus (USDA-NRCS)

ROUGH LEAF DOGWOOD (***Cornus drummondii***) is also known as Cornel Dogwood, Small Flower Dogwood, and White Cornel. The specific epithet, "*drummondii*", honors Thomas Drummond (1780-1835), a Scottish botanical explorer. Rough-leaf Dogwood is a shrub to 3 m high that is loosely branched, with twigs that are reddish brown to brown. The bark on young branches and trunks are rather smooth, pale gray to brown. On old trunks, it is gray with narrow ridges and fissures and the scales are small. The pith is brown. The blades are ovate to elliptic-lanceolate and mostly narrower than 1". The upper surface is scabrous while the lower surface is pubescent. The cream colored flowers are produced in a cyme from May to August with the leaves. The drupes are round, white, closely grouped, and have a rancid odor during development. Quail, turkey, and over 40 species of birds and several mammals eat the fruit. Whitetail deer browses the twigs and foliage. It has been known in cultivation since 1836. The wood is pale brown, with sapwood paler, heavy, hard, strong, durable, and close-grained. It is used for small woodenware articles, especially shuttle-blocks and charcoal. This species is widely distributed in Louisiana from the Inland Prairie Forests to bottomlands (FAC). The current distribution includes 57 parishes.

FLOWERING DOGWOOD ***(Cornus florida)*** is also known as Virginia Dogwood, Arrow Wood, Box Wood, American Box Wood, Dogwood, Cornelius Tree, False Box, Nature's Mistake, Cornel, White Cornel, and Florida Dogwood. The specific epithet "*florida*" refers to the large petal-like bracts surrounding the flower head. Flowering Dogwood is a shrub or small tree that reaches up to 40 ft. The bark is grayish brown or black with squarish rough checks. The leaves are smooth on both surfaces, shiny, elliptical to broadly oval, 3-6" long, 2-3" wide and light on the undersurface. The flower buds form in the fall as conspicuous gray-green buttons at the tips of branches. The flowers are small, but are surrounded by four large white rarely pink bracts. The bracts are often mistakenly called petals. The flowers are produced from March to April before the leaves come out. When, in full bloom, the tree is a sheet of white, the most conspicuous feature of the spring landscape with some towns having, i.e. Pollock hosting a dogwood celebration every year. The drupe is bright red drupe and borne in close small clusters. The strong, heavy, very close-grained, hard, splinter-free wood weighs 51 lb per cu ft. The wood is brown, sometimes changing to shades of green and red. It is used for small woodenware articles, tool handles, wheel hubs, pulleys, shuttles, bobbins, mallets, golf club heads, wood engraving, weaving spindles, and barrel hoops. The powdered bark is used as toothpaste, black ink (when mixed with iron sulfate) and a quinine substitute. The dried bark of the roots is used for intermittent fever, scarlet dye, astringent, tonic, antiperiodic, and malaria. The very bitter inner bark was mixed with cherry bark in a remedy for jaundice. Flowering dogwood is used by 28 species of birds and is the 21st on the list of preferred plants by quail, turkey, squirrels, and white-tailed deer. It is the state tree of Virginia and Missouri and is the state flower of North Carolina. There is a story that the cross on which Jesus Christ was crucified was made from dogwood. It is said that the reddish tip of the flower petals was colored with his blood. Flowering Dogwood was first cultivated in 1731. It is found in the uplands in rich, well-drained soils (FACU) of Louisiana. It is reported from 48 parishes and is absent from the large floodplains and the coastal marsh regions.

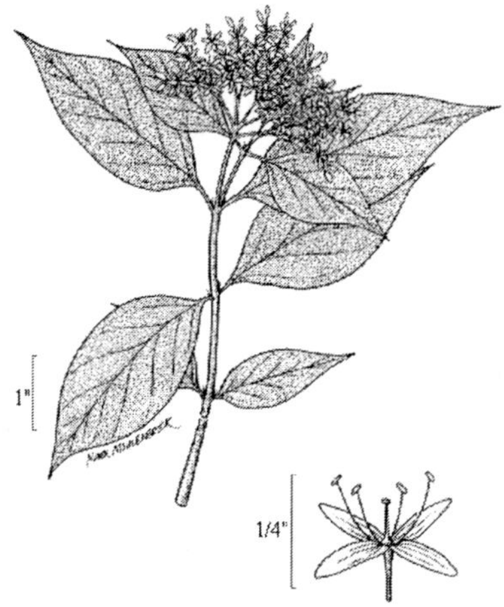

Cornus drummondii (USDA-NRCS)

Cornus florida (USDA-NRCS)

SWAMP DOGWOOD (***Cornus foemina***) is also known as Gray Dogwood, Stiff Cornel Dogwood, Stiff Dogwood, and Blue Fruited Dogwood. The specific epithet, "*formina*", is an obscure reference to "female", perhaps because the plant is smaller and more slender than other dogwoods. It is a shrub or small tree up to 15 ft tall. The bark is smooth. The twigs are red to gray with white pith. The blades are lanceolate to elliptic 1-3" long and mostly narrower than 1". Both surfaces are smooth and the upper surface is dark green while the lower surface is pale green. The flowers are small, cream-colored, and borne in spreading clusters from May to June. The fruits are globular, blue and mature from August to October. Swamp Dogwood has been in cultivation since 1758. Widely distributed (49 parishes) in wet areas (FACW-) throughout the state. (syn. *Cornus stricta)*

HAZEL NUT - *Corylus americana* - Betulaceae

Hazel Nut is also known as Filbert and American Filbert. The genus name, "*Corylus*", is the classical Greek name, which probably originated from the hood-like or helmet-like appearance of the involucre. The specific epithet, "americana", denotes its origin in America. The name Filbert is of French origin, the European Filbert becoming ripe about the time of St. Philbert's Day (August 22). Hazelnut is a thicket forming or colonial shrub that reaches a height of 10 ft with pubescent or glandular pubescent twigs. The bark is smooth and light gray. The leaves are alternate, simple, and somewhat heart-shaped and double toothed. The flowers are found from March to May. The light green to pale brownish yellow male catkins are up to 8" long, straight, slender and regularly spaced along the upper stems in spring. About the time the leaves are shed in the fall, the new catkins are pale green and short, becoming brown in the winter. The red female flowers are tiny and almost completely enclosed by bracts. They are located near the ends of the twigs. Flowers develop before the leaves. The fruit is an edible nut enclosed in thin flattened, hairy, ragged-edged husks. Several mammals and birds eat the nuts and deer browse the foliage. Beavers eat the bark. It has been cultivated since 1798. The sweet nuts are sold on the market and may be eaten raw, or ground and made into cake-like bread. Hazelnut is rare in Louisiana with wild populations in Morehouse and Union parishes near the Arkansas border on well-drained soils (FACU).

HAWTHORN - *Crataegus* - Rosaceae

The genus name, "*Crataegus*", is of Greek origin and translates as "flowering thorn" or "strong wood". Hawthorns are shrubs to small trees with alternate, simple, deciduous leaves with pinnate major veins. Most are armed with thorns. The flowers are produced in corymbs that may be simple or compound. The flowers have five green sepals, five white petals, five styles, and many (sixteen to twenty) pale yellow stamens. The fruit is a pome. The fruits of most hawthorn species are eaten by several species of birds. The fruit of hawthorns, though not usually good raw, is used for making tea, jam, and jelly, with Mayhaw jelly being the most common in Louisiana. Steeping the fruit with peppermint leaves makes a pleasant tea.

BAR BERRY HAWTHORN (***Crataegus berberifolia***) is also known as Sabine Hawthorn. The specific epithet, "*berberifolia*", refers to the Bar Berry-like leaves. This is a small tree or large shrub to 10 ft and 6" diameter. The bark is light gray, becoming furrowed with age. The twigs are brown, slender, with white pubescence when young, glabrous with age, spineless or with slender red-brown to black spines that are 1-1.5" long. The blades are oblong-cuneate to spathulate, 1.5-2.5" long and 0.75-1.25" wide, coarsely pubescent above, glabrous with age, pubescent beneath, particularly along the veins, firm, dark green above, paler beneath, cuneate at base, rounded or obtuse at apex, serrate above the middle. The petioles are short and densely villose when young. The flowers are produced from March to April and the sepals and petals are

Cornus foemina (USDA-NRCS)

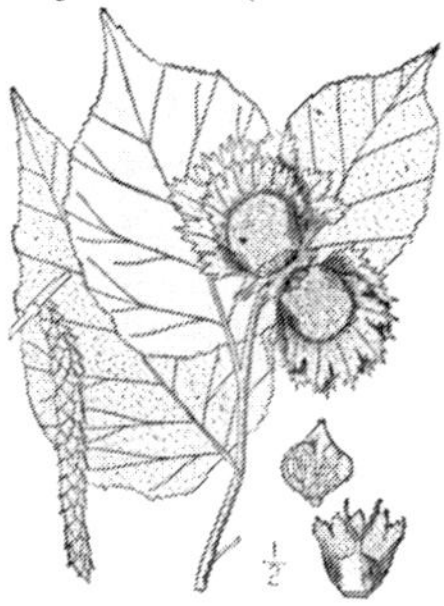

Corylus americana (Britton and Brown 1943)

Crataegus berberifolia (Vines 1960)

pubescent. The yellow-orange pome is subglobose, about 0.5" in diameter, and has 2-3 small ridges. It is found infrequent in Louisiana with reports from 19 parishes only west of the Mississippi River (FAC). The species is considered to be closely related to ***Crataegus crus-galli*** and perhaps could be described as a Southern pubescent variety of it. (syn. *Crataegus crocina, Crataegus crus-galli* var. *berberifolia, Crataegus edura, Crataegus fera, Crataegus tersa*)

BLUE BERRY HAWTHORN (***Crataegus brachyacantha***) is also called Blue Haw and Pamett Bleu, in French. The specific epithet, "*brachyacantha*", refers to the short thorns. It is a shrub or small tree to 18 ft and 6" diameter. The inner bark is reddish-brown and the outer bark is gray to brownish and with loose scales. The twigs are reddish brown to gray, glabrous, short, with curved spines approximately 0.8" long. The blades are elliptical, 1-2" long and 0.5-1" wide, glabrous above, glabrous to pubescent below, dark green above, paler below, narrow at base, blunt or rounded at apex, wavy, saw-toothed and rounded with gland-tipped teeth. The leaf veins of blue haw extend from the midrib to the lobes and sinuses and the larger leaves are 3-lobed. The flowers are glabrous and produced March to April. The pomes are on short, straight, stubby side twigs and are bright blue to black, globose, and about 0.5" in diameter. They begin to drop soon after ripening in late summer. The wood is hard, strong, heavy, and not large enough to be commercially usable. It is found in wet clay soils (OBL) throughout the state with reports from 26 parishes.

COCKSPUR HAWTHORN (***Crataegus crus-galli***) is also called Red Haw. The specific epithet, "*crus-galli*", means "cock's spur", referring to the long thorns that resemble the spurs of a rooster. It is a large shrub to small tree that may reach 20 ft. The bark is dark gray to brown and oak-like on large plants. The twigs are reddish brown when young, graying with age, glabrous, with many slender, brown straight spines up to 4" long. The blades are oblanceolate to narrowly elliptical, about 1" long and 0.6" wide; slightly thick and leathery, widest above the middle and tapering to a narrow base, short-pointed or rounded at apex, serrate above the middle with gland-tipped teeth; glabrous, dark shiny green above, and pale green below with raised veins. The petiole is 0.5-0.75" long, stout and winged. Some branches or entire plants may be without thorns. The glabrous flowers are produced from April to May. The pomes are dull red to yellow or green, subglobose or ovoid, and about 0.5" in diameter. The pomes remain attached throughout the winter. The wood is heavy, hard, fine-grained, and suitable for tool handles. Cockspur Hawthorn is an important plant as a wildlife food source. Several birds including bluejay, fox sparrow, cedar waxwing, quail, ruffed grouse, bear, raccoon, deer, flicker, and smaller mammals feed upon the fruits. The thorniness, dense branching and heavy foliage makes Cockspur Hawthorn favorite nesting sites for many birds. Beavers eat the bark of Cockspur Hawthorn. It is widely distributed on well drained to moderately drained soils (FAC-) with reports from 28 parishes west of the Mississippi River. (syn. *Crataegus algens*, *Crataegus bushii*, *Crataegus cherokeensis, Crataegus sabineana*)

FAN LEAF HAWTHORN (***Crataegus flabellata***) is more common east and north of Louisiana and perhaps does not reach the state. The specific epithet "*flabellata*" means "fan like" and refers to the shape of the blades. It is a shrub or small tree with spines 1-2" long. The twigs are glabrous. The blades are ovate, 1.5-3" long, 1-2" wide, margins doubly toothed and usually with 4-6 shallow lobes, glabrous or pubescent above, glabrous below, the base truncate to widely cuneate, and the apex is acute. The pedicels are glabrous as are the sepals. The pome is globose, 0.5-0.75" in diameter, and red. It is rare; one report from St. Tammany Parish.

Crataegus brachyacantha (Vines 1960)

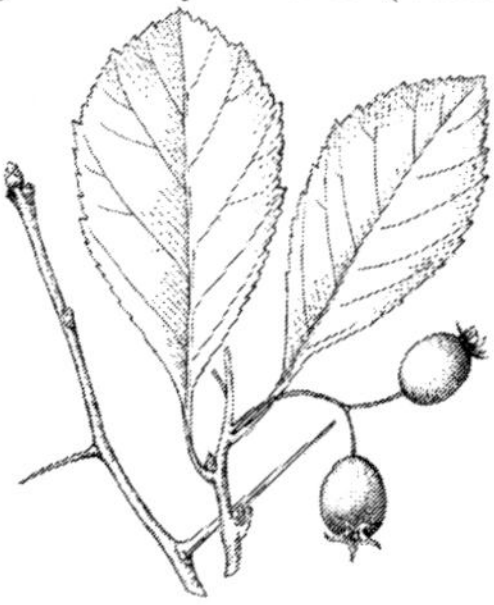

Crataegus crus-galli (USDA-FS)

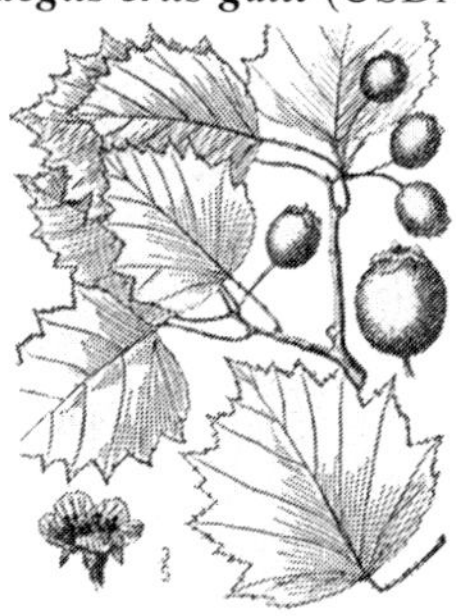

Crataegus flabellata (Britton and Brown 1943)

BIG FRUIT HAWTHORN ***(Crataegus macrosperma)*** is also known as Roan Mountain Haw and Variable Haw. The specific epithet, "*macrosperma*", refers to the large seeds. It is a small tree to 25 ft with numerous, stout, short, 0.5-1" long, dark reddish brown to purplish or black spines. The bark is grayish brown and broken into small oblong scales. The young twigs are reddish brown to purplish with grayish lenticels and the older twigs become gray. The blades are deltoid or broadly oval, the base subcordate to cordate, the apex obtuse, and the margins serrate with the teeth gland-tipped, and often with 3-5 pairs of shallow lobes above. The flowers are produced from April to May and the pedicels are glabrous. The pomes are 0.5-1" in diameter and dark bright red. Rare in Louisiana in well drained soil with reports from Ouachita and Winn parishes.

PARSLEY HAWTHORN ***(Crataegus marshallii)*** is also known as Parsley Haw. The specific epithet, "*marshallii*", honors the botanist Humphrey Marshall (1722 - 1801), a Pennsylvania botanist and father of American dendrology, who first described Parsley Hawthorn. Parsley Hawthorn is a shrub or small tree that reaches up to 20 ft and 4" diameter. The bark is smooth and gray to brown, scaling off in thin large plates and exposing the reddish-brown underbark. The twigs are light brown, pubescent with small spines about 1-1.5" long. The blades are broadly ovate, 1-2" long and 0.8-1.5" wide, deeply divided into 5-7 narrow coarsely toothed lobes; glabrous, lustrous green above, paler beneath; truncate, cuneate, or subcordate at base, acute at apex, margins serrate, and the petiole is slender, 1-2" long and tomentose. The name parsley is descriptive of the parsley-like margin of the leaves. The white flowers are borne in clusters of 3 to 12, having 5 petals per flower, 20 stamens, and rose-colored anthers and appear before the leaves from March to April. The flowers have a foul odor that is attractive to flies. The pomes are red, about 0.3" diameter and clings all winter. The heavy, hard, strong wood weighs 46 lb. per cu ft and is reddish brown. Game birds and Mammals, including deer, use Parsley Hawthorn for food during the winter. It is widely distributed in well drained to somewhat moist sites (FAC) throughout the state with records from 43 parishes.

MAY HAW ***(Crataegus opaca)*** is also known as May Hawthorn and Riverflat Hawthorn. The specific epithet, "*opaca*", refers to the dull fruit. It is a rounded shrub or small tree. It has stout, very sharp thorns scattered over the branches that are more numerous on the older wood. The blades are oblong to obovate or elliptic, 1-2.5" long, 0.5-1.3" wide, apex acute or rounded, base cuneate, upper surface dark green and glabrous, and lower surface dense rusty brown pubescent. The margins are crenate serrate to three lobed. The large white flowers, up to 1.5" wide, bloom before the leaves appear, are in groups on green pedicels and have greenish centers, dark brown anthers, and 5-6 cupped petals. The fruits are 0.5" or more in diameter, greenish to reddish, and shiny. This is the famous May Haw of the South, from which preserves are made. The large size and acid character of the pomes make it particularly desirable for that purpose. The wood is heavy, hard, strong, and not large enough for commercial use. Widely planted but natural populations are found in wet to moist (OBL) areas especially flats throughout the state with records from 31 parishes.

PEARSON HAWTHORN ***(Crataegus pearsonii)*** is a small tree to about 15 ft and 4" diameter. The specific epithet "*pearsonii*" honors Professor Gilbert Person of Guilford College, NC. The twigs are light to dark gray or brown, glabrous and with slender chestnut brown spines to 3" long. The blades are about 2.5" long, obovate to orbicular, truncate at base, and obtuse to broadly rounded apically. The margins are singly or doubly serrate with glandular teeth andoften lobed above the middle. The flowers are on glabrous to pubescent pedicels, with

Crataegus macrosperma (Britton and Brown 1943)

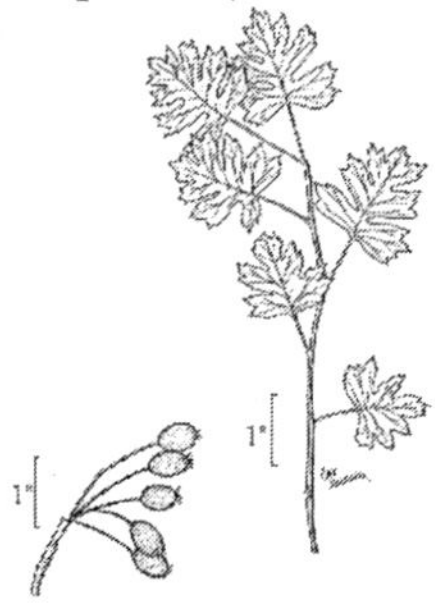

Crataegus marshallii (USDA-NRCS)

Crataegus opaca (USDA-FS)

Crataegus pearsonii (Vines 1960)

glandular sepals, and 20 stamens with yellow anthers. The pomes are bright red, subglobose, and about 0.5" in diameter. Infrequent in well drained hardwood slope forests sporadically throughout the state with records from Claiborne, Jackson, Lincoln, and Washington parishes.

FROSTED HAWTHORN ***(Crataegus pruinosa)*** is also known as Waxy Fruit Thorn. The specific epithet, "*pruinosa*", means "frosty" and refers to the waxy bloom of the fruit. It is a shrub or small tree to 20 ft and 4" in diameter. The bark is gray and scaly and the twigs are brown, and glabrous with straight spines about 1.5" long. The blades are deltoid or ovate, 1-2" long and 0.8-1.5" wide; glabrous, bluish-green color at maturity and pruinose (having a whitish-powdery secretion on the surface); blunt at base, margins doubly serrate and with 6-8 pairs of shallow lobes often present. The flowers are glabrous with 20 rose colored stamens and are produced from April to May. The pomes are reddish purple and pruinose, globose, and 0.4-0.6" diameter. The fruits of hawthorn species are eaten by several species of birds. It is infrequent to rare in well drained upland areas in north Louisiana with records from Caldwell, Catahoula, Ouachita, and Union parishes.

LITTLE HIP HAWTHORN (***Crataegus spathulata***) is also known as Pasture Hawthorn. The specific epithet, "*spathulata*", refers to the spathe- or spoon-shaped leaves. It is a shrub or small tree to 20 ft. The bark is light brown to dark gray and in large flakes that peel back to expose the bright reddish orange inner bark. The twigs are red-brown, glabrous with straight spines to 1.5". The spines are sparse and may be completely absent from some plants. The blades are spathulate to oblanceolate, 0.8-1.5" long and 0.4-1" wide, hairy when young, glabrous at maturity, shiny dark green above, paler beneath; tapering at base, rounded at apex, margins crenate-serrate, sometimes three-lobed and with winged petioles. The wood is reddish brown, heavy, hard, strong, weighing 45 lb per cu ft, not large enough for commercial use. The flowers are small, glabrous, numerous, with 20 pale yellow stamens, and are produced from March to May. The pomes are bright red, globose, small, about 0.25" in diameter. If fruits are not present, Little Hip Hawthorn can be separated from Blue Haw (***Crataegus brachyacantha***) by the veins: those of Blue Haw extend from the midrib to the lobes and sinuses while those of Little Hip Hawthorn extend only to the lobes. Common in well drained sandy sites (FAC) and clay soils throughout Louisiana but absent from the coastal marsh and large floodplains. There are 27 parishes currently recorded for this species.

THREE FLOWER HAWTHORN ***(Crataegus triflora)*** is similar to ***Crataegus crus-gallii*** but with the flowers and fruits in threes. The specific epithet, "*triflora*", refers to the flowers in threes. It is very rare in Louisiana with records only from Caldwell Parish. (Not Iillustrated)

ONE FLOWER HAWTHORN ***(Crataegus uniflora)*** is also called Dwarf Thorn. The specific epithet, "*uniflora*", means "one-flowered". It is a small shrub to 12 ft with spreading branches. The twigs are red-brown, turning to gray with age, densely pubescent, with slender spines 0.4-2.4" long. The blades are obovate, elliptic or ovate with sunken veins, 0.4-2.4" long and 0.3-1.3" wide, the upper surface with scattered hairs to glabrous and dark green and lustrous, lower surface pubescent, and paler, cuneate at base, serrate with tiny black glands, and with a short petiole. The flowers are solitary on tomentose pedicels with sepals that are glandular-serrate and persist on the fruit. The 20 stamens are white or pale yellow and the flowers are produced from April to May with the leaves. The pomes are subglobose, green-yellow to dull red, and about 0.5" diameter. It is rare in Louisiana in dry sandy soils with reports from Bienville, Caddo, Vernon, and Winn parishes.

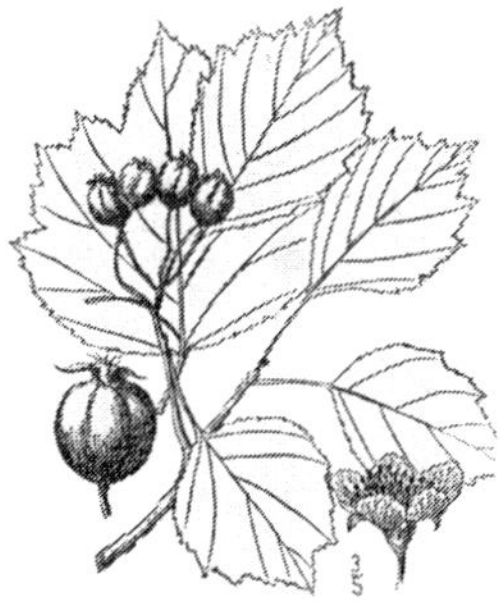

Crataegus pruinosa (Britton and Brown 1943)

Crataegus spathulata (USDA-FS)

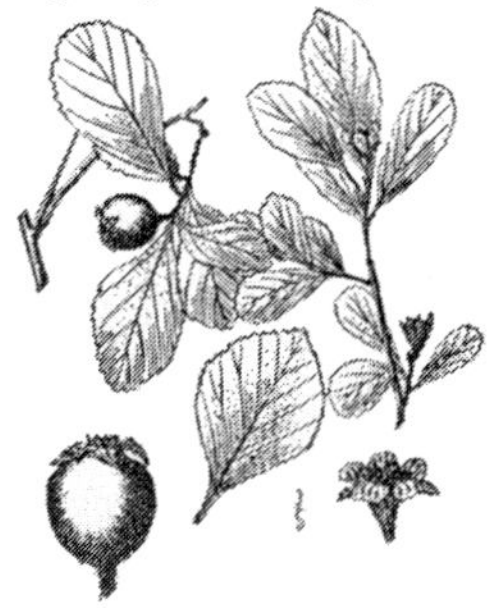

Crataegus uniflora (Britton and Brown 1943)

GREEN HAWTHORN (***Crataegus viridis***) is also known as Green Thorn and Southern Thorn. The specific epithet, "*viridis*", means green. It is a small tree to 35 ft and 6" diameter that often has a fluted base and forms thickets. The crown is dense and rounded. The bark is pale gray, scaly, with orange inner bark. The twigs are brown, glabrous and spines are often absent. The blades are ovate-oblong, 1-2.5" long, 0.5-1.5" wide, glabrous when mature; cuneate or rounded at base, acute to acuminate at apex, margins serrate or doubly serrate, often with shallow lobes toward the base, tufts of hairs present where veins join, dark green above, paler beneath. The petioles are slender and glabrous. The wood is reddish brown, heavy, hard, tough, weighing about 40 lb per cu ft, not large enough to be commercially valuable. The flowers are glabrous, with 20 pale yellow stamens and appear from March to April. The pomes are red or orange, globose, and small, 0.25" diameter. These pomes cling all winter, adding a bright touch to winter woods at the edge of swamps. The fruit is used to make jam, jelly, and tea. It is very common in bottomland hardwood forests (FACW) throughout the state with reports from 46 parishes.

SILVER LEAF CROTON - *Croton argyranthemus* - Euphorbiaceae

Silver Leaf Croton is also known as Healing Croton. The genus name, "*Croton*", is the Greek word for tick, referring to the fact that the seeds of some plants of this genus resemble a tick. The specific epithet, "*argyranthemus*", refers to the silvery scales of the flowers. It is a plant that is 12-24" tall and often sprawling. The leaves are simple, alternate, evergreen, and with pinnate major veins. The margins are entire. The upper surface is scaly or glabrate and the lower surface is clothed with conspicuous silvery scales, often with a central mass of rusty brown scales. The crushed leaves have a strong odor. The flowers are greenish racemes. The fruit is an oval-globose capsule. It is rare in very sandy soil in western Louisiana with records from Caddo, Sabine, and Vernon parishes.

CRYPTOMERIA - *Cryptomeria japonica* - Taxodiaceae

Cryptomeria is Greek for hidden part and the specific epithet "*japonica*" is for Japan. Cryptomeria is also known as Japanese Cedar and Sugi. Cryptomeria is an introduced evergreen from Japan and China that reaches a height of 80 ft. The bark is reddish-brown, fibrous, fissured, and peeling off in long strips. The leaves are needlelike 0.25-0.75" long, and with an odor when crushed. The leaves curve forward and inward with the base extending toward the twig, slightly stiff and blunt, somewhat flattened, 4-angled, and dull blue-green. The cones are 1-2" long, 0.5-1" diameter, rounded, dull, short-stalked, and opening at maturity in autumn but remaining attached. Many cone-scales are with bristly point in center and 3-5 points at end. Under the cone scale, there are 2-5 slightly 3-angled, narrowly winged seeds. Cultivated in Louisiana and persisting with records from Iberia, Lafayette, Ouachita, and Washington parishes.

CHINESE FIR - *Cunninghamia lanceolata* - Taxodiaceae

Chinese Fir is an evergreen tree native to China that reaches 80 ft. The genus, "*Cunninghamia*", honors James Cunningham, who discovered the tree in 1702. The specific epithet "*lanceolata*" refers to the shape of the needles. The twigs are green the first year with the young branches whorled and later becoming more irregular. The bark is gray-brown on the outside, irregularly scaly and peeling to reveal the reddish brown inner bark. Needles are arranged in a spiral around the stem, and have a pleasant upward arch. The needles are linear to lanceolate, about 3" long, sharp pointed, with a broad decurrent base, finely serrate, and with two whitish bands beneath. The crushed leaf has an odor. The cones are roundish to ovate with leathery serrate and pointed scales. The seeds are narrowly winged. In its native habitat in China, the wood is widely used for construction of houses and shipbuilding, cabinet making, and coffins. The Chinese valued

Crataegus viridis (USDA-NRCS)

Croton argyranthemus (Vines 1960)

Cryptomeria japonica (Bailey 1949)

this tree for its many medicinal properties. A wood decoction was used as a bath for fetid (smelly) feet. The decoctions were ingested for lacquer poisoning, to help chronic ulcers, cholera, and even alleviate flatulence. An essential oil was also made from the stem and used for bruises, pain from rheumatism, and wounds. Similarly, the ash from old bark was applied to burns, scalds, and wounds. A decoction from the cones was also taken as a cough remedy. Cultivated throughout Louisiana and persisting with 13 parish records. (Not Illustrated)

TITI - *Cyrilla racemiflora* - Cyrillaceae

Titi is also called American Cyrilla, Black Titi, Burn Wood Bark, Leather Wood, Swamp Cyrilla, Red Titi, White Titi, and He Huckle Berry. The genus name, "*Cyrilla*", honors Cyrillo Dominico, professor of medicine at Naples (1734-1799), and the specific epithet, *"racemiflora"*, refers to the racemose flowers. Titi is a small, often twisted, thicket forming shrub or small tree up to 30 ft that is tardily deciduous and appearing evergreen. The bark is thin, with bright red-brown scales. The twigs have a ridge that runs downward from just below the base of the petiole. The leaves are alternate, simple, and with pinnate major veins. The blades are elliptic to narrowly obovate, entire, and the veins on the lower surface are visible. The white flowers appear in late spring and early summer in distinct racemes. The fruit is a small, inconspicuous, dry drupe. The wood is reddish brown, close-grained, heavy, hard, not strong, and has no commercial value. The flowers are considered good bee food. The spongy lower bark was used as a styptic. It is an important deer browse when the shoots are tender. It is found in baygalls and along the edges of swamps and streams preferring acid, sandy, or peaty soils (FACW). It is found in 12 parishes in southeastern and southwestern Louisiana.

WATER WILLOW - *Decodon verticillatus* - Lythraceae

Water Willow is also known as Swamp Loosestrife, Willow Herb, Water Oleander, and Peat Weed. The genus "*Decodon*" is from Greek "deca" for ten and "odon" for the teeth on the calyx. The specific epithet "*verticillatus*" refers to the whorled or verticillate leaves. It is an aquatic shrub. The twigs are recurved, 4-6 angled, and tan to brown. The leaves are simple, opposite or whorled, deciduous, and with pinnate major veins. The blades are lanceolate to elliptic, 1.5-8" long, apex acuminate, base cuneate to obtuse, upper surface dark green, lower surface pale green, and margins entire. The purple to pink flowers are in axillary cymes. The fruit is a subglobose capsule. In wet areas such as edges of ponds and lakes (OBL) throughout the state with almost all records west of the Mississippi River. The current records include nine parishes.

.

DECUMARIA VINE - *Decumaria barbara* - Saxifragaceae

Decumaria Vine is also known as Wood Vamp. The genus name, *"Decumaria"*, is from *decum* ("ten"), referring to the sometimes 10-numerous flowers. The specific epithet, *"barbara"*, ("of Barbary"), comes from the fact that the species was originally thought to be African. The glabrous vine climbs by aerial roots. The leaves are opposite, simple, deciduous, entire or toothed, and with pinnate major veins. The blades are ovate to elliptic, entire to slightly denticulate, and 2-4" long. The blades are glossy above and glabrous or minutely pubescent beneath. The fragrant flowers are in terminal, flat-topped corymbs, 4-10 cm broad. The flowers are small and white and there are 7-10 petals. The fruiting capsules are urn shaped, strongly but delicately ribbed, and are 4-5 mm long. The seeds are yellow, 2 mm long, and terete. This vine is found in moist areas (FACW) along streams especially in southeastern Louisiana. It is reported from 17 parishes.

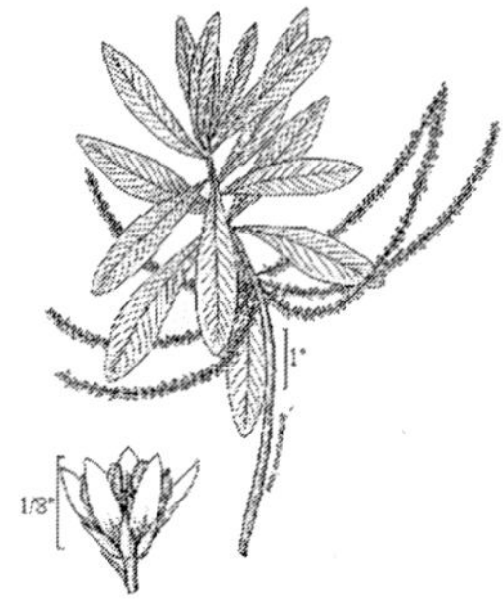

Cyrilla racemiflora (USDA-NRCS)

Decodon verticillatus (USDA-NRCS)

Decumaria barbara (USDA-NRCS)

ILLINOIS BUNDLE FLOWER - *Desmanthus illinoensis* - Fabaceae

Illinois Bundle Flower is also known as Prairie Mimosa, Prickle Weed, and Illinois Desmanthus. The genus "*Desmanthus*" is from Greek "desme", a bundle, and "anthos", a flower. The specific epithet "*illinoensis*" is for the state of Illinois. It is a small shrub to 4 ft. The stems are green to brown, striate angled, and pubescent to glabrous. The leaves are alternate, deciduous, and two even pinnately compound. The leaves are 2.5-4" long with numerous leaflets (6-14 pairs of pinnae with 20-30 pairs of leaflets per pinnae). The leaflets are small, lanceolate to linear, shorter than 0.2" long, apex acute and often apicuate, base truncate and asymmetrical. The greenish white flowers are in peduncled axillary heads. The fruit is a flattened, oblong to lanceolate, curved brown legume. In a range of habitats (FAC) but especially on clay soils throughout the state. The current records include 57 parishes.

PERSIMMON - *Diospyros virginiana* - Ebenaceae

Persimmon is also known as Jove's Fruit, Eastern Persimmon, Plaqueminer, Winter Plum, Possum Wood, Seed Plum, and Date Plum. It is a medium to large tree that reaches up to 100 ft but usually less than 40 ft. The genus, "*Diospyros*", is from the Greek means "divine fruit of the gods". The specific epithet"*virginiana*", refers to the state of Virginia. The name of Persimmon is old and is from the Lenape Indians who referred it as pasimenan. The bark is brown to black, fissures with deep ridges broken into rectangular checked sections (alligator skin). The twigs are thin, slightly zigzag, brownish or gray. The leaves are alternate, simple, deciduous, ovate to elliptic, entire, 2-6" long, 1.6-3" wide, usually somewhat acuminate, rounded at base, shiny glabrous dark green above, whitish and glabrous to densely hairy below. Persimmon is dioecious. The flowers are stiff, greenish yellow, bell shaped, and produced in the spring. The fruit is a yellow to orange or dark red berry and is found from fall to frost. The seeds are large, flat and shiny brown. The strong, hard wood weighs 52 lb per cu ft. In very old trees, the small heart is almost black, very hard, dense, and strong. The thick sapwood is cream-colored. The wood is used for woodenware, turney, plane stocks, handles, shoe lasts, golf clubs, shuttles, and billiard cues. The orange-colored, hard, immature fruit will pucker the mouth, but the fully ripen purplish-red or brownish-red, soft fruit can be eaten raw or prepared in different ways; some say that the fruits have to have been exposed to frost to remove the puckering quality. The fruit was known and appreciated by early settlers and explorers, being mentioned in writings of De Soto in 1539, Jan de Laet in 1558, and John Smith in the seventeenth century. The fruits have been used for jam, nut bread, wine, puddings, and vinegar. Confederate soldiers boiled persimmon seeds for coffee. The dried leaves make an excellent tea with a high percentage of vitamin C. Some herbal uses include an infusion of syrup from the green persimmon fruits used to treat diarrhea, chronic dysentery and hemorrhage of the womb. A decoction was made from the persimmon bark and used to gargle in ulcerated sore throat and also for diarrhea. The fruits are used by 16 species of birds, skunks, raccoons, opossums, hogs, grey and fox squirrels, and whitetail deer. The leaves are browsed by whitetail deer. The flowers are a valuable source of nectar. Persimmon is widely distributed in Louisiana from dry sandy soils to bottomland areas (FAC) but is absent from the acid pH baygalls of west central Louisiana. It occurs in all 64 parishes with current reports from 62.

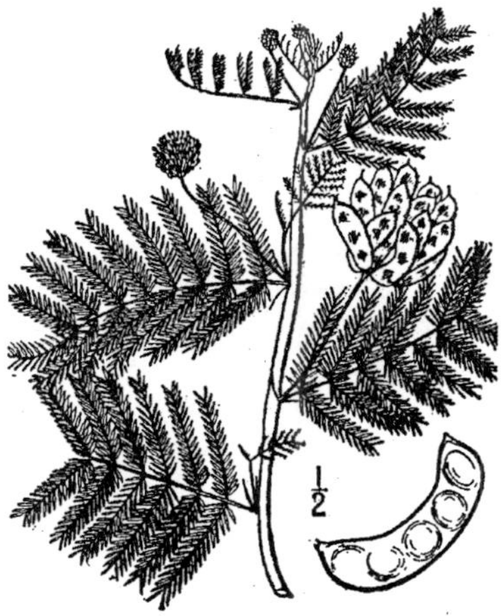

Desmanthus illinoensis (Britton and Brown 1943)

Diospyros virginiana (USDA-FS)

EASTERN LEATHER WOOD - *Dirca palustris* - Thymelaeaceae

Eastern Leather Wood is also known as Atlantic Leather Wood, Wicopy, Wickup, Moose Wood, Rope Bark, Swamp Wood, Leather Bark, Leather Bush, and American Mezereon. The genus name, "*Dirca*", is from Dirke, mythological name of a spring near Thebes. The specific epithet, "*palustris*", refers to the plant's habitat in low wet soil. It is shrub to 7 ft tall and 4" in diameter. The bark is gray, tough and leathery. The twigs are yellow to green and glabrous. The leaves are alternate, simple, deciduous, and with pinnate major veins. The blades are obovate to oval, 2-4" long, 1.4-2.8" wide, membranous, pubescent when young, becoming glabrous above, but remaining pubescent beneath, narrow or rounded at base, acute to subacute at apex, and the margins are entire. The petioles are 1-2 mm long. In the fall, the leaves turn an attractive yellow color. The flowers are in fascicles of 2-4 with short pedicels. The yellow calyx is funneled shaped and the petals are absent. The flowers are produced from March to April before the leaves. The fruit is a green to yellow, oval to oblong drupe that is 0.5-1" long, with a solitary dark brown seed that is streaked. The twigs and bark, being used for baskets and cordage by the Indians, has given it some of its names. It is also said to cause inflammation and pustulation on the skin of some people. The drupes are reported to be narcotic. The plant was introduced into cultivation about 1750. It is rare in well drained areas (FACU-) with reports from Caldwell and Catahoula parishes only.

THORNY OR AUTUMN OLIVE - *Elaeagnus* - Elaeagnaceae

The genus, "*Elaeagnus*", is from the Greek word "ealai" for olive and the Latin word "*agnus*" meaning lamb for the stellate pubescence on the underside of the leaves. The plants are shrubs to small trees that are often pruned into hedges. The leaves are alternate, simple, and with pinnate major veins. The lower surface is copiously covered with silvery scales. The flowers are borne in the leaf axils and are silvery white but quickly turning brown. The fruit is a red maturing drupe that is speckled with the silvery scales. Elaeagnus plants are one of the few non-legumes whose roots support nitrogen-fixing nodules.

THORNY OLIVE (***Elaeagnus pungens)*** is also known as Silver Thorn, Spotted Elaeagnus, Russian Olive, and Thorny Elaeagnus. The specific epithet "*pungens*" refers to the thorns on the twigs. It is a shrub to small tree that can reach heights of 15 ft. The branches are limber and armed with thorns in the axils. The leaves are evergreen with elliptic to linear-ovate blades that have entire to slightly undulate margins. The upper surface is lustrous dark green with a deep yellow blotch of variable size near the central portion; the major veins on the underside are brown. The fruit is oblong, and green to pinkish brown or red and covered with fine scales. It is native to Japan. Widely cultivated across the state and persisting with records from eleven parishes.

AUTUMN OLIVE (***Elaeagnus umbellata)*** is also known as Cardinal Olive and Autumn Elaeagnus. The specific epithet "*umbellata*", refers to the umbel arrangement of the flowers and fruits. It is a deciduous, much branched shrub, rarely over 12 ft. The bark is smooth and gray when young, becoming split and furrowed later. It occasionally has thorns in the axils of the leaves. The leaves are deciduous with oval to lanceolate blades. The flowers are yellow, fragrant, and borne in umbels in the leaf axils. The drupes are good grouse and turkey food. It was introduced to the United States from East Asia in the 1830's. Planted for wildlife in the state and persisting with records from Concordia, Franklin, Natchitoches, Ouachita, and Vernon parishes.

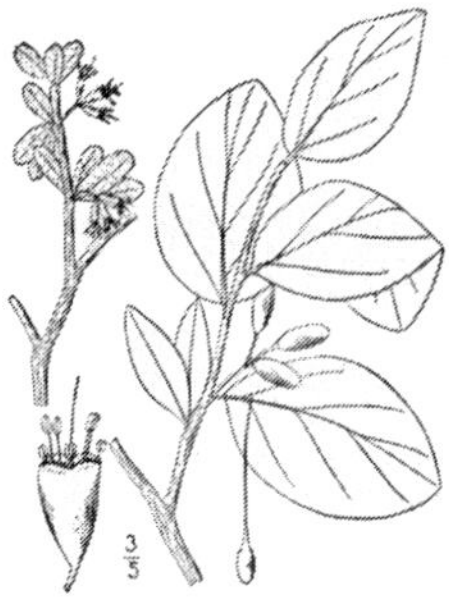

Dirca palustris (Britton and Brown 1943)

Elaeagnus pungens (Nelson 1996)

Elaeagnus umbellata (Godfrey 1988)

LOQUAT - *Eriobotrya japonica* - Rosaceae

The genus name, "*Eriobotrya*", is the Greek word for "woolly cluster," referring to the hairy panicles. The specific epithet, "*japonica*", is for Japan, where is has been cultivated for a long time. Other common names include Japanese Plum and China Plum. It is a large shrub or small tree to 30 ft. The new twigs are woolly pubescent. The leaves are alternate, simple, evergreen, and with pinnate major veins. The blades are elliptical-lanceolate, 5-12" long, 3-4" wide, thick and stiff, and with conspicuous parallel, oblique veins. The upper surface is glossy dark green while the lower surface is whitish or rusty-hairy. The flowers are in panicles at the ends of branches and produced in fall or early winter. The five petals are small, white, and sweetly fragrant while the calyx is brown and pubescent. The fruit is a pear shaped yellow-orange 1-2" long pome with three to five large brown seeds. It is indigenous to southeastern China and has been cultivated in Japan for over 1,000 years. Cultivated throughout the state and persisting with records from 12 parishes.

CORAL BEAN, CORAL TREE - *Erythrina* - Fabaceae

The genus, "*Erythrina*", is from the Greek "erythros" (red) referring to the color of the flowers. The genus includes shrubs to small trees that are armed on the twigs and leaves. The twigs are green to reddish brown, glabrous, angled, and finely grooved. The leaves are alternate, deciduous, and pinnate compound with three leaflets. The blades are dull green to yellowish-green. The flowers are typical legume with the large petal the standard and the smaller wings and keel. The fruit is a legume with typical "bean" seeds. The seeds contain the alkaloid erythroidine, which strongly affects the motor nerves.

CORAL TREE (***Erythrina crista-gallii***) is also known as Cockspur Coral Tree, Crybaby Tree, Fireman's Cap, Dragon's Teeth, and Immortelle. It is a medium to large tree to 30 ft. The leaflets are elliptic to ovate. The flowers are in terminal racemes that are in groups of two to three or single. The legume is 3-10" long and the seeds vary from yellow to orange or red. The wood is soft, light, and used for corks or carvings. This species is native to South America. It is cultivated in Louisiana and persists with records from East Baton Rouge, East Carroll, Lafourche, Ouachita, and St. Charles parishes.

CORAL BEAN ***(Erythrina herbacea)*** is also known as Mamou, Eastern Coral Bean, Cardinal Spear, Cherokee Bean, and Red Cardinal Flower. The specific epithet, "h*erbacea*", refers to the herbaceous character of the plant in temperate regions. The name "Mamou" is said to derive from an Acadian French term meaning "large root". Coral Bean is a shrub; but in our area is killed back to the ground during most winters. The blades are deltoid to hastate with a truncate to cuneate base and an acuminate or acute apex, 2.5-3.5" long, and 1-2" wide. The flowers are in a distinct raceme, large, dark red, and produced from March to May before the leaves unfold. The flowers are borne on stems up to 5 ft tall making a massive display of red. The legumes are linear, often falcate, 2-4" long, and with a stiff apiculate apex. The seeds are scarlet, hard, bony, lustrous, with a distinct black hilum, and persist on the open legume halves. The persistent seeds add to the fall and winter attractiveness of the species. The seeds are used in Mexico to kill dogs and rats. A fish poison has also been made from the seed. The seeds are strung into necklaces and used by the Mexican Indians in a game called "patol." The Indians used the roots as a diaphoretic. The bark is used for a yellow dye. This species is widely distributed in Louisiana with 41 parish records and is found mostly in well drained, often sandy habitats.

Eriobotrya japonica (Vines 1960)

Erythrina crista-gallii (USDA-FS)

Erythrina herbacea (Thomas and Allen 1998)

STRAW BERRY BUSH AND WAHOO - *Euonymus* - Celastraceae

The genus name, "*Euonymus*", is the Greek name meaning "true name". The genus includes shrubs to small trees. The leaves are opposite, simple, and with pinnate major veins. The fruit is a capsule with red seeds.

STRAW BERRY BUSH ***(Euonymus americana)*** is also known as Wahoo, Bursting Heart, Fish Wood, Burning Bush, Hearts-A-Bustin-With-Love, and Brook Euonymus. The specific epithet, "*americana*", is for America. It is a sprawling shrub with green, squarish stems that reaches a height of 8 ft. The leaves are tardily deciduous. The blades are lanceolate to elliptic, 0.75-4" long, and 0.5-1" wide, with finely serrate margins, a rounded to cuneate base, and acute apex. The blades are sessile or with a petiole shorter than 0.5". The flowers are yellowish green, borne at the end of long stalks, and appear from March to June. The petals are 5 in number and nearly triangular in shape. The fruits are somewhat rounded capsules with a warty surface. The capsules split open at maturity to expose bright red seeds and takes on the appearance of a strawberry. The seeds are eaten by several species of birds. The plants are heavily browsed by deer, sometimes to the point of reducing the stand of plants. Strawberry Bush was introduced into cultivation in 1697. It is found on well drained to moist soils (FAC-) throughout the state but is absent from the marsh and the large floodplains. The current distribution includes 41 parishes.

WAHOO ***(Euonymus atropurpurea)*** is also known as Eastern Wahoo, Spindle Tree, Bleeding Heart, Arrow Wood, Indian Arrow, Bitter Oak, Straw Berry Tree, and Burning Bush. The specific epithet, "*atropurpurea*", refers to the purple flowers and fruits. It is a shrub or small erect tree that reaches 25 ft. The leaves are deciduous. The bark is smoothish, gray often mottled, cracking on old stems into shallow ridges. The twigs are squarish and green. The blades are ovate-elliptic with finely serrate margins, 2-5" long, 1-2" wide, acute or cuneate at base, and with an acute to acuminate apex. The petiole is 0.5-1" long. The flowers are maroon to purple in color and borne in few-flowered clusters on long, slender stalks. The petals are four in number. The flowers appear from May to July. The fruit is a pinkish to tan, four-winged capsule that splits at maturity. The seeds are brownish to bright red, sometimes inconspicuous and remaining hidden in the capsule until falling, other times they are conspicuous and dangling on a threadlike aril. A few birds eat the fruits. The wood is almost white, close grained, heavy, hard, tough, and weighs 41 lb per cu ft. The twigs are browsed by whitetail deer. It is known to be a purgative for livestock. The bark was once ground into a powder and used as a purgative, especially by the American Indians and early colonists. However, the internal damage that sometimes results from high dosages and overdose led the U.S. Food and Drug Administration to label the plant as unsafe for medicinal purposes. Wahoo is widely planted as an ornamental and has been in cultivation since 1756. It is rare with only collections from Copenhagen in Caldwell Parish (FAC-).

JAPANESE SPINDLE TREE ***(Euonymus japonica)*** is also known as Euonymus. It is a shrub to 15 ft tall. The twigs are green but not distinctly square. The leaves are evergreen. The blades are obovate to narrow oval, 1-3" long, lustrous dark green but with many variegated, often yellow forms, leathery, tapered at base, serrated except at base. The flowers are greenish white. The fruits are pinkish-red. It is currently reported from 12 parishes. (Not Illustrated)

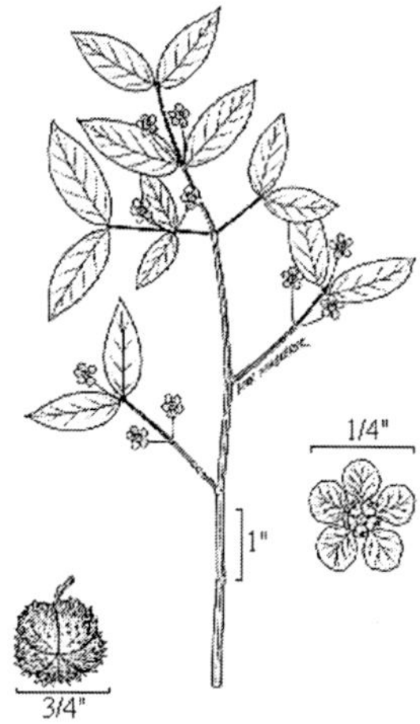

Euonymus americana (USDA-NRCS)

Euonymus atropurpurea (Britton and Brown 1943)

AMERICAN BEECH - *Fagus grandifolia* - Fagaceae

American Beech is also known as Beech, Carolina Beech, Red Beech, Ridge Beech, White Beech, and Beech Nut. The genus, "*Fagus*", is an old Greek word referring to its edible nuts and the specific epithet, "*grandifolia*", refers to the grand or large foliage. It is a large tree that can reach 120 ft. The bark is with smooth, light gray, sometimes mottled; the smooth bark is an attractant to humans who carve their names and/or initials into the bark leading to another name of Initial Tree. The twigs are slender, brown, and slightly zigzag, often with numerous side branches or spurs and the buds are very long, sharp-pointed, glabrous, with numerous shiny red-brown scales and are often described as cigar shaped. The leaves are alternate, simple, deciduous, and with pinnate major veins. The blades are elliptical or ovate, 2.5-5" long, 1.2-2.5" wide, with numerous straight parallel side veins that end in a tooth on the coarsely serrate margins. The upper surface is dark blue-green above while the lower surface is pale green. Beech trees retain their golden brown leaves following frost much longer than most deciduous trees. The flowers are green, often in clusters and produced from March to May when the leaves are half grown. Male and female flowers appear on the same tree. The fruit is a short stalked prickly nut with two seeds per involucre and is produced from September to November. The wood is strong, hard, tough, not durable, and weighs 43 lb per cu ft. The wood has been used for chairs, tool-handles, shoe lasts, flooring, cooperage, crates, spools, brush racks, clothespins, toys, cross-ties, pulpwood, boxes, veneer, laundry appliances, general millwork, and for fuel. The fruit is used by many species of birds, squirrels, raccoons, opossums, porcupines, gray and red foxes, white tailed deer and even passenger pigeons. It is used for food (nuts, flour, and coffee) and vegetable oil by humans. It was an important food source for Indians and settlers. The fruit was eaten whole, ground into flour, roasted for coffee, and used medicinally. Extractant from wood has been used medicinally as a stimulant, expectorant, antiseptic, and antipyretic. A decoction made from the leaves of the American Beech is used for diabetes, bed-wetting, obstinate ulcers, skin diseases, headaches, and low spirits. It has been used for vomiting, seasickness, chronic bronchitis, and pulmonary tuberculosis. It is extremely shade-tolerant and is part of the northern hardwood climax forest. It is found on well drained to mesic soils, especially along streams, throughout the state but is absent from the marsh and the large floodplains (FACU). It is currently reported from 36 parishes.

FIG - *Ficus carica* - Moraceae

The genus, "*Ficus*", is an ancient Latin name while the specific epithet, "*carica*", is Caria, an ancient country in Asia Minor, from where it is thought to originate. It is a shrub or small tree to 30 ft. The twigs are stout, smooth, gray, and pubescent when young but becoming glabrous with age. The sap is milky and thick. The leaves are alternate, simple, deciduous, and with palmate major veins. The blade is ovate, about 4-8" long and wide, 3-7 lobed, with a cordate to truncate base, and scabrous and pubescent on both surfaces. The minute flowers are enclosed in a pear-shaped receptacle. The fruit is a pear shaped synconium with seedless achenes inside. The fig fruit is actually a hollow receptacle with hundreds of small fleshy flowers facing each other on the inside. In their native habitat, figs are pollinated by a tiny gall wasp that enters the flower cluster through a small opening in the apex. Each flower then produces a small fruit containing the seeds. The wasp is not present in most of North America, and so seeds are not produced. Since ancient times, infusions of figs have been taken as an aperient. They are reportedly useful as a hot poultice for ulcers. It is cultivated in the state and persisting with records from 38 parishes.

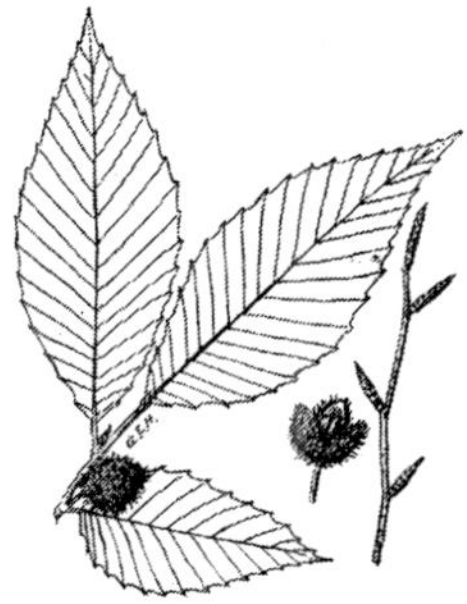

Fagus grandifolia (USDA-FS)

Ficus carica (Vines 1960)

Firmiana simplex (Moore 1960)

CHINESE PARASOL TREE - *Firmiana simplex* - Sterculiaceae

The genus name, "*Firmiana*", honors Karl Joseph von Firmian (1718-1782), at one time governor of Lombardy. The specific epithet, "*simplex*", refers to the lobed but simple leaves. Other common names include Varnish Tree, Phoenix Tree, and Bottle Tree. It is a shrub to small tree to 35 ft. The bark is smooth and green and the twigs are stout, smooth, and grayish green. The leaves are alternate, simple, deciduous, and with palmate major veins. The blades are ovate to orbicular, 4-12" long and wide, 3-5 lobed, and the lobes are acuminate. The base is cordate and the upper surface is dull green, the lower surface pale green, and both surfaces are glabrous to pubescent. The greenish white flowers are in a terminal panicle. The fruit is a capsule that opens into 5 leaf-like bracts with seeds along the margins. The seeds are globular and brown to reddish brown. The tree has been cultivated since 1757 and is native to Japan and China. Planted throughout the state and persisting and spreading with records from 12 parishes. (syn. *Firmiana platinifolia*)

SWAMP PRIVET - *Forestiera* - Oleaceae

The genus name, "*Forestiera*", honors the French botanist and physician, Charles Le Forestier. The genus includes shrubs to small trees with opposite, simple, deciduous leaves that have pinnate major veins. The margins are slightly toothed but appear entire due to the small size of the teeth. Flowers are inconspicuous and produced before the leaves. The fruit is a drupe that is ovoid to oblong, dark purple to black, and about 0.3" long.

EASTERN SWAM P PRIVET (***Forestiera acuminata***) is also known as Swamp Privet, Texas Forestiera, Forestiera, Swamp Ash, and Adelia. The specific epithet, "*acuminata*", refers to the acuminate apex and base of the blades. It is a shrub or small tree rarely over 40 ft. The bark is dark brown and ridged. The twigs are light brown, slender with numerous lenticels, and glabrous. The blades are elliptical to oblong-ovate; 2-4.5" long, 1-2" wide; acuminate at apex and base, serrulate above the middle, glabrous, occasional with a few hairs on veins beneath, yellowish green above, and paler beneath. The petioles are slender and ½" or longer. The flowers are in fascicles, subtended by yellow bracts. The flowers are produced from late February to March with the drupes maturing in May. The drupes are considered to be an excellent food for wild ducks and birds. The wood is soft, medium heavy (40 lb per cu ft), coarse grained, and yellow brown. Deer browses the foliage and twigs. It is widespread in Louisiana but always in wet habitats (OBL). It is currently reported from 46 parishes.

UPLAND SWAMP PRIVET (***Forestiera ligustrina***) is a shrub to about 15 ft and often with leaning trunks. The specifiic epithet, "*ligustrina*", means "Ligustrum or privet like". The blades are elliptic to obovate, 1-2" long, 3/4-1" wide, with an obtuse to acute tip and base, the margins toothed from base to apex, and sparsely to densely pubescent below. The petioles are shorter than 1/4". The flowers are produced in late summer or very early fall with fruits maturing in the fall. Rare in the state in upland well drained habitats (FAC) or somewhat mesic areas but not real wet. There are records from Bossier, Caddo, and DeSoto parishes in northwest Louisiana.

GOLDEN BELLS - *Forsythia suspensa* - Oleaceae

The genus name, "*Forsythia*", honors William Forsyth (1737-1804) a Scottish gardener who became superintendent of the Royal Garden of Kensington Palace. The specific epithet, "*suspensa*", means "hanging", refers to the flowers. It is a shrub to 10 ft. The leaves are opposite, simple, deciduous, and with pinnate major veins. The blades are elliptic to

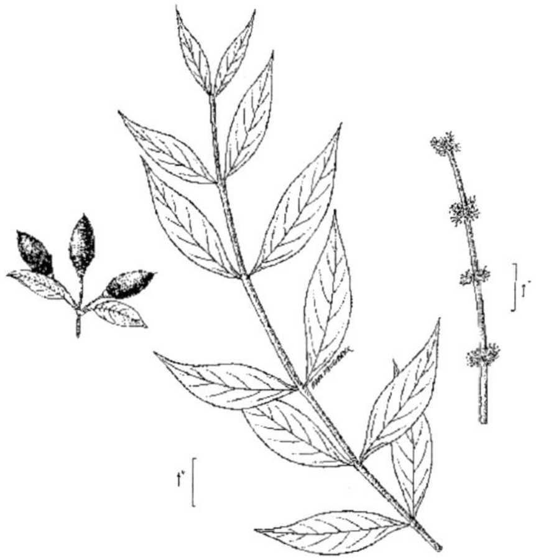

Forestiera acuminata (USDA-FS and USDA-NRCS)

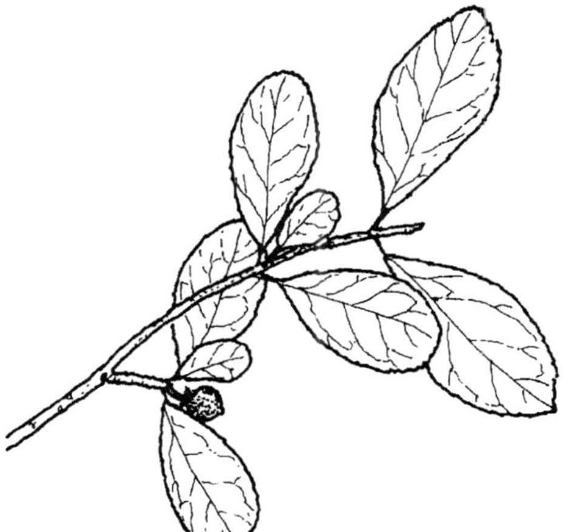

Forestiera ligustrina (Nelson 1996)

oblanceolate, 2-4" long, 0.5-1.5" wide, the base and apex acute to acuminate, with irregularly serrate margins. The blades are dark green above and paler yellow green below. The flowers appear before the leaves in February to March. The flowers are bright golden yellow, bell shaped with 4 lobes. The fruit is a capsule with winged seeds. Cultivated throughout the state and persisting with reports from 11 parishes. (Not Illustrated)

CAROLINA BUCKTHORN - *Frangula caroliniana* - Rhamnaceae

The genus, "*Frangula*", is from the Latin, "frangere = to break", because the wood of some species breaks easily. The specific epithet, "*caroliniana*", is for the Carolinas, North and South Carolina. Other common names include Indian Cherry, Yellow Buckthorn, Bog Birch, Alder Leaf Buckthorn, and Polecat Tree. It is a large shrub or small tree to 30 ft and 6" diameter. The bark is thin, grayish, and slightly fissured. The twigs are thin, gray, soon becoming glabrous. The leaves are alternate, simple, deciduous, and with pinnate major veins. The blades are elliptical, 2-5" long, 0.8-1.5" wide, with serrate margins. The lateral veins are very distinct and straight. The blades are hairy when young, dark shiny green above, and pale green below. The greenish yellow flowers are in small clusters in the leaf axils. The fruit are spherical drupes that are red but change to a lustrous black at maturity. The drupes are about 0.4" diameter with 2-4 reddish brown seeds. The fruits are eaten by several species of birds. The wood is light brown, sapwood yellow, close-grained, fairly hard, rather weak, and weighs 34 lb per cu ft. In mesic to well drained habitats (FACU) throughout the state with reports from 41 parishes. (syn. *Rhamnus caroliniana*)

ASH - *Fraxinus* - Oleaceae

The genus name, "*Fraxinus*", is the ancient Latin name for a species of European Ash. The genus includes small to large trees. The leaves are opposite, deciduous, and one-pinnately compound with an odd number of leaflets. The flowers are imperfect, small, green to purple, and are produced with or before the leaves from April to May. The male flowers are in dense clusters and the female flowers are in short, more open, drooping branches. The fruit is a yellow to light brown samara. Some superstitions connected to the Ash are: An ash twig around a cow's horn will keep witches from milking her; Lightning never strikes an ash; Serpents fear to approach an ash; and If ash is out before the oak, twill be a summer of fire and smoke but if oak is out before the ash, twill be a summer of wet and splash.

WHITE ASH (***Fraxinus americana***) is also known as American Ash, Cane Ash, Biltmore Cane, Biltmore Ash, Biltmore White Ash, and Small Seed White Ash. . The specific epithet, "*americana*", refers to America. It reaches to 60 ft and 18" diameter. The bark is light gray to dark brown with ridges that are narrow and separated by deep fissures and looks diamond-shaped. The twigs are thick, gray-brown, with raised crescent-shaped leaf scars. The leaves are 8-12" long with 5 to 9 leaflets. The leaflets are ovate lanceolate, pointed, toothed or with smooth margin, dark lustrous green above and paler below, 2.4-5" long and 1.2-2.4" wide. The samaras are yellow to brown 1-2" long, about 1/4" wide, and with a narrow wing that does not extend down the cylindrical body. The strong, hard, stiff, heavy (41 lb per cu ft), light brown with thick, lighter colored sapwood, close-grained, tough, durable, and shock resistant wood is a very important timber species because of its blond colored wood, attractive grain figure, and workability. The wood is used for tanks, silos, toys, musical instruments, cabinets, refrigerators, millwork, sash, doors, frames, vehicle parts, farm utensils, woodenware, butter tube, veneer, cross ties, sporting goods, furniture, cooperage, handles, ships, railroad cars, frame parts of airplanes, interior finish, oars, carriages, tennis rackets, baseball bats, snowshoes, and skis. The bark from the trunk, branches, and roots are used for a bitter tonic, astringent, cathartic diuretic,

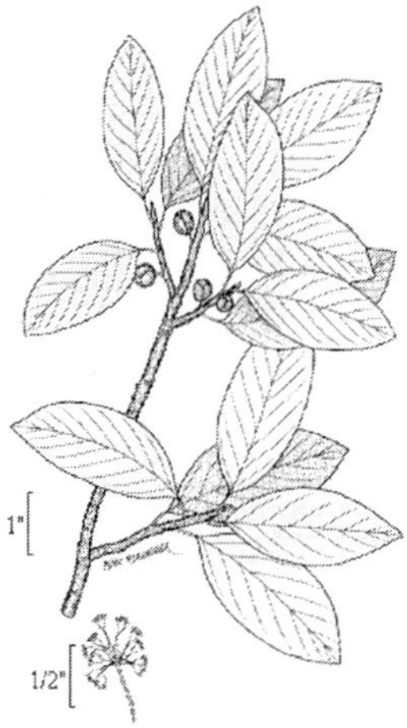

Frangula caroliniana (USDA-NRCS)

Fraxinus americana (USDA-FS)

a febrifuge, diaphoretic, anti-arthritic, alterative fever sores of the lips and constipation. Chewed bark is used as a poultice for sores. Tea from the buds is used to treat snakebite. The fruit is used by a number of birds including purple finch and pine grosbeak. Rabbit, porcupine, and white-tailed deer use the foliage. Indians and early settlers believed this tree to have special snake repelling qualities. It has been known in cultivation since 1724. On well drained soils of slope forests and in the Inland Prairie Forests (FACU). It is currently reported from 45 parishes. (syn. *Fraxinus biltmoreana*)

ARIZONA ASH (***Fraxinus berlandieriana***) is also known as Plumero, Fresno, and Mexican Ash. The specific epithet, "*berlandieriana*", honors the Swiss botanist, Jean Louis Berlandier (1805-1851), who collected extensively in Mexico and Texas. Arizona Ash is a small tree to 30 ft with gray or reddish bark with shallow fissures and narrow ridges. The twigs are gray, red, or green with scattered lenticels and small oval raised leaf scars. The leaves are 3-10" long with 3-5 leaflets. The leaflets are 3-4" long, 0.5-1.5" wide, elliptic to lancelate or obovate, acuminate to acute at the apex, cuneate or rounded at base, entire or remotely serrate, dark green and glabrous above, and lighter green and glabrous or pubescent with a few hairs below. The samaras are 1-1.5" long, about 0.5" wide, and with a wing that extends almost to the base of the achene. The wood is light brown, close-grained, light, soft, and with lighter sapwood. Native to the southwestern US and Mexico and widely planted in Louisiana. Persisting with records from East Baton Rouge, Orleans, Ouachita, and St. Charles parishes but much more widely distributed.

CAROLINA ASH (***Fraxinus caroliniana***) is also called Poppy Ash, Pop Ash, and Water Ash. The specific epithet, "*caroliniana*", refers to the states of North or South Carolina. It is a small tree to 40 ft. The bark is gray, often blotched, smoothish, with small scales, and tends to have a pinkish color when young. The twigs are green and pubescent when young but turn gray to brown and become glabrous with age. The leaves are 7-12" long with 5-9 (usually 7) leaflets. The leaflets are 2-6" long, 0.5-3" wide, oblong-ovate, acuminate to acute at the apex, cuneate or rounded at base, with serrate or sometimes entire margins, dark green, lustrous, and glabrous above, and lighter green and glabrous or pubescent below. The samaras are 1-3" long, 0.5-0.75" wide, and with a wing that extends to the base of the achene and is elliptic in shape. There is often a third wing on the samaras. The wood is yellowish white, close-grained, soft, weak, and weighs 22 lb per cu ft. This ash is found along the edges of streams, lakes and other water bodies (OBL). Carolina Ash is widely distributed with records from 45 parishes.

GREEN ASH ***(Fraxinus pennsylvanica)*** is also known as Water Ash, River Ash, Red Ash, Swamp Ash, and Martian Butt (Created by Charles Allen in the mid 1990's). The specific epithet, "*pennsylvanica*", refers to Pennsylvania. It is a tree to 60 ft tall and 2 ft diameter. The bark is gray-brown, divided into narrow scaly ridges, with reddish-brown inner layer. The twigs are gray, thick, and with raised leaf scars nearly straight across the top. The 5-9 leaflets are Lancelot or ovate, acuminate or acute at apex, cuneate at base, margins usually toothed but sometimes entire, 2-5" long, 1-1.6" wide, glabrous shiny green above, and paler and usually slightly hairy below. The samaras are 1.2-2" long, spatulate or oblanceolate and the end of the wing is square, notched, rounded, or acute, and the extending nearly to the base of the narrow body of the achene. The wood is heavy (40 lb per cu ft), hard, strong, coarse-grained, light brown with thick lighter colored sapwood much like the White Ash. Many birds and mammals use the fruits. Green Ash is a common tree that is found in bottomland hardwood sites (FACW) throughout the state. Probably in all parishes with records from 57 currently recorded.

Fraxinus berlandieriana (USDA-FS)

Fraxinus caroliniana (USDA-NRCS)

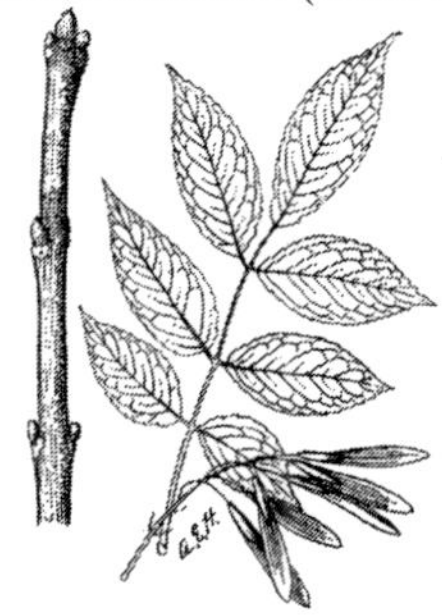

Fraxinus pennsylvanica (USDA-FS)

PUMPKIN ASH ***(Fraxinus profunda)*** is also known as Red Ash and Swell Butt Ash. The specific epithet, "*profunda*", refers to the thick hairs on the young leaves. It is a large tree to 120 ft and 3 ft diameter; the base is often buttressed. The bark is smooth and light gray on young trees, turning darker and more furrowed on older trees. The young twigs are velvety pubescent but turning glabrous with age, both are light gray, and the leaf scars are large. The leaves are 9-18" long with 7 (3-9) leaflets. The leaflets are 5-10" long, 1.5-5" wide, lancelate to elliptic, with entire to serrulate margins, apex acuminate, base cuneate to rounded, sometimes asymmetrical, young leaves densely pubescent, and older leaves glabrous and dark yellowish green above and paler and pubescent below. The samaras are 2-3" long, 0.5" wide, straight or falcate, apex rounded or emarginate, often apiculate, and the wing decurrent nearly to base of achene or sometimes only 0.5-0.75 of length of achene. The fruit rarely has three wings. Pumpkin Ash is similar to Green Ash but in wetter sites (OBL) throughout the state. The current records include 26 parishes from across the state with many records from the Mississippi River floodplain. *(*syn. *Fraxinus tomentosa)*

HUCKLE BERRY - *Gaylussacia* - Ericaceae

The genus name, "*Gaylussacia*", honors J.L. Gay-Lussac (1778-1850), a French chemist. The plants are low shrubs with alternate, simple, deciduous leaves with pinnate major veins. The margins are entire and the apex is obtuse and mucronate, and the base cuneate. The flowers are white to pinkish and bell shaped. The fruit is a globose black false berry with persistent calyx lobes.

DWARF HUCKLE BERRY ***(Gaylussacia dumosa)*** is also known as Bush Huckle Berry and Gopher Berry. The specific epithet, "*dumosa*", refers to its bushy habit. It reaches 2 ft tall with terete pubescent twigs. The blades are oval to oblanceolate to spatulate. The upper surface of older leaves is deep green and lustrous with the lower surface paler and with amber glandular dots. The young blades are pubescent and glandular dotted. The flowers have short curly hairs. The berry is 0.25-0.5" diameter, shiny, and covered with short hairs. Pine forests (FAC) in southeastern Louisiana with reports from St. Helena, St. Tammany, Tangipahoa, and Washington parishes.

HAIRY HUCKLE BERRY (***Gaylussacia mosieri***) is also known as Mosier's Huckle Berry and Wooly Huckle Berry. It is a shrub to 5 ft with gland tipped hairs on twigs and leaves. The leaves are elliptic to oblanceolate. The flowers have long straight hairs. The false berry is hairier than Dwarf Huckleberry and larger, 0.3-0.75" diameter. Hairy Huckleberry is rare in wet areas near bogs and baygalls (FACW) in southeastern Louisiana with records from St. Tammany, Tangipahoa, and Washington parishes.

YELLOW JESSAMINE - *Gelsemium* - Loganiaceae

The genus name, "*Gelsemium*", refers to the Italian name for Jessamine, "gelsomino". The genus includes vines that climb by twining or lie prostrate on the ground. The twigs are smooth and green or reddish. The leaves are opposite, simple, evergreen, and with pinnate major veins. The blades are lanceolate to ovate, 1-3" long, 0.5-1" wide, apex acuminate, base cuneate, margins entire, the upper surface dark shiny green, and the lower surface paler. The flowers are large, tubular, and yellow. The fruit is flattened 2-valved capsule with a beak. The numerous seeds are flattened.

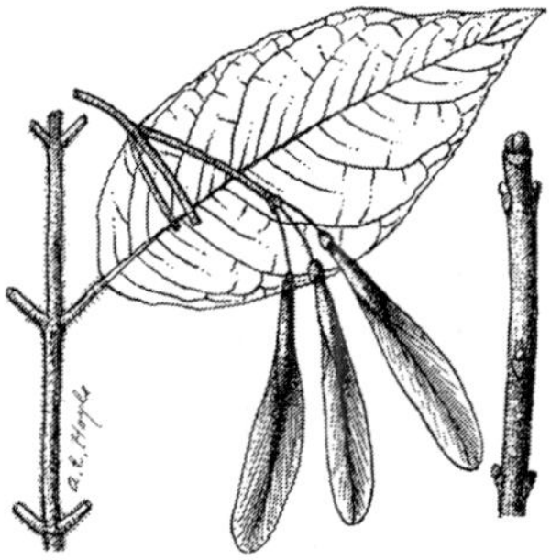

Fraxinus profunda (USDA-FS)

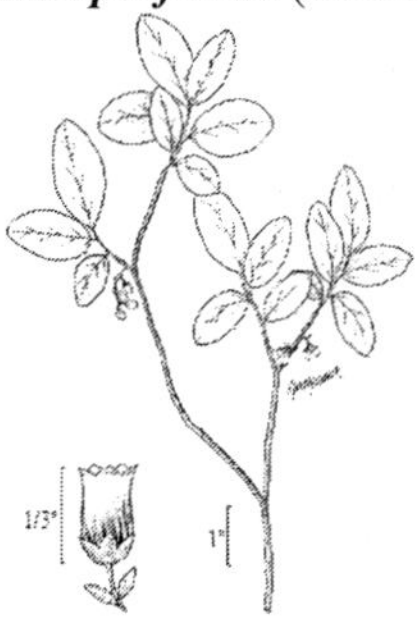

Gaylussacia dumosa (USDA-NRCS)

Gaylussacia mosieri (Godfrey 1988)

RANKIN'S YELLOW JESSAMINE (*Gelsemium rankinii*) is very similar to Common Yellow Jasmine but with flowers with no fragrance. The calyx lobes are acute to acuminate. The capsule is ellipsoid, 0.4-0.6" long, smaller than 0.3" wide and the beak is more than 2 mm long. The seeds are wingless and shorter than 4 mm. Rare in moist hardwood forests (FACW+) in southeastern Louisiana with reports from Livingston, St. Helena, St. Tammany, Tangipahoa, and Washington parishes

COMMON YELLOW JESSAMINE *(Gelsemium sempervirens)* is also known as Caroline Jessamine, Jasmine, Gelsemimi, Gelsemin, Wild Jessamine, Carolina Wild Woodbine, and Evening Trumpet Flower. The specific epithet, "*sempervirens*", is for its evergreen habit. The flowers are fragrant and the calyx lobes obtuse. The capsule is oblong, 0.5-0.8" long, and 0.3-0.5" wide and the beak is shorter than 2 mm. The seeds are longer than 7 mm and have wings. The dried rhizome is a poisonous alkaline gelsemine but is used medicinally in the form of a tincture to treat facial neuralgia, rheumatism and gonorrhea. It can cause vertigo and even death. The dried roots are used as a nervous system depressant, febrifuge, anodyne, antispasmodic, nervine, sedative, and mydriatic. It is also often used as a mail box plant or on trellises and as a ground cover. It is the state flower of South Carolina. Bobwhite quail eats the seeds and marsh rabbit and white-tailed deer eat the leaves. It is found in dry to moist habitats throughout the state but absent from the large floodplains and the marsh (FAC). It is currently reported from 45 parishes. (syn. *Gelsemium nitidum*)

GINKGO - *Ginkgo biloba*- Ginkgoaceae

Ginkgo is also known as Maiden Hair Tree. The genus "*Ginkgo*" is from the Chinese word meaning silver fruit. The specific epithet "*biloba*" refers to the two lobed leaves. It is a dioecious tree to 100 ft and 6 ft diameter. The bark is dark brown or gray and smooth when young but becoming fissured with age. The stout twigs are lustrous, glabrous, and green to yellow or brown. The leaves are alternate but often clustered on spur shoots, simple, deciduous, and with dichotomous veins. The blades are wedge shaped, 2-4" wide and long, apex cleft and wavy, base cuneate, and glabrous. The male cones are catkin like and the ovules are solitary or paired on long stalks. The seeds are covered with a strong smelling aril. The seeds are eaten in the Orient and the leaves are used as an herb to improve memory, I think. It is a native of China that was first introduced into the United States in 1784. Widely planted and persisting throughout the state.

LOCUST - *Gleditsia* - Fabaceae

The genus name *Gleditsia* is for Gottlieb Gleditsch, a German botanist who died in 1786. The genus includes trees with large usually branched thorns and thin twigs that are usually swollen at the nodes. Leaves are alternate, 1 or 2 even pinnate compound, and deciduous. The 12 to 30 leaflets are subsessile and oblong-lanceolate with crenulate margins. The flowers are in racemes, greenish to yellowish, and produced April-May. The fruit is a flattened legume. The two species are best separated by the fruit which is sessile, very long (10-18"), linear, and pubescent in Honey Locust and stalked, short 1-3" long, ovate, and glabrous in Water Locust. The number of seeds per fruit is several to many in Honey Locust with the seeds surrounded by pulp. In Water Locust, the number of seeds per fruit is mostly one but rarely up to three and the seeds are not surrounded by pulp. The leaflets in Honey Locust are pubescent at least along the midvein and entirely glabrous in Water Locust. The habitat is also different with Water Locust being in very wet sites (OBL) and Honey Locust in well drained to somewhat wet sites (FAC-).

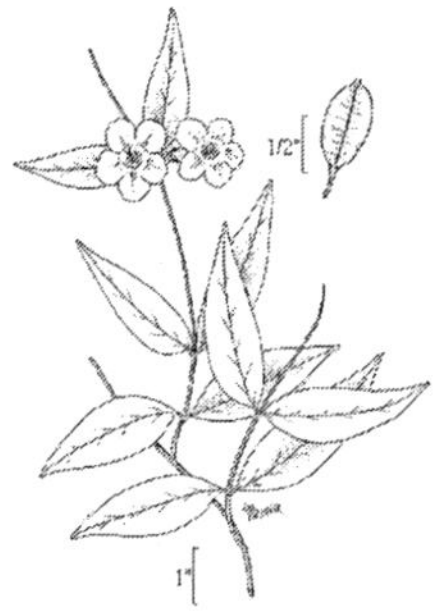

Gelsemium rankinii (USDA-NRCS)

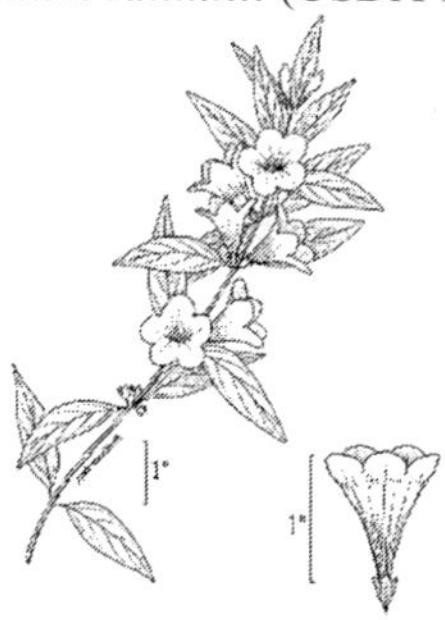

Gelsemium sempervirens (USDA-NRCS)

Ginkgo biloba (USDA-FS)

WATER LOCUST (***Gleditsia aquatica***) is also known as Swamp Locust and Thorny Acacia. The specific epithet, *aquatica*, refers to the plant's swampy habitat. It is a tree that reaches up to 40-60 ft high with glabrous leaflets. It resembles Honey Locust but has larger, oblong leaflets that are more widely spaced, more branched thorns on the trunk, and smaller, thin-walled pods. The bark is smooth, reddish brown or dull gray, broken into small plate-like scales on older parts. The fruit is a thin chestnut brown legume, 1-3" long, about 1" in width, long-stalked, without pulp, and containing 1 or rarely 2 or 3 seeds. The wood is similar to Honeylocust and weighs 46 lb per cu ft. It is light reddish brown, sapwood yellow, strong, heavy, hard, and close-grained. The wood is used for posts and fuel. Water Locust is a preferred beaver food. Along the edges of lakes, streams and other water bodies (OBL) throughout the state with records from 43 parishes.

HONEY LOCUST (***Gleditsia triacanthos***) is also known as Confederate Pin Tree, Sweet Locust, Sweet Bean, Thorn Tree, Thorny Locust, Black Locust, Thorny Acacia, Thorn, Three Thorned Acacia, and Honeyshucks. The specific epithet, "*triacanthos*", refers to the commonly three branched thorns. It is a tree that reaches up to 120 ft and 20" diameter. The bark is grayish brown to black and with deep fissures in older trees. The leaflets are pubescent, sometimes only along the midrib. The reddish to purple-brown fruit is a large flattened pubescent legume that may be 10-18" long and 1-1.5" wide. The mature legume contains several to many dark brown oval seeds that are surrounded by pulp. The wood weighs 44 lb per cu ft and is strong in bending, stiff, high shock resistance, close-grained, fairly durable in the soil, rather hard, red or bright red brown with thin pale sapwood, and moderately durable. It is used for farm implements, fuel, lumber, posts, vehicles, furniture, crossties, and hubs for wheels. The thorns are used for pins, spear points, animal traps, carding wool, and pinning up the mouths of woolsacks. Indians ate the legumes. During the Civil War, the spines were used to pin together the tattered uniforms of the soldiers (hence, the name Confederate Pin Tree). The Cherokees made bows from the wood. Humans, cattle, white-tailed deer, gray squirrel, and cottontail eat the pods. The flowers are said to be good bee pasture. It has been widely planted ornamentally and can be trimmed to form an impenetrable hedge. A thornless form is available. Honeylocust has been in cultivation since 1700. Moist to well drained sites (FAC-) and the clay soils of Inland Prairie Forests throughout the state in all parishes with current records from 55 parishes.

SILVER BELL - *Halesia diptera* - Styracaceae

Silver Bell is also known as Halesia, Snowdrop Tree, Snow Bell, Cowlicks, and Two Winged Silver Bell. The genus, "*Halesia*", honors Dr. Stephen Hales (1677-1761), English scientist and inventor. The specific epithet, "*diptera*", means "two-winged (the fruit)". It is a large shrub or a slow growing tree to 30 ft with a diameter to 12". The bark is reddish-brown to gray with irregular fissures separating on the surface into small flakes. A good winter identification is the twig's bark, which forms long, loose brown fibers. The broken twigs have an odor. The leaves are simple, alternate, deciduous, and with pinnate major veins. The blades are 2-3" wide, 3-5" long and elliptical to oval. The blades have a light green, smooth to hairy, upper surface and a paler, hairy, lower surface. The petiole is longer than Snow Bell, ***Styrax***. The medium sized cluster of 2 to 6 white flowers occur during February and April on branches of the previous year's growth. The young leaves and white flowers appear simultaneously. The flowers droop and are very attractive. They are 1" across and consist of 4 waxy, white petals surrounding a tight bundle of stamens. The fruit is an oblong woody dry flattened drupe, 1-2" long, with two broad thin wings with 2 narrower ones in between and is sour when green but edible. Hummingbirds are attracted to the bell shaped flowers. Gray and fox squirrels sometimes eat the

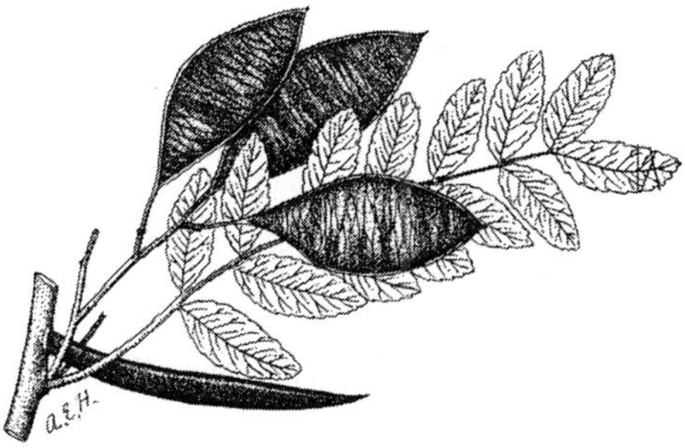

Gleditsia aquatica (USDA-FS)

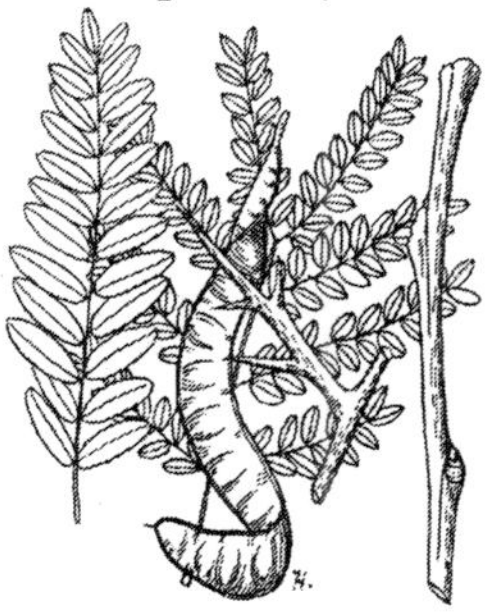

Gleditsia triacanthos (USDA-FS)

Halesia diptera (USDA-NRCS)

fruit. It has been cultivated since 1758, and is occasionally grown in Europe. The wood is light brown, close-grained, light, soft, brittle, and the sapwood is lighter. In moist to well drained sites (FAC) but absent from the marsh and large floodplains. The current records include 36 parishes.

WITCH HAZEL - *Hamamelis virginiana* -Hamamelidaceae

The genus name, "*Hamamelis*", is from the Greek word "*hama*" ("at the same time") and "*melon*" ("apple"), possibly because of the presence of both fruit and flower simultaneously. The specific epithet, "*virginiana*", refers to the state of Virginia. It is also known as Winter Hazel, Hazel Nut, Snapping Hazel, Spotted Alder, Striped Alder, Winter Bloom, Tobacco Wood, Pistachio, and Witch Elm. It is a large shrub or small tree that reaches up to 30 ft. The twigs are glabrous or sparsely pubescent. The bark is light brown and relatively smooth or scaly. The leaves are alternate, simple, deciduous, and with pinnate major veins. The blades are obovate to almost round; 2-6" long and almost as wide; wavy-toothed; with 5-7 prominent veins; glabrous above; and sometimes with hairs on veins below. The blades are a dull green above and paler below and tend to develop yellow spots in the summer. The inconspicuous, refreshing fragrant, yellow flowers have long, strap-shaped petals and are borne in clusters along the stem, appearing in late October through December. The fruit is a 2 valved woody brown hairy capsule with two recurved beaks. The seeds are lustrous black oblong and about 0.3" long. The capsule ejects the seeds with great force and the seeds are reported to travel as much as 25 ft. The capsules remain on the plant and are a good winter characteristic. The hard, close-grained wood weighs 43 lb per cu ft. It has a thick white sapwood and light reddish brown heartwood. There is no value as a timber tree because it does not grow large enough. The branches are used as divining rods to find water and to locate buried treasure and ore deposits. The leaves and twigs are used for medicinal extractions and lotions and salves. It contains tannic acid, which lessens the inflammation, stops bleeding, check secretions of mucous membranes. The Indians made tea from the leaves. The seeds are edible by humans. The bark and leaves contain essential oil to make astringent, hemoclatic used for hemorrhoids, varicose veins, and bruises. The Indians used decoction from the bark to heal wounds. One may place Witch Hazel twigs in water and add hot rocks in an enclosed room to create a sauna for sore muscles. The fruits and young twigs are used by 5 species of birds, squirrels, cottontail, white-tail deer, and bear. Witch Hazel is usually said to have its name from confusion by the settlers between this plant and the true European Hazel, and there is some resemblance between the leaves and the appearance of the young fruits. The Hazel of Europe was famous for its magical properties; some legends say that one can find witches by means of it, and others that with its help witches can find water, or gold or other desirable subterranean things. The origin of Witch Hazel is related to Anglo-Saxon wicken, meaning to bend, or Old English wick, meaning quick or living, or possibly even to the modern word, switch. It was introduced into cultivation in 1736. In mesic to well drained soils (FACU) throughout the state but absent from the marsh and large floodplains. The current records include 40 parishes. (syn. *Hamamelis macrophylla)*

ENGLISH IVY - *Hedera helix* - Araliaceae

The genus name, "*Hedera*", is the ancient Latin name of ivy. The specific epithet," *helix*", is an old generic name meaning "twining". It is also known as Common Ivy. It is a vine climbing with aerial rootlets up to a height of 90 ft or creeping on the ground. The stems are stiff and rope-like, green to brown, glabrous, and with or without aerial roots. The leaves are alternate, simple, evergreen, and with palmate major veins. The blades are entire to 3-5 lobed, ovate to round, dark green above, paler dull green below, and 1.5-4" long. The flowers are borne in terminal, globose or somewhat paniculate umbels, usually forming a raceme. The flowers have five greenish petals. The fruit is a fleshy, black, globose or angled, 1/4" in diameter, berry-like

Hamamelis virginiana (USDA-FS)

Hedera helix (Vines 1960)

drupe. The fresh, bitter leaves were once used in domestic medicine as a local application in skin diseases and old ulcers. Both the leaves and the berries are poisonous when eaten, and in some persons contact with the leaves causes dermatitis. Resin from the bark of old plants has been used in medicine under the name of Ivy Gum. It was at one time used to plug carious teeth for the relief of toothache. In medieval Europe, it was believed to be a cure for the plague. English Ivy has been cultivated in gardens in both hemispheres since ancient times. It is planted as a ground cover in shady places, as a wall cover, and for borders. It persists and spreads from old home sites and often climbs and covers trees. It is native to Europe and currently reported from 26 parishes.

SHRUB ALTHEA - *Hibiscus syriacus* - Malvaceae

The genus name, "*Hibiscus*", is the ancient name of the European Marsh Mallow. The specific epithet, "*syriacus*", is for Syria, where it was once supposed to be native. However, more recent investigations prove it to be originally from China and India. Other common names include Rose-of-Sharon and Rose Mallow. It is shrub to 18 ft. The bark is gray to brown and somewhat roughened. The twigs are green to brown. The leaves are simple, alternate, deciduous, and with palmate major veins. The blade is triangular to rhombic-ovate or elliptic to oval, 1.5-5" long, cuneate at base, and margins crenate to lobed. The younger leaves are pubescent but become glabrous with age. The calyx is densely but finely stellate pubescent. The petals are white or rose-purple and are darker at the base. The fruit is a capsule and the seeds are ciliate. It was introduced into cultivation about 1600. Cultivated throughout the state and persisting at old home sites. The current records include 25 parishes.

HYDRANGEA - *Hydrangea* - Hydrangeaceae

The genus "Hydrangea" is from the Greek hydor (water) and aggos (a jar) referring to the cup-shaped fruits. The species are shrubs. The leaves are simple, opposite, deciduous, and with pinnate major veins. Two kinds of flowers are produced; the sterile are large and showy with a white four-parted calyx and the fertile ones small and inconspicuous with five white petals and ten stamens. The fruit is a brown ribbed capsule with many small seeds.

SMOOTH HYDRANGEA ***(Hydrangea arborescens)*** is also known as Hill-of-Snow, Mountain Hydrangea, Wild Hydrangea, and Sevenbark; this name comes from the fact that the bark peels off in layers of different colors. The specific epithet, "*arborescens*", refers to its likeness to a small tree. Hydrangea reaches a height of up to 6 ft with glabrous smooth twigs but loose and peeling on older stems. The blades are ovate, acuminate, coarsely serrate, 2.5-7" long and 1.5-4.5" wide, somewhat rough above, and glabrous or hairy below. The inflorescence is a flat topped compound cyme. The blades are thin, ovate, pointed, toothed, base rounded to cuneate, glabrate or slightly downy to densely tomentose beneath. There are five white petals and ten stamens. The flowers and fruits are eaten by some wildlife species, including wild turkey and white-tailed deer. The foliage is reported to be poisonous to livestock. It has been in cultivation since 1736. A drug called hydrangin was made from the reddish yellow rhizome. It was used in the treatment of bladder stones and cystitis. It is reported to be a favorite medicine of the Cherokee Indians. A strong purgative, but in lesser doses it is nephritic and removes gravel and stones from the bladder. Rare in well drained slopes (FACU) mostly in southeastern Louisiana with reports from East Baton Rouge, East Feliciana, St. Helena, and West Feliciana parishes; it is also reported from one parish (Union) west of the Mississippi River.

Hibiscus syriacus (Britton and Brown 1943)

Hydrangea arborescens (Britton and Brown 1943)

OAK LEAF HYDRANGEA ***(Hydrangea quercifolia)*** is also known as Sevenbark, Gray Beard, and Old Man's Beard. The specific epithet, "*quercifolia*", refers to the oaklike leaves; the common name has a similar origin. It is a spreading, straggly shrub that reaches up to 12 ft tall. The bark is thin reddish brown and exfoliating in shaggy strips. The blades are ovate to suborbicular or oblong, 4-8" wide and long, deeply 3 to 7 lobed, the lobes with small lobes or serrate, apex of lobes acute to acuminate, base rounded to truncate and often decurrent on the petiole, upper surface dull green and glabrous except for a few white hairs along the impressed veins, and the lower surface densely white pubescent. The inflorescence is a panicle. The flowers of the clusters turn reddish brown when the white fades in mid-summer and remain showy until the fall. It is widely planted as an ornamental. Infrequent in well drained slope forests and absent from the floodplains and marsh. The current records include 15 parishes with no records from extreme western Louisiana.

ST. JOHN'S WORT, ST. ANDREW'S CROSS - *Hypericum* - Clusiaceae

The genus name, *"Hypericum",* is a Latinization of the ancient Greek word "huperikon", referring to a European species of the genus or from the Greek hyper (above) and eikon (a picture); the plant was hung above pictures to ward off evil spirits. The genus includes shrubs and also some herbaceous species. The leaves are opposite, simple, evergreen, and with pinnate major veins. The blades are sessile or sub-sessile with entire margins and a punctate lower surface. The two, four, or five sepals are green and persistent. The four or five petals are yellow. The fruit is a capsule with numerous small lustrous seeds.

SHORT LEAF ST. JOHN'S WORT ***(Hypericum brachyphyllum)*** is also known as Pineland St. John's Wort and Coastal Plain St. John's Wort. The specific epithet "*brachyphyllum*" is for the short leaves. It is an erect, single-stemmed, winged shrub to 3 ft. The leaves are in conspicuous axillary clusters and have a distinct articulation joint at the base. The blades are needle-like, less than 1 cm long and 1 mm wide. The margins are revolute and the new leaves are lustrous above. The flowers have five petals and are typically borne individually in the leaf axils. Just before opening, the flower buds are often bright red and offer an attractive contrast to the deep green foliage. The capsule is narrow and nearly cylindrical. It is found in rare in wet areas (FACW) in southern Louisiana with reports from Calcasieu, St. Tammany, and Washington parishes. (syn. *Hypericum aspalathoides*)

ROUND POD ST. JOHN'S WORT ***(Hypericum cistifolium)*** is also known as Rockrose Leaved St. John's Wort. The specific epithet, "*cistifolium*", means "leaves like the Rockrose". It is a single-stemmed (sometimes a few branches) shrub up to 3 ft tall with winged stems. The leaves do not have the articulation at base. The blades are lanceolate to linear-lanceolate with revolute margins. The flowers have five petals. The capsule is globose to ovoid. It is found infrequent in wet areas (FACW) mostly in southern Louisiana with reports from nine parishes. (syn. *Hypericum opacum*)

ST. PETER'S WORT (***Hypericum crux-andreae***) is also known as Atlantic St. Peter's Cross. The specific epithet "*crux-andreae*" is translated as "crux = cross and andreaea = Andrew". It is a shrub with reddish, flaking bark that reaches 3 ft. The leaves have the articulation at the base. The blades are lanceolate to oblong or oblanceolate. The flowers have four sepals (two large and two small) and four petals. The fruit is an egg-shaped capsule. It has been in cultivation since 1816. It is widely distributed in moist to somewhat well-drained soil

Hydrangea quercifolia (Vines 1960)

Hypericum brachyphyllum (Vines 1960 as ***Hypericum aspalathoides***)

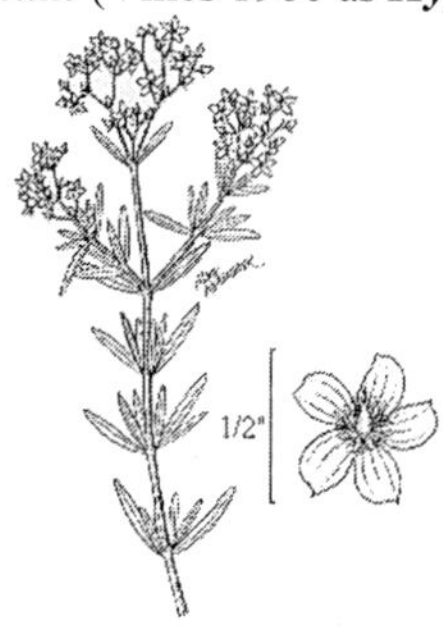

Hypericum cistifolium (USDA-NRCS)

Hypericum crux-andreae (USDA-NRCS)

(FACW-) and the Cajun Prairie. The current reports include 36 parishes. (syn. *Ascyrum cunefolium*, *Ascyrum stans, Hypericum stans*)

SHRUBBY ST. JOHN'S WORT ***(Hypericum densiflorum)*** is also known as Bushy St. John's Wort and Dense St. John's Wort. The specific epithet, "*densiflorum*", is for the dense flowers. It is a branched shrub to 7 ft with winged twigs. The leaves have the articulation at the base and often have smaller leaves in the axils of larger ones. The blades are linear to oblanceolate or oblong, 1- 2" long, about 1/4" wide, apex acute or obtuse, the base attenuate, and with revolute margins. Several of the leaves are in the axils of the main leaves. The sepals and petals are in fives and the flowers are numerous in corymb-like panicles. The capsules are ovoid. It is found throughout the state in wet to moist areas (FACW-). The current records include 36 parishes. (syn. *Hypericum glomeratum*, *Hypericum prolificum* var. *densiflorum)*

SAND BUSH ST. JOHN'S WORT ***(Hypericum fasciculatum)*** is also known as Peel Bark St. John's Wort. The specific epithet, "*fasciculatum*", refers to the clustered (fascicled) leaves. It is a tall shrub with green to brown twigs that are sharply angled. The bark is buff, burnt orange, or cinnamon colored. The leaves are fascicled and have the articulation at base. The blades are linear-filiform, 1 mm wide, 0.5-2" long, the apex is acute, and margins are distinctly revolute so as to create two distinct longitudinal grooves along the lower surface. The flowers have five petals and five linear (similar to the leaves) sepals. The capsule is ovoid. Frequent in wet to moist areas (FACW+) especially bogs and baygalls in southeastern and southwestern Louisiana. The current records include 13 parishes.

SHOWY ST. JOHN'S WORT (***Hypericum frondosum***) is also known as Golden St. John's Wort and Cedarglade St. John's Wort. The specific epithet "*frondosa*" is for the leaves. It is a shrub to 4 ft with slightly winged stems. The leaves have the articulation at base. The blades are elliptic to oblong, 1-3" long, 0.3-0.75" wide, apex acute to obtuse or rounded, base attenuate, and margins slightly revolute. The flowers have five parts with the petals 0.5-1" long. The flowers are solitary or in groups of three. The capsule is conic shaped. It is found in well drained soils often on slopes throughout the state with records from 15 parishes. (syn. *Hypericum aureum*, *Hypericum splendens*)

BEDSTRAW ST. JOHN'S WORT (***Hypericum galioides***) is a shrub to 6 ft with reddish brown twigs that are angled to terete. The specific epithet, "*galioides*", means "like *Galium,"* or appearing like a bedstraw plant. The leaves are often clustered in the leaf axils with linear-oblong or oblanceolate blades. The blades are 0.5-2.5" long, narrower than 1/8 ', apex obtuse or acute, and the margins entire and revolute. The petals are five and the capsules are 0.25" long or shorter. In wet areas along edges of bogs, baygalls, and other wet areas (OBL) mainly in the southwestern, southeastern, and west central parts of the state. The current records include 17 parishes.

Hypericum densiflorum (USDA-NRCS)

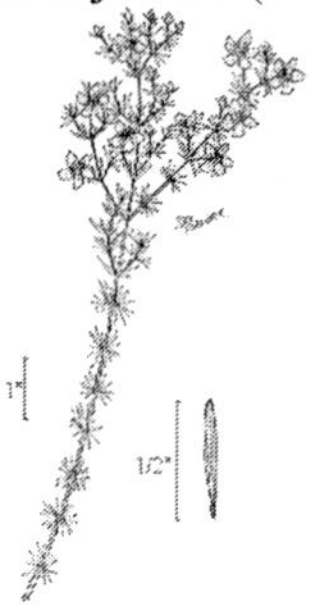

Hypericum fasciculatum (USDA-NRCS)

Hypericum frondosum (Britton and Brown 1943)

Hypericum galioides (USDA-NRCS)

ST. ANDREW'S CROSS ***(Hypericum hypericoides*** *ssp.* ***hypericoides)*** is also known as Arrayanalla. The specific epithet, "*hypericoides*", is for its resemblance to certain species of Hypericum. It is a shrub to 3 ft that is rather short-lived. The bark is reddish-brown and sheds in thin strips. The leaves have the articulation at base and often have clumps of smaller leaves in the axils of larger leaves. The blades are narrowly oblanceolate to linear, 0.5-1.5" long, 0.15-0.3" wide, and apex obtuse or rounded. The flowers are erect on stalks shorter than 6 mm. The sepals
and petals are in fours. The sepals include two small and two large ones. The capsule is erect, flattened, and ovoid to spindle-shaped and about 1/3" long. The extract of leaves is used as an astringent and resolutive. The seeds are reported to have purgative properties; it is used for kidney ailments and stones. The plant is browsed by deer. It is widely distributed in Louisiana on well drained to mesic sites (FAC) with reports from 58 parishes. (syn. *Ascyrum hypericoides, Ascyrum linifolium*) The ssp. ***multicaule*** (syn. *Hypericum stragulum*) differs in the plant being decumbent instead of erect and thus mostly shorter than 1 ft. The specific epithet "*multicaule*" means many stemmed. This variety is reported from six parishes in southeastern and southwestern Louisiana.

FIVE LOBE ST. JOHN'S WORT ***(Hypericum lobocarpum)*** is also known as Dense St. John's Wort. The specific epithet "*lobocarpum*" is for the lobed fruit. It is similar to ***Hypericum densiflorum*** but has a capsule that is deeply 4 or 5 lobed. It is found in similar habitats as ***Hypericum densiflorum*** with reports from 13 parishes. (syn. *Hypericum densiflorum* var. *lobocarpum, Hypericum oklahomense*) (Not Illustrated*)*

DOGBANE ST. JOHN'S WORT ***(Hypericum nudiflorum)*** is also known as Early St. John's Wort and Naked St. John's Wort. The specific epithet, "*nudiflorum*", refers to the fact that the inflorescence has only small subtending leaves. It is a branching shrub that reaches 6 ft with 2-angled twigs that are green when young and red with age. The leaves do not have the articulation at base. The blades are linear-oblong to narrowly elliptic or oval-lanceolate with bracted subulate leaves in their axils. The blades are 1-2.5" long, apex obtuse to rounded, and the base clasping the stem. The flowers are in an open cyme and have five sepals and petals. The sepals are less than half as long as the petals. The capsule is conic-ovoid and shorter than 0.25" long. It is found in wet areas (FACW) throughout the state with reports from 30 parishes. *(*syn. *Hypericum apocynifolium*)

BROOM BRUSH (***Hypericum prolificum***) is also called Paint Brush St. John's Wort, Rock Rose St. John's Wort, and Shrubby St. John's Wort. The specific epithet, "*prolificum*", means "abundant", probably in reference to the flowers. It is a shrub to 8 ft with two-winged branches. The bark is papery, whitish, and exfoliating. The twigs are 2-angled, red when young, and turning gray to stramineous with age. The leaves have the articulation at base and sometimes have the fascicles of small leaves in the axils of the larger or regular leaves. The blades are linear to oblong or narrowly elliptic, 0.5-3.5" long, 1/6-3/5" wide, apex acute to acuminate, and base narrowed to a short winged petiole. The flowers are in terminal cymes with five sepals and five petals. The capsule is conic to narrowly ovoid and 0.3-0.6" long. It is browsed by deer. It is found in well drained slope forests (FACU) throughout the state with reports from 16 parishes currently. *(*syn. *Hypericum spathulatum)*

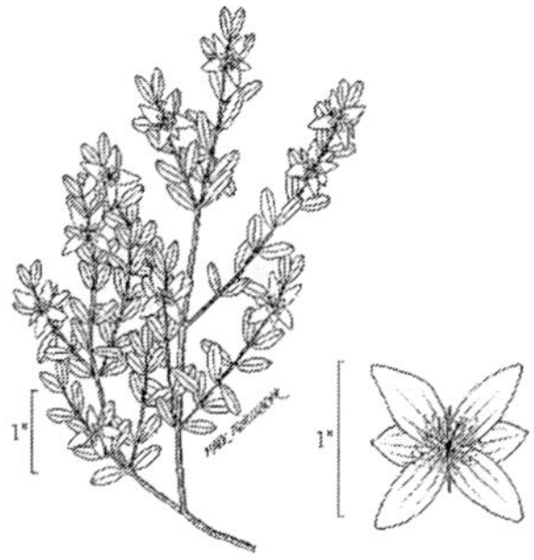

Hypericum hypericoides ssp. ***hypericoides*** (USDA-NRCS)

Hypericum nudiflorum (Vines 1960)

Hypericum prolificum (USDA-NRCS)

PINELAND ST. ANDREW'S CROSS (***Hypericum suffruticosum***) is similar to ***Hypericum hypericoides***. The specific epithet "*suffruticosum*" is for the somewhat shrubby habit. It differs in being a low growing often decumbent shrub to 0.5 ft. The flowers are on stalks longer than 6 mm and the stalks are reflexed. Rare with only one report from pine forests in St. Tammany Parish. (syn. *Ascyrum pumilum*)

HOLLY - *Ilex* - Aquifoliaceae

The genus name, "*Ilex*", is the old name for Holly Oak, *Quercus ilex*, of Europe. The genus includes mostly hollies with other common names like gallberry, yaupon, and dahoon. All are shrubs to small trees with alternate, simple leaves with pinnate major veins. Some have leaves with spinose margins and all have a pair of stipules in the form of small black dots at the base of the leaf. The flowers are white, imperfect, axillary, solitary or fascicled, and produced from March to June. Most hollies are dioecious with a pyrene fruit with 4-7 seeds that are grooved or ribbed; the fruit is often called a berry but is closer to a drupe than a berry and is often termed a pyrene since it has more than one seed as in drupes.

CAROLINA HOLLY (***Ilex ambigua***) is also known as Sand Holly and Possum Holly. The specific epithet, "*ambigua*", refers to its vague similarity to other species. It reaches a height of 20 ft. The trunk and branches are crooked and the bark is gray. The twigs are slender, reddish brown with lenticels. The leaves are deciduous or nearly evergreen. The blades are 2-3" long, elliptic or oval, thin, fine-toothed, usually somewhat hairy beneath, light green with slightly indented veins above, and cuneate bases. The small white flowers have 4 petals, sometimes 5, rarely 6. The white petals are often tinted with light purple and the pedicels are very short. The shiny red fruits are about 0.25" diameter, sessile, and have grooved seed. Carolina Holly furnishes food for a number of wildlife species. This species is widely distributed in Louisiana primarily on well drained, often sandy soils but is absent from the marsh and the Mississippi and Red River floodplains. The current records include 23 parishes.

SERVICE BERRY HOLLY (***Ilex amelanchier***) is also known as Sarvis Holly and June Berry Holly. The specific epithet, "*amelanchier*", refers to its superficial resemblance to a Serviceberry (*Amelanchier*). It is typically a shrub but sometimes treelike and reaches 15 ft. The leaves are deciduous and the blades are oblong, 2-4" long, entire to minutely serrate, and the lower surfaces are copiously shaggy pubescent. It has staminate flowers that are clustered. The fruit is dull red and 0.3-0.4" diameter. It has been cultivated since 1880. This species is restricted to moist to wet habitats (OBL) in extreme southeastern Louisiana with records from St. Tammany, Tangipahoa, and Washington parishes.

DAHOON (***Ilex cassine***) is also known as Christmas Berry, Yaupon, Cassine Holly, and Henderson Wood. The specific epithet, "*cassine*, is from the Indian name, Cassena, that was applied to a drink prepared from this species and/or Yaupon. Although the leaves of dahoon were sometimes mixed with those of Yaupon in teas, Dahoon contains no caffeine. It is a slow growing and short-lived shrub or tree up to 30 ft tall that sprouts profusely. The bark is gray, covered by roughened lenticels. The young twigs are coated with silky pubescence. The leaves are evergreen and the blades are oblong to elliptic to oblanceolate, pointed or less commonly rounded at the tip, tapering to the base, usually toothless. The leaves are also usually smooth on the upper surface, smooth or hairy on the lower surface, up to 6" long, and up to 2" wide. The fruits are red or orange red, spherical, 1/4-1/3" diameter. Pre-Columbian Creek Indians roasted the leaves and made them into a decoction. Only men were permitted to drink this tea, to which

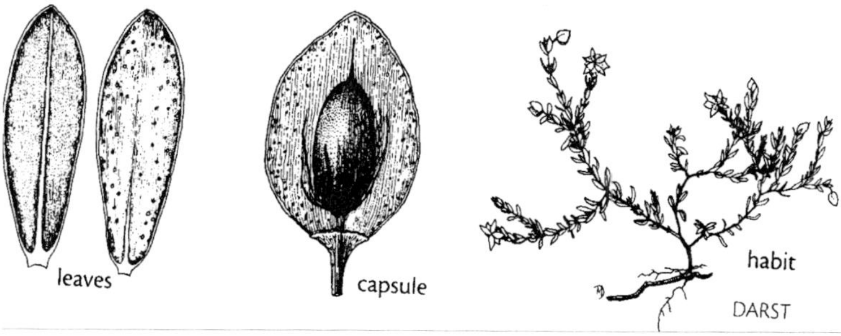

Hypericum suffruticosum (Tobe et. al. 1998)

Ilex ambigua (Vines 1960)

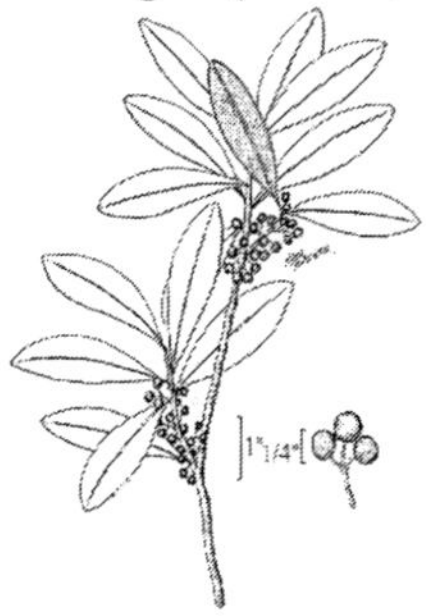

Ilex amelanchier (USDA-NRCS)

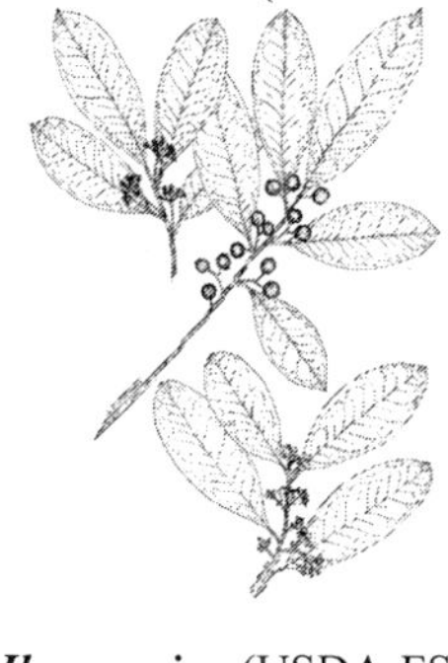

Ilex cassine (USDA-FS)

many virtues were attributed. The wood is pale brown, close-grained, light, soft, not strong, easily worked, and weighing 30 lb per cu ft. Dahoon is highly susceptible to fire. It provides an important food supply for wintering songbirds. Several animals depend on the berries. Deer browse on the twigs and foliage. The plant has been cultivated since 1726. Dahoon varies greatly because it hybridizes naturally with other hollies. Dahoon is found naturally in Louisiana along streams, low pinelands, pond margins, flatwoods, and bogs (FACW) near the coast but has been planted inland. The current records include 13 parishes.

LARGE GALL BERRY (***Ilex coriacea***) is also known as Baygall Bush, Large Sweet Gall Berry, Gall Berry, Ink Berry, Shining Ink Berry, and Tall Ink Berry Holly. The specific epithet, "*coriacea*", refers to the thick, leathery leaves. It is a tree or shrub to 15 ft and is fast growing but short-lived. The bark is gray with blackish streaks. The leaves are evergreen and the blades may have sharp teeth near the pointed tips or may lack teeth, 2-3" long, and glaucous beneath. The leaves are often marked with fine black dots beneath. Large Gallberry has black fruit. It is one of the most preferred browse plants within its range and furnishes palatable browse for deer throughout the year. Game birds and animals, as well as songbirds use the fruits less than the other hollies. Large gallberry is a common plant of the acid pH baygalls (FACW) of southeastern and southwestern Louisiana. The current records include seven parishes.

BUFORD HOLLY (***Ilex cornuta***) is also known as Buford Chinese Holly. It is a shrub or small tree to 25 ft. The leaves are evergreen and the blades are 2-4" long, lustrous dark green, often with a single terminal spine or occasionally with two lateral spines, and with a tough texture. The fruit is red and 0.25-0.5" diameter. Widely planted and persisting and escaping throughout the state. (Not Illustrated)

DECIDUOUS HOLLY (***Ilex decidua***) is also known as Possum Haw Holly, Swamp Holly, Meadow Holly, Prairie Holly, Welk Holly, Bear Berry, Christmas Berry, and Winter Berry. The specific epithet, "*decidua*", refers to the fact that the leaves are shed in the autumn. It is a medium to large bushy shrub or small tree that reaches a height of 30 ft. The bark is thin, smooth and mottled gray to brown and is often covered with numerous small, warty bumps. The leaves are deciduous and often clustered on spur branches. The blades are 1.5-3" long, 0.4-1.2" wide, obovate, light green above, pale and slightly hairy beneath, and serrate with gland-tipped teeth. The flowers have 4 rounded, white petals, 4 sepals and 4 stamens, with the male and female usually on separate trees. The red (sometimes yellow to orange) 1/4" fruits are clustered along the bare branches and remain after the leaves fall. The wood is creamy white, heavy (47 lb per cu ft), hard, and close-grained. Nine species of birds including bobwhite quail and several mammals use the fruit. Deer browse the foliage. This species is widely distributed throughout the state and a wide range of habitats from lowlands and swamps (FACW-) to ravines and rich, upland habitats. The current records include 63 but it grows in all 64 parishes.

GALL BERRY (***Ilex glabra***) is also known as Appalachian Tea, Canadian Winter Berry, Evergreen Winter Berry, Dye Leaves, Ink Berry and Low Gall Berry Holly. The specific epithet, "*glabra*", refers to the smooth glabrous leaves. The common name Gallberry refers to the bitter taste. It is an evergreen shrub up to 12 ft tall, with extensive rhizomes, often forming colonies. The leaves are evergreen, leathery, obovate to elliptic, pointed at the tip, tapering to the base, smooth, usually with a few teeth in the upper half. It is often reddish and dotted on the lower surface, shiny on the upper surface, up to 3" long, and up to 3/4" wide. The fruits are black at maturity, spherical, up to 1/4" diameter, and contain 5-8 nutlets. The fruits are food for many songbirds, bobwhite, and wild turkey. The marsh rabbit browses the leaves. The berry is

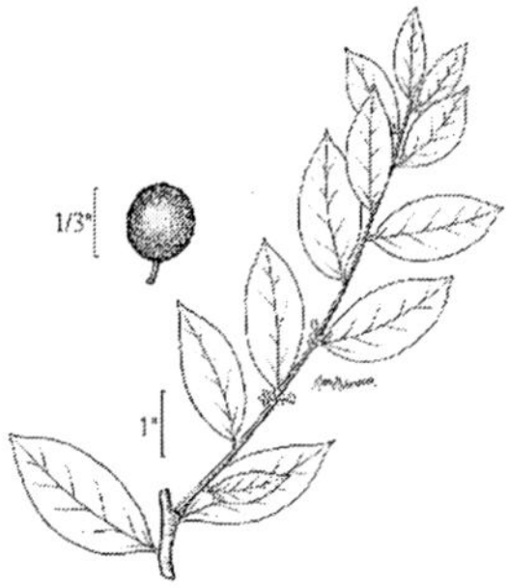

Ilex coriacea (USDA-NRCS)

Ilex decidua (USDA-NRCS)

Ilex glabra (USDA-NRCS)

considered a good source of honey. In the past, the leaves were boiled for a tea; however, it contains no caffeine. The berries, which persist during the winter, have been used as a source of black ink. It has been cultivated since 1759. It is restricted to savannahs, low woods, and bogs (FACW) of southeastern Louisiana. The current records include eight parishes with the non cultivated records all east of the Mississippi River.

GEORGIA HOLLY ***(Ilex longipes)*** is also known as Chapman's Holly, Long Stem Holly, and Yaupon Holly. The specific epithet, "*longipes*", refers to the long-stalked fruits, which is a conspicuous feature. It is a large shrub up to 25 ft tall and is similar to Deciduous Holly and often considered a variety of that species. The leaves tend not be on spur shoots as much as in Deciduous Holly. The main difference is in fruit which is on a stalk that can be up to 1" long. The habitat is also different with Georgia Holly being restricted to better drained upland habitats (FAC) of southeastern, central, and northwestern Louisiana. The current records include 16 parishes.

MYRTLE HOLLY (***Ilex myrtifolia***) is also known as Myrtle Dahoon Holly, Small Leaf Dahoon, Cypress Holly, and Myrtle Leaved Holly. The specific epithet, "*myrtifolia*", refers to the small, myrtle-like leaves. It is a shrub that reaches 30 ft and is similar to Dahoon and often considered a Variety of that species. The bark is pale gray or almost white, and roughened by warty protuberances. The leaves are evergreen with blades that are 0.5-2" long, stiff, entire, leathery, short, and very narrowly elliptic. The apices are with a small, but distinctive point. The margins are mostly entire and revolute or with a few spinescent teeth apically. The fruit is typically red but sometimes orange or yellow. Like Dahoon, it is found naturally only near the coast (FACW) and is also widely planted inland. The non cultivated records are from St. Tammany and Tangipahoa parishes with an additional cultivated record from Lafayette Parish. (syn. *Ilex cassine var. myrtifolia)*

AMERICAN HOLLY ***(Ilex opaca***) is also known as Christmas Holly, Prickly Holly, Evergreen Holly, Holly, Yule Holly, and White Holly. The specific epithet, "*opaca*", refers to the dull green leaf. It is the state tree of Delaware. American Holly is the largest holly and can become a small tree to a height of 40-70 ft and 1 ft diameter. The bark is light or dark gray and smooth, with occasional wart-like protuberances on old trees. The leaves are evergreen (persist for about 3 years), and the blades are oval, tough and leathery, 2-4" long, and 0.8-1.6" wide. The blades are dark shining green above, yellow green below, have wavy margins with sharp stiff spines (rarely absent), with a rounded base, and a spinose apex. Male and female flowers are on separate trees and have 4 white, rounded petals. The male flower has one white stamen at each notch between the petals and a hole in the center. The female flower has a small, green, rounded pistil in the center and sterile stamens. The fruit is red (sometimes orange or more rarely yellow), 3/8" diameter and appears from November to December and clings until spring. The wood is tough, high shock resistant, close-grained, not durable under exposure, and rather heavy (46 lb per cu ft). The wood is nearly white when first cut, turning brown with age and exposure, with thick, and rather lighter colored sapwood. It is used for cabinets, interior finish, novelties, handles, fixtures, scientific instruments, tannery, piano keys, ship models, and inlays. The leaves are used for medicinal teas (diuretic and purgative), while the fruits induce vomiting. The branches are used for Christmas decorations. At least eighteen species of birds eat the fruit. Wild turkeys sometime feed on the berries in treetops as well as on the ground. The flowers are a valuable source of nectar. The leaves are browsed by cattle. American Holly can be found on rich, moist woods (FAC-) along streams, borders of swamps and dry, well-protected slopes

Ilex longipes (Vines 1960)

Ilex myrtifolia (Vines 1960)

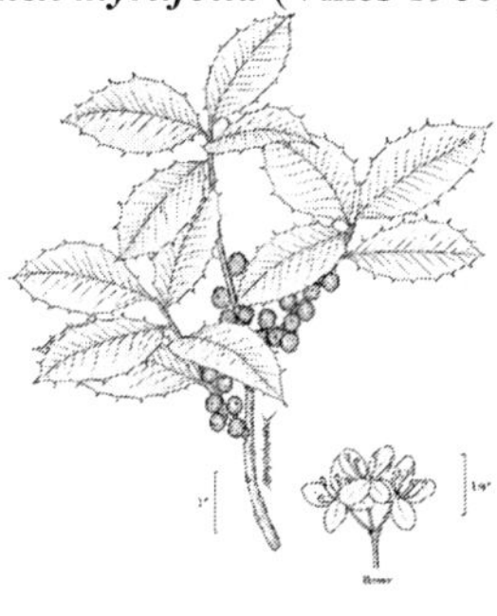

Ilex opaca (USDA-NRCS)

except for the floodplain of the Red and Mississippi Rivers. The current records include 45 parishes.

ROUGH WINTER BERRY (***Ilex verticillata***) is also known as Winter Berry, Striped Alder, False Alder, Fever Bush, Virginia Winter Berry, Common Winter Berry, and Black Alder. The specific epithet, "*verticillata*", refers to the axillary clusters of flowers. It is a variable small to large shrub that grows to a height of 25 ft. The leaves are deciduous and the blades ovate to elliptic, 1.5-4" long, 0.5-2" wide, dull green above with a network of irregular veins, and with lower veins prominent on a pubescent surface. The apex is acuminate to acute, the base acute to acuminate or rarely rounded, and the margins are sharply serrate. The leaves turn very dark in the fall. The red, rarely yellow fruit is about 1/3" diameter. The fruit and bark have been used in domestic medicine as an astringent, internally for diarrhea, skin lesions and is reported to make a good gargle for sore throat and pharyngitis. It can be mixed with boric acid it makes a good hemostatic against hemorrhage. A decoction made with vinegar, instead of water, is a good external wash for lice, scabies and scabs. This same liquid is used for washing the teeth. The leaves are made into a tea and used for breast inflammations, diarrhea, dropsy, and similar swellings. The fruit is valuable as a wildlife food, being eaten by at least 20 species of birds. The bright red fruit is also used for winter dry bouquets. It has been cultivated since 1763, and is used for ornamental planting in moist places. This species is found along the edges of lakes, ponds, and other moist areas (FACW), sometimes on old stumps in lakes throughout the state. The current records include 16 parishes.

YAUPON (***Ilex vomitoria***) is also known as Cassena, Cassine, Cassio Berry Bush, Evergreen Cassene, Yapon, Yopon, Youpon, Yaupon Holly, Emetic Holly, Evergreen Holly, South Sea Tea, Carolina Tea, Appalachian Tea, Yopon del Indio, Chocolate del Indio, Indian Black Drink, Black Drink Plant, and Christmas Berry. The specific epithet, "*vomitoria*", refers to its use as a medicine as an emetic. It is a shrub to small tree that reaches up to 25 ft. The bark is brownish to mottled grey almost black with thin, small scales on old trunks. The leaves are evergreen and the blades are elliptic to oval, 0.5-2" long, 0.5-1" wide, margins crenate, apex rounded or obtuse, sometimes emarginate and mucronate, and the upper surface is lustrous dark green with the lower surface paler. The red (rarely yellow) fruit is glistening, ca 1/4" diameter, and borne in profusion close to the stem and cling until spring. The heavy, hard, nearly white turning yellow when exposed to air, close-grained, strong wood weighs 46 lb per cu ft and is used for turnery, inlay and woodenware. The branches are used for decoration for Christmas. The leaves are dried to make tea (black drink) which contains caffeine and possesses emetic and purgative qualities. The fruit is eaten by at least 7 species of birds and raccoons. It is good browse for white-tail deer. Cabeza de Vaca, the early Spanish explorer, noted that the east Texas Indians used the leaves in make a medicinal tea called "Black Drink" which possessed emetic and purgative qualities. Some sources state the Indians drank it to prove their suitability for the warpath. A strong decoction cause vomiting, hence the specific epithet "*vomitoria*". Yaupon thrives in a variety of soil conditions and ranges throughout the state from the marsh and prairie to the dry pine forests (FAC). The current records include 55 parishes.

FLORIDA ANISE - *Illicium floridanum* - Magnoliaceae

The genus name, "*Illicium*, means "allurement," and is in reference to the flowers. The specific epithet, "*floridanum*", means "of Florida." Other common names include Polecat Tree, Poison Bay, Sweet Laurel, Swamp Anise, Star Anise, Star Bush, Stink Bush, Purple Anise Tree, Matacaballos (horsekiller), and Ixcapantl. It is a shrub or small tree to 25 ft with dark brown to black or gray bark that is broken into small fissures. The twigs are smooth and green when

Ilex verticillata (USDA-NRCS)

Ilex vomitoria (USDA-FS)

Illicium floridanum (Vines 1960)

young but turning reddish brown to gray with age. The leaves are alternate, simple, evergreen, and with pinnate major veins. The blades are elliptic to oblong or lanceolate, 2.5-6" long, apex acute to acuminate, base acute or cuneate, and the margins are entire. The blades are leathery and somewhat fleshy, dark green and glabrous above, paler below with inconspicuous veins, and with a strong unpleasant odor to the crushed blade. The flowers are nodding with 20-30 linear dark red petals. The sepals are more than 10 mm long. The fruits are a drooping star-shaped cluster of follicles. The leaves are reported as being poisonous to stock. Locally abundant along the banks of streams (FACW) in southeastern Louisiana and cultivated in other areas. The current uncultivated reports include East Baton Rouge, Livingston, St. Helena, St. Tammany, Tangipahoa, and Washington parishes.

ANIL DE PASTO - *Indigofera suffruticosa* - Fabaceae

The genus name, "*Indigofera*", is for the yield of indigo from some species. The specific epithet, "*suffruticosa*", refers to its suffrutescent manner of growth. Anil de Pasto is also known as Anil Indigo and Indigo. It is a sparingly branched perennial plant that reaches a height of 6 ft. The stems are rather erect and herbaceous above, but become woody near the base. The angled stems are hairy green or gray when young but becoming glabrous and gray to brown with age. The leaves are alternate and pinnately compound. The 7-15 leaflets are subsessile, oblong to elliptic or oblanceolate to obovate, 0.3-1.25" long, apex mucronate, base acute, upper surface sparsely pubescent when young but becoming glabrous with age, lower surface pubescent throughout, and margins entire. The blade hairs are blond and two branched. The flowers are reddish to purplish, small and borne in spike-like clusters in the leaf axils. The fruits are linear curved legumes with a long slender beak. It is found rare in sandy soils (FACU) sporadically throughout the western part of the state. The current reports include eight parishes

VIRGINIA SWEET SPIRE - *Itea virginica* - Grossulariaceae

The genus name, "*Itea*", is the Greek name for willow. The specific epithet, "*virginica*", refers to the state of Virginia. Other common names include Virginia Willow, Sweet Spire, and Tassel White. It is a shrub to 10 ft with brown to gray smooth scaly bark. The young twigs are green but turn brown with age and the pith is chambered. The leaves are alternate, simple, deciduous, and with pinnate major veins. The blades are oblong, elliptic or oval, 0.8-3.5" long, 0.4-1.5" wide, glabrous and dull green above, paler green and pubescent below, cuneate at base; acute or acuminate at apex, and the margins are serrate. The blades turn dull green to yellow or red in the fall. The white (rarely pink) flowers are in terminal recurved racemes. The fruit is a capsule. It is widespread in moist to wet areas along streams (FACW+) throughout the state. The current reports include 52 parishes.

MARSH ELDER - *Iva* - Asteraceae

The genus name, "*Iva*", is a medieval name for a medicinal plant, *Ajuga iva*. The genus includes woody shrub and herbaceous species. The leaves are simple, deciduous, and with pinnate major veins. The flowers are small, green, and in head inflorescences. The fruits are obovoid, flattened, achenes with no pappus.

SALT MARSH ELDER (***Iva frutescens***) is also known as Sump Weed, Jesuit Bark, False Jesuit Bark, Big Leaf Sump Weed, and High Water Shrub. The specific epithet, "*frutescens*", refers to the shrubby character. A much branched partly woody plant that reaches 12 ft. The leaves are opposite and the blades somewhat thickened. The blades are three-veined at the base, lanceolate-oblong to linear, 2-6" long, 0.5-2" wide, irregularly serrate with large teeth, both surfaces pale green, the apex acute or acuminate, and the base cuneate. The achenes

Indigofera suffruticosa (Vines 1960)

Itea virginica (USDA-NRCS)

Iva frutescens (USDA-NRCS)

are about 1/8" long. It is found in marshes (FACW+) of southern Louisiana with 13 parish records.

SEA COAST MARSH ELDER ***(Iva imbricata)*** is also known as Dune Sump Weed, and Beach Elder. The specific epithet, "*imbricata*", refers to the imbricate bracts of the receptacle. A much branched partly woody plant that reaches 10 ft. The leaves are alternate or opposite below. The blades are fleshy, linear-elliptic to spatulate-oblong, 0.25-2" long, about 0.3" wide, entire or remotely serrate with large teeth, surfaces pale green, the apex acute to obtuse and some mucronate, and the base attenuate. The achenes are about 1/12" long. In marshes (FACW+) of southern Louisiana with six parish records.

PRIMROSE JASMINE - *Jasminum mesnyi* - Oleaceae

The genus "*Jasminum*" is from Yasmin, the Persian name. The specific epithet "*mesnyi*", honors William Mesny (1842-1919), a Major General in the Chinese National Army who collected in China. It is also known as Japanese Jasmine. A shrub with arching branches to 10 ft. The twigs are green and square in cross section. The leaves are opposite, pinnate compound, and evergreen. The 3 leaflets are oblong lanceolate, entire, subsessile, medium green above, paler below, apex acute to acuminate, terminal leaflet 1.5-3" long, and lateral leaflets 0.5-2" long. The flowers are large, yellow, and with a dark center. The fruit is a drupe. Persisting near old home sites with reports from 11 parishes. (syn. *Jasminum primulinum*)

BLACK WALNUT - *Juglans nigra* - Juglandaceae

The genus name, "*Juglans*", is from the Latin *Jovis glans,* meaning "acorn [or any nut of similar shape] of Jove." The specific epithet, "*nigra*", refers to the dark wood or the nuts. Black Walnut is also known as American Black Walnut and Eastern Black Walnut. The bark is gray to dark brown, becoming deeply furrowed into scaly ridges. The twigs are thick, gray, finely hairy when young, with raised 3-lobed leaf scars, and chambered brown pith. The terminal bud is single, narrow, and with paired gray hairy scales. The leaves are alternate, even-pinnately compound, 12-24" long with 9-21 leaflets. The leaflets are broadly lanceolate, uneven at base, acuminate, the lateral veins meshing near edge and not running to the edge, finely serrate, 2.4-5.1" long, dark green, glabrous above and soft-hairy below. The plants are monoecious with staminate catkins and single or paired pistillate flowers. The fruit is a nut, often paired, 1.6-2.4" diameter, and with a thick pulpy husk. The wood is dark rich brown, durable, strong, heavy, hard, close-grained, is easily worked, glues well, does not warp, shrink, or swell much, takes a good polish, and weighs 38 lb per cu ft. The whitish sapwood is sometimes stained to match the color of the heartwood to bring a better price. The wood is used in making superior furniture, cabinets, veneers, musical instruments, interior finish, sewing machines, caskets, coffins, posts, railroad crossties, and fuel. Large amounts were used for gunstocks during the Civil War and World War I. The trees are about 12 years old before bearing nuts. Confections and cakes are made from the nuts, which were also a favorite with the American Indians. Squirrels and birds are fond of the large nuts. The squirrels often develop dark "moustaches" from the staining juice of the husks. The hulls are used to treat athlete's foot. This is the plant from which allelopathy was first reported and the fruits were thought to be good for curing brain ailments during the 15th Century when the Doctrine of Signatures was believed. It has been cultivated since 1686. Sporadically throughout the state on well drained to mesic soils (FACU) and widely planted. The current records include 38 parishes.

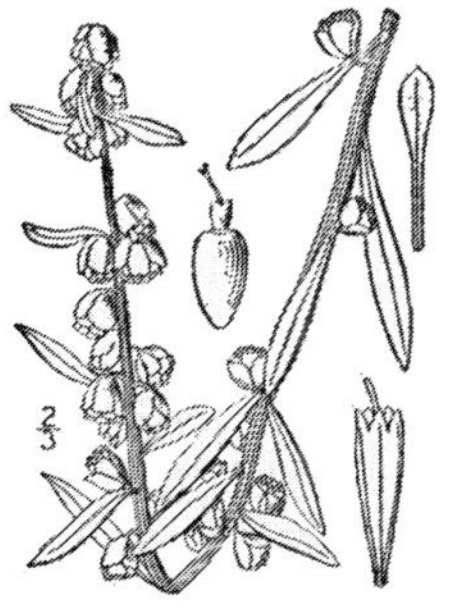

Iva imbricata (Britton and Brown 1943)

Jasminium mesnyi (Hall 1993)

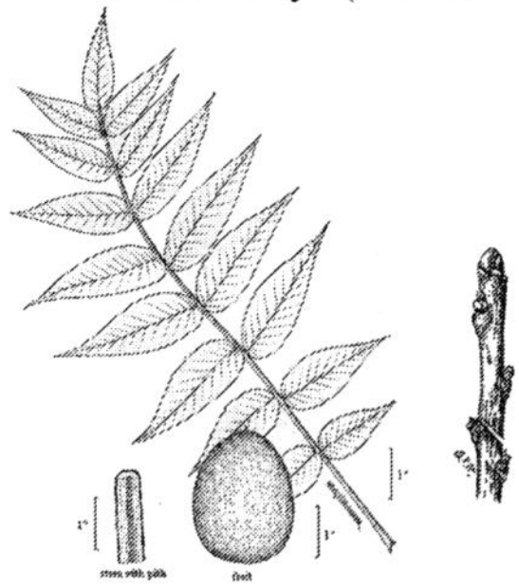

Juglans nigra (USDA-FS and USDA-NRCS)

RED CEDAR - ***Juniperus*** - Cupressaceae

The genus name, "*Juniperus*", is the ancient classical name for the old-world junipers. The genus includes shrubs or trees with evergreen needle-like or scale-like or awl shaped leaves. The plants are dioecious with separate male and female trees. The male cones are small, terminal, solitary, and oblong-ovoid. The female cones are fleshy, berry like, and aromatic. The seeds are bony, ovoid, and 1 to 3 per cone. Gin derives its characteristic taste and odor from juniper female cones. The female cones also are used as a spice and after roasting are used as a very poor substitute for coffee. The wood is quite resistant to decay when in contact with the soil.

COASTAL JUNIPER (***Juniperus communis***) is also known as Dwarf Juniper, Ground Juniper, Horse Savin, Aiten, Ground Cedar, Prostrate Juniper, Melmot Tree, Low Savin, Gin Juniper, Hackmatack, Gorst, Fairy Circle, and Pasture Juniper. The specific epithet, "*communis*", means "common or widespread". It is native to both Europe and North America. Common Juniper is a tree (up to 40 ft), or more commonly a shrub. The needles are in whorls of three, sharp, hollowed, 3-sided, whitened above and 1/4-7/8" long. The young bark is yellow to red, the older bark is dark reddish brown to gray and shreds into papery scales. The female cones blue-black, with a white powder, and are nearly spherical. The wood is close-grained, durable, hard, the heartwood light brown, and the sapwood paler. The oil from the leaves and wood is used in perfumery, and the aromatic foliage is burned as incense in India. The female cones are sometimes used therapeutically in India and Europe. The plant supplies food for several birds and mammals. The wood is used for fuel in Europe, burned for incense in India, and used as a fumigant elsewhere. Cultivated in Louisiana and persisting and perhaps escaping with records from Livingston, Natchitoches, Ouachita, Richland, and St. Tammany.

COASTAL RED CEDAR (***Juniperus virginiana*** var. ***silicicola***) is also known as Southern Red Cedar, Red Cedar, Eastern Red Cedar, Sand Cedar, and Coast Juniper. The specific epithet, "*silicicola*", refers to growing in the sand. It is similar to Eastern Red Cedar but with twigs more slender and drooping. It is a smaller tree up to 50 ft. The female cones are usually bright blue, and only 1/8-3/16" diameter. Mostly along the coast in somewhat wetter sites (FAC) than Eastern Red Cedar. The current records include Natchitoches, Orleans, St. Bernard, St. Tammany, and West Feliciana parishes.

EASTERN RED CEDAR (***Juniperus virginiana*** var. ***virginiana***) is also known as Pencil Wood, Juniper Bush, Pencil Cedar, Virginia Juniper, Red Cedar, Red Savin, Carolina Cedar, and Red Juniper. The specific epithet, "*virginiana*", means "of Virginia" and refers to the general area from which this species was first described. It is a small to medium-sized tree to 70 ft. The bark is light reddish brown, separating into large fibrous strips. There are two kinds of leaves, both usually found on the same tree. The more common kind is dark green, minute, scale-like and clasping the stem in four ranks so that the stem appears square. The other kind, usually appearing on vigorous young shoots, is awl-shaped, quite sharp-pointed, spreading and whitened. Very young trees bear this kind of leaf only, which gives them a different appearance from that of the old trees. The female cone is globular, 0.25-0.4" diameter, glaucous, and dark blue to purplish. Some people say Baton Rouge (Red Stick) is named after the Eastern Red Cedar because the Indians were carrying the pole made from it. Eastern Red Cedar has been often used as a Christmas tree. The light, brittle, soft wood weighs 30-33 lb per cu ft. The heart is wine red and the sapwood pure white. It is used for cabinets, pails, novelties, posts, poles, woodenware, millwork, paneling, sills, closets, shavings, wardrobe chests, and pencils. The dry bark is used as the tinder in flint and steel or sunglass fire sets. The leaves are used as a stimulant, emmenagogue taeniafuge, and the Indians used the leaves as incense. The female

Juniperus communis (USDA-FS)

Juniperus virginiana var. silicicola (Vines 1960)

Juniperus virginiana var. virginiana (USDA-FS)

cones, leaves, and twigs are boiled and inhaled to treat bronchitis. The boiled mixture of bark and water is used to treat skin rash. The extract of cedar oil is used as perfume, moth repellent, insecticide, and immersion oil in microscopes, scenting soaps, and causing abortions. It has been cultivated since 1664. Twenty bird species and numerous mammals eat the fruit. Beaver occasionally eats the bark. Eastern Red Cedar is a favorite in old country graveyards, where, to the imagination of our forebears, perhaps, its finger seemed to point to heaven and its evergreen boughs spoke symbolically of life eternal. But another superstition in parts of Louisiana is "Do not plant a cedar tree because when it gets big enough to shade your grave, you will die". The cedar is the host for cedar apple rust. Twig-laden bagworm cocoons are also frequent on the branches. Eastern Red Cedar is the state tree of Tennessee. Common throughout Louisiana in open well drained (FACU-) areas but disappears from an area with the onset of shaded conditions. Very common in the Inland Prairie Forests. Probably in all 64 with current records from 56 parishes.

MOUNTAIN LAUREL - *Kalmia latifolia* - Ericaceae

The genus name, "*Kalmia*", honors Peter Kalm (1715-1779), a Swedish botanist, who collected, 1748-1751, in North America. The specific epithet, "*latifolia*", means broad-leaved. Mountain Laurel is also known as Big Leaf Ivy, Ivy Leaf Laurel, Spoonhunt, Spoon Wood, American Laurel, Wood Laurel, Small Laurel, Poison Laurel, Broad Leaf Kalmia, Calico Bush, Sheep Laurel, Poison Laurel, Leather Leaf, Calico Tree, Calico Flower, Poison Ivy, Mountain Ivy, Lambkill, Ivy Wood, Clamoun, and Southern Mountain Laurel. It is a thicket-forming shrub or small tree to 35 ft. The bark is reddish brown to dark gray, narrowly and deeply fissured longitudinally, about 1/16" thick. The leaves are very closely alternate, simple, evergreen, and with pinnate major veins. The blades are thick, dark green above, and yellowish-green beneath. The flowers are white, pink or rose, cup-shaped, and borne in many flowered corymbs. The fruit is a five-lobed, woody capsule with many seeds. The wood is reddish brown, hard, sapwood lighter color, heavy, strong, brittle, fine-grained, and weighs 44 lb per cu ft. The wood is used for making small woodenware articles, pipes, handles, and fuel. The leaves have been used medicinally, in internal treatment of diarrhea and syphilis, and in external treatment of skin diseases. The leaves are poisonous to sheep or cattle; seem to be eaten by white-tail deer in small quantities with impunity, but are toxic in large amounts. Records also exist concerning human fatalities from ***Kalmia*** honey. At one time, the burls were tried as a replacement for Italian briar for pipes but the taste was not favorable and the attempt was dropped. The native range in Louisiana is only along stream banks (FACU) in extreme southeastern Louisiana with uncultivated records only from Washington Parish.

GOLDEN RAIN TREE - *Koelreuteria paniculata* - Sapindaceae

The genus "Koelreuteria" honors Joseph Gottleib Koelreuter (1733-1806) a German Professor of Botany. The specific epithet "*paniculata*" is for the panicle flowers and fruits. Other common names include Panicled Golden Rain Tree and Pride of China. It is a medium tree to 35 ft. The bark is rough and light brown. The twigs are light brown, somewhat zig-zag, with prominent leaf scars and lenticels. The leaves are alternate, deciduous, and pinnate to two odd pinnate compound. The 7-15 leaflets are 1.5-3" long, 1-2" wide, ovate to ovate lanceolate, sparsely pubescent, and the margins coarsely dentate or the terminal leaflet lobed. The yellow flowers are numerous in a terminal panicle. The fruit is an inflated papery capsule. It is native to China and Japan and was introduced into the U.S. by Thomas Jefferson in 1809. Cultivated throughout the state and persisting and spreading with records from 18 parishes.

Kalmia latifolia (USDA-FS)

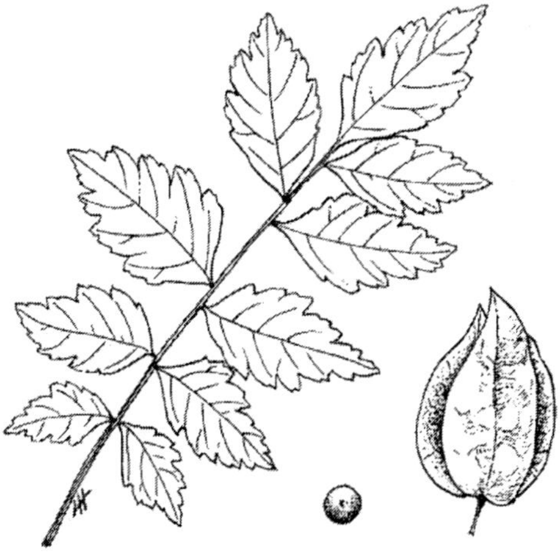

Koelreuteria paniculata (USDA-FS)

CREPE MYRTLE - *Lagerstroemia indica* - Lythraceae

The genus name, "*Lagerstroemia*", honors Magnus Lagerstroem (1696-1759), a Swedish friend of Linnaeus, and the specific epithet, "*indica*", is for India. Crepe Myrtle is also known as Ladies' Streamer. It is a small tree or shrub due to pruning that reaches a height of 35 ft, with a very smooth fluted trunk. The bark is smooth, brown, and exfoliates to expose a smooth, often convoluted, pale surface. The twigs are glabrous, pale, and 4-angled. The leaves are simple, opposite or alternate, deciduous, and with pinnate major veins. The blades are obovate-oval, entire, acute or obtuse, broad-cuneate or rounded at the base, 0.5-2" long, glabrous and lustrous above, and paler and glabrous or pubescent along veins below. The flowers are bright pink, purple, or white in showy terminal panicles. The fruits are oval-globose capsules about 0.5" long. It was introduced in the United States from China in 1747. Widely planted and persisting but apparently does not produce viable seeds. The current reports include 44 parishes.

LANTANA - *Lantana* - Verbenaceae

The genus name, "*Lantana*", is the old name once applied to the Viburnum for the similar inflorescence. The genus includes small shrubs to 6 ft that may be herbaceous during colder winters. The stems are hairy, rarely glabrous, and often with prickles and the twigs are 4-angled. The leaves are opposite, simple, deciduous to evergreen, and with pinnate major veins. The crushed leaves have a strong pungent odor. The flowers are bracted in axillary spikes but appear flattened with the petal color commonly yellow but varying to orange, red, lavender, or blue. The fruit is a two-seeded berry.

COMMON LANTANA (***Lantana camara***) is also called Lantana. The specific epithet, "*camara*", is derived from a Latin word "to arch" referring to the arching stems. The blades are ovate to oblong, 0.75-4.75" long, 0.5-0.75" wide, apex acute to short-acuminate or rarely obtuse, base abruptly rounded to acutely narrowed or subcuneate, margins crenate-serrate, upper surface reticulate-veined, rough and scabrous, and lower surface densely short-pubescent, especially thick on the veins. The calyx lobes are acute and the berry is black. Widely planted as a butterfly nectar plant and persisting and escaping (FACU). It is currently reported from 44 parishes.

TEXAS LANTANA (***Lantana urticoides***) is also called West Indian Shrub Verbena, Bunch Berry, Calico Bush, and many Hispanic and Native American names. The specific epithet, "*urticoides*", refers to the leaves like the genus ***Urtica*** (stinging nettle). The blades are ovate to broadly ovate, 0.33-1.75" long, 0.5-0.75" wide, apex acute, base truncate or semicordate, margins coarsely serrate, upper surface dark green, rugose, scabrous, and distinctly veined, and lower surface paler, and hairy, especially on the veins. The calyx lobes are obtuse and the berry is dark blue. Widely planted as a butterfly nectar plant and persisting and escaping (FACU). The current reports include 9 parishes. (syn. *Lantana horrida)*

LESPEDEZA - *Lespedeza* - Fabaceae

The genus *Lespedeza* honors Vincente Manuel de Cespedes, who was the Spanish governor of Florida and patron of Michaux. In Louisiana, it includes mostly herbaceous species but two shrub species. The shrubs are small to 10 ft tall, usually branched basally, and are woody near the base only. The leaves are alternate, deciduous, and one-pinnate compound with three leaflets. The leaflets are entire with lateral veins at right angles to the midrib and distinctly parallel to each other. The purple flowers are typical legume shape and in racemes in the leaf axils. The fruit is a flattened legume. Deer, bobwhite quail, turkey and other game birds use both species for food. Both are native to Japan, are widely planted, persist, and escape.

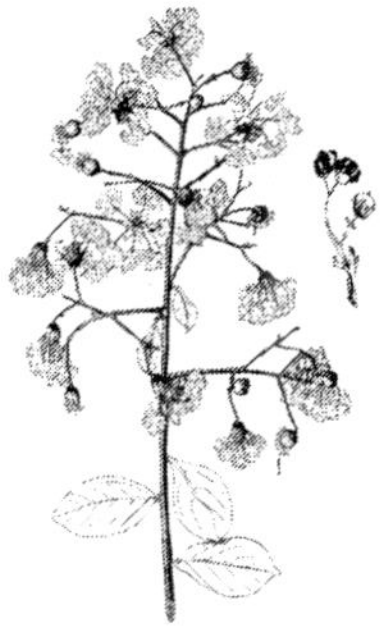

Lagerstroemia indica (USDA-FS)

Lantana camara (Vines 1960)

Lantana urticoides (Vines 1960 as ***Lantana horrida***)

SHRUBBY LESPEDEZA (***Lespedeza bicolor***) is also known as Japanese Bush Clover and Bicolor Lespedeza. The specific epithet, "*bicolor*", refers to two colors. The leaflets are elliptic to oval, 1-2" long, 0.5-1" wide, lower surface pubescent, and upper surface glabrous to slightly pubescent. The calyx tube is longer than the lobes. The seeds are green or splotched with purple. The current reports include six parishes. (Not Illustrated)

THUNBERG'S LESPEDEZA (***Lespedeza thunbergii***) is also known as Bush Clover. The meaning of the specific epithet, "*thunbergii*", honors Carl Peter Thunberg (1743-1828), Dutch physician and botanist who introduced many Japanese plants. The leaflets are elliptic to ovate or lanceolate, 1.5-2.5" long, 0.75-1.5" wide, and pubescent on both surfaces. The calyx tube is shorter than the lobes. The seeds are purple. It is currently reported from Bossier, Washington, and Webster parishes.

DOGHOBBLE - *Leucothoe* - Ericaceae

The genus name, "*Leucothoe*", honors Leucothoe, daughter of Orchamus, king of Babylonia who is said to have been changed into a shrub by her lover, Apollo, after he was buried alive by his rival, Clytia. It includes shrubs with alternate, simple, evergreen to tardily deciduous leaves with pinnate major veins. The white to pink urn shaped flowers are in axillary racemes. The fruit is a capsule with many seeds.

COASTAL DOGHOBBLE (***Leucothoe axillaris***) is also known as Doghobble and Coastal Leucothoe. The specific epithet, "*axillaris*", refers to the axillary inflorescence. It is a shrub up to 6 ft with arching branchlets. The twigs are terete, puberulent when young and glabrous later. The leaves are evergreen and leathery with blades that are oval to elliptic to lanceolate, with entire to serrate margins that are revolute, pointed at the tip, tapering and rounded at the base. The blades are dark green on the upper surface, paler on the lower surface, usually short-hairy, up to 5" long, and 2" wide. It has been in cultivation since 1765. Moist to wet areas along streams (FACW) in extreme southeastern Louisiana with non-cultivated reports from St. Tammany and Washington parishes.

SWAMP DOGHOBBLE (***Leucothoe racemosa***) is also known as Fetter Bush, White Osier, Swamp Sweet Bells, Pepper Bush, and Sweet Bells Leucothoe. The specific epithet, "*racemosa*", refers to the inflorescence. It is a shrub that reaches 12 ft tall. The twigs are erect or divergent, puberulent at first, and glabrous later. The leaves are tardily deciduous and the blades are elliptic to oval, 0.5-2" long, and 0.25-1" wide. Both surfaces are somewhat pubescent and the margins are obscurely crenate to serrate. It has been in cultivation since 1736. The leaves are reported as being poisonous to livestock and a powder of the leaves and bark is considered to be a powerful errhine. Moist to wet areas along streams (FACW) throughout the pine regions of the state. The current reports include nine parishes. (syn. *Leucothoe elongata*)

Lespedeza thunbergii (Nelson 1996)

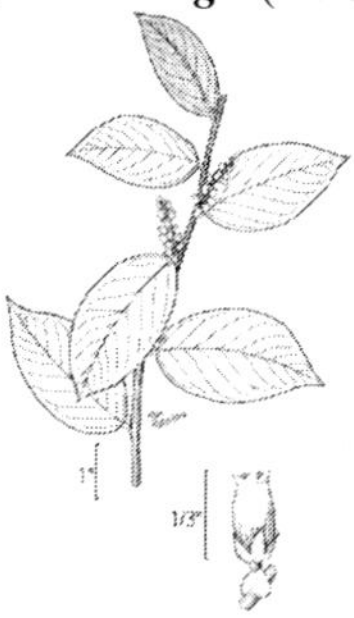

Leucothoe axillaris (USDA-NRCS)

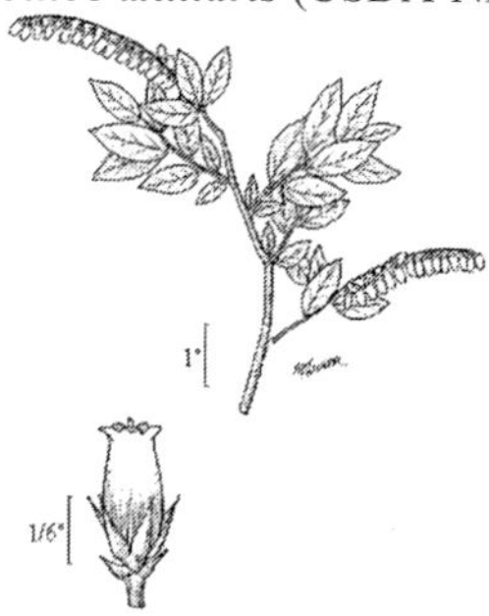

Leucothoe racemosa (USDA-NRCS)

GOPHER APPLE - ***Licania michauxii*** - Chrysobalanaceae

Gopher Apple is also known as Ground Oak and Ground Plum. The genus *Licania* is assumed to be an anagram of "Caligni", the local name in French Guyana for *Licania incana*, the first species described in this genus. The specific epithet, "*michauxii*", honors Andre Michaux (1746-1803), French traveller and plant collector. It is a very low evergreen shrub up to 4 ft that spreads by underground stems. The leaves are alternate, simple, elliptic to almost spatulate, oak-like in appearance, and with pinnate major veins. The leaves of the midstem are larger than the leaves at the top and bottom. The margins are entire to slightly wavy. The white flowers are borne in terminal clusters that commonly extend beyond the uppermost leaves and appear in the spring; sepals pubescent. The fruit is an elliptical red drupe. Very common east of Louisiana but only one old report from Washington Parish. (syn. *Chrysobalanus incanus, Chrysobalanus oblongifolius, Geobalanus oblongifolius)*

PRIVET - ***Ligustrum*** - Oleaceae

The genus name, "*Ligustrum*", is the ancient name credited to Linnaeus. The genus includes shrubs to small trees with opposite, simple, entire leaves with pinnate major veins. The flowers are white and the fruit is an oblong dark blue to black drupe.

JAPANESE PRIVET ***(Ligustrum japonicum)*** is also known as Privet Berry and Prim. The specific epithet, "*japonicum*", refers to Japan, its native province. It reaches a height of 20 ft and has glabrous branches. The leaves are evergreen and the blades are leathery, ovate, 1-2.5" long, 1-1.5" wide, apex and base acute to rounded, and lower surface yellowish. Widely cultivated throughout the state and persisting with a few escaping populations. It is currently reported from 35 parishes.

GLOSSY PRIVET (***Ligustrum lucidum***) is also known as Privet. The specific epithet, "*lucidum*", refers to the lucid, or shining, leaf. It reaches a height of 20 ft and has glabrous twigs. The leaves are evergreen and the blades are leathery, ovate, 2-6" long, 1-2.5" wide, base acute to acuminate, apex acuminate, and lower surface light green. It has been cultivated since 1794. Widely cultivated throughout the state and persisting with a few escaping populations. It is currently reported from 25 parishes.

CHINESE PRIVET (***Ligustrum sinense***) is also known as Hedge Privet. The specific epithet, "*sinense*", refers to its Chinese origin. It is a shrub to 10 ft tall with pubescent twigs. The leaves are deciduous to tardily deciduous and the blades oval to elliptic, 0.5-2" long, 0.3-0.75" wide, and light green on lower surface. Widely cultivated and escaping and forming dense thickets; this species is a serious pest in many parts of the state (FAC). (syn. *Ligustrum villosum)*

SPICE BUSH - ***Lindera*** - Lauraceae

The genus name, "*Lindera*", honors John Linder, a Swedish physician (1676-1723). It includes shrubs to small trees. The leaves are alternate, simple, deciduous, and with pinnate major veins. The crushed leaves and twigs have an odor. The flowers are yellow and produced before the leaves appear. The fruit is an oblong red (rarely yellow) drupe.

Licania michauxii (Vines 1960 as ***Geobalanus oblongifolius***)

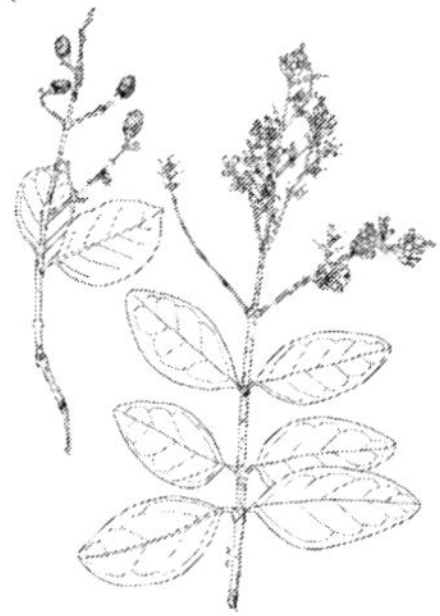

Ligustrum japonicum (USDA-FS)

Ligustrum lucidum (Vines 1960)

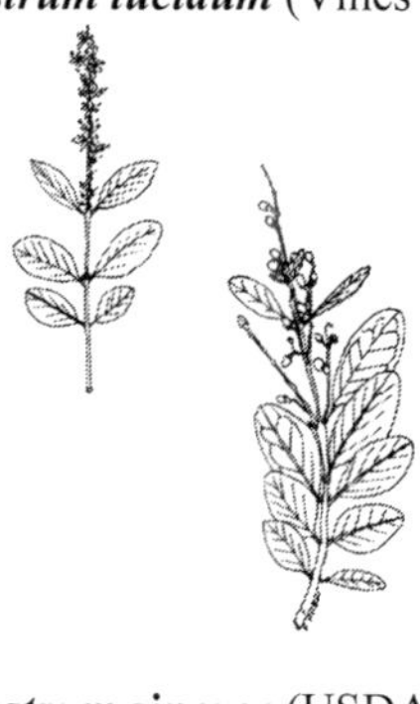

Ligustrum sinense (USDA-FS)

COMMON SPICE BUSH (***Lindera benzoin***) is also known as Benjamin Bush, Spice Wood, Fever Bush, Snap Bush, and Wild Allspice. The specific epithet, "*benzoin*", denotes its similarity, in odor, to the true balsamic resin of ***Styrax benzoin***, an Asiatic tree. It reaches 20 ft tall with light grayish brown bark and gray glabrous twigs. The blades are oval to obovate, thin, 2.5-5" long, 1-2.5" wide, apex acuminate, margins ciliate, bright green and glaucous above, and glabrous and more rarely pubescent underneath. The flowers are small, yellowish, and highly fragrant. They are produced in the spring before the leaves appear. The fruits are shiny, red or occasionally yellow drupes with one large seed and are fragrant when crushed. The leaves, twigs, bark and fruit contain an aromatic oil which was made into a fragrant tea by the pioneers. The bark is aromatic, tonic, astringent, stimulant, and pleasant to chew. A substitute for allspice was once made from the dry, powdered drupes. Twenty-four species of birds are known to feed upon the fruit, also rabbit and white-tailed deer nibble the leaves. It has been in cultivation since 1683. Moist to mesic habitats (FACW) throughout the state with current records from 34 parishes. (syn. *Benzoin aestivale*, *Laurus benzoin*)

POND BERRY (***Lindera melissaefolia***) is also known as Southern Spice Bush, Jove's Fruit, Spice Wood, Benjamin Bush, and Hairy Spice Bush. The specific epithet, "*melissaefolia*", means melissa-like leaves. It reaches a height of 10 ft with hairy green twigs. The firm blades are broadly lanceolate to narrowly ovate, with rounded bases, minutely pilose to densely pubescent on both surfaces, and when crushed yields a sassafras-like odor. The twigs are sometimes brewed into tea and used as a spring tonic. The plant is sometimes nibbled by marsh rabbit, and the fruit eaten by a number of species of birds. No current records from Louisiana but one very old one from the Ouachita River at the Arkansas border. In Mississippi, it is found around the edges of small ponds in the Mississippi floodplain (OBL). This species is federally listed as a rare species.

BOG SPICE BUSH (***Lindera subcoriacea***) is a relatively new species that was first described from collections in Mississippi in 1983. The specific epithet "*subcoriacea*" refers to the somewhat leathery texture of the leaves. It reaches a height of 7 ft with light gray and pubescent young twigs that turn darker gray and glabrous with age. The blades are subcoriaceous, pubescent, elliptic to oblanceolate, 1.5-3.5" long, 1-1.5" wide, apex obtuse to rounded, base cuneate, upper surface dark green, and lower surface pale green. The crushed leaves and twigs have a lemon odor. Rare; wet edge of streams (OBL) in southeast Louisiana with the only report from Washington Parish. (Not Illustrated)

SWEET GUM - *Liquidambar styraciflua* - Hamamelidaceae

Sweetgum is also known as Red Gum, Gum, American Sweet Gum, Sap Gum, Star Leaved Gum, Gum Tree, Liquidamber, Copalm, Liquid Storax, Styrax, Alligator Wood, Alligator Tree, White Gum, Opossum Tree, Bilsted, Satin Walnut, Gum Wood, and California Red Gum. The genus, "*Liquidambar*", is from the Latin "liquidus" (liquid) and "ambar" (amber), referring to the resin obtained from the bark. The specific epithet, "*styraciflua*", means flowing with styrax (storax, an aromatic basalm used in medicine and perfumery). It is a rapidly growing large tree that reaches up to 150 ft and 3 ft diameter. The bark is brown to gray, rough and is deeply furrowed creating round-topped ridges. The young twigs are green but turning brown or gray with age. Many twigs are winged. A cross section of the twigs reveals a star-shaped darker pith. The leaves are alternate, simple, deciduous, long petioled, and with palmate major veins. The blades are palmately 5-lobed and serrate, almost star-shaped, 3-7" long and wide, shiny dark green above, and light green with tufts of rusty hairs below. The crushed leaves have a pleasant odor. In fall, the leaf coloring is strikingly beautiful, the colors ranging from pale yellow

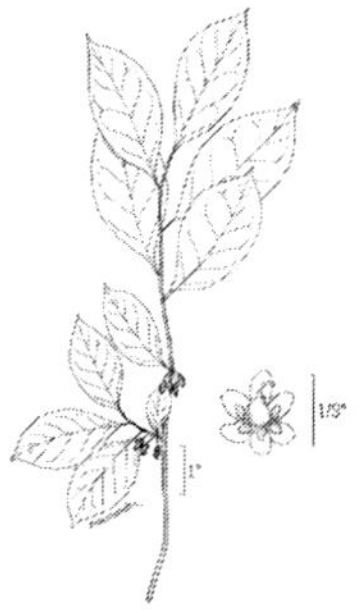

Lindera benzoin (USDA-NRCS)

Lindera melissaefolia (USDA-NRCS)

Liquidambar styraciflua (USDA-FS)

through orange and all shades of red to bronze and purple. The flowers are numerous in green ball-like clusters. The natural interlocking grain of this wood. It is usually made into veneer, often being sold for Circassian Walnut, Mahogany, or Cherry. The wood is difficult to season and should be carefully kiln-dried. The extract is called American storax or styrax (a basalm) and is used as a stimulant, expectorant, antiseptic, applied externally for scabies, fumigating powders and pastilles, as an adhesive, for healing wounds, to treat dysentery, desharpener for tobacco and treating diarrhea. Oil of styrax is used for scenting soap and perfumes. When damaged, Sweetgum exudes an amber-colored sap (hence the name Liquidambar, referring to the amber-colored liquid sap) that was chewed by youngsters before the advent of "store-bought" chewing gum. The twigs are used to clean teeth. The gumballs (fruit) persist in the winter providing food for 25 species of birds, gray squirrels, and eastern chipmunks. Also the fruits are used for decorations. It has been in cultivation since 1681. Widely distributed, perhaps our most common and widespread species, throughout the state in wet to dry habitats (FAC+). It probably occurs in all with current reports from 62 parishes.

YELLOW POPLAR - *Liriodendron tulipifera* - Magnoliaceae

Yellow Poplar is also known as Tulip Poplar, Tulip Tree, White Poplar, Blue Poplar, Hickory Poplar, Popple, White Wood, Yellow Wood, Canoe Wood, Cucumber Tree, Lynn Tree, Saddle Tree, and Saddle Leaf. The genus, "*Liriodendron*, is from the Greek "leiron" (a lily) and "dendron" (a tree). The specific epithet "*tulipifera*" means tulip-bearing, referring to the tulip-like flowers or the silhouette of the leaves. The use of the name "Poplar" is misleading, because the tree is not related to the genus, "***Populus***", which contains the cottonwood trees. It is the state tree of Indiana, Kentucky, and Tennessee. It is large, fast growing to a height of 200 ft and reported to be the tallest and fastest growing hardwood tree in North America. On young trees, the bark is thin, mottled and smooth, and becoming brown and furrowed on old trees. The pith is chambered and the twigs have stipular ring scars. The leaves are simple, alternate, deciduous, and with palmate major veins. The blade is broad, 4-6" long and wide, truncate base, apex truncate or notched, and with two large lateral lobes below the apex. The upper surface is dark green and the lower surface is paler. The blades are aromatic when crushed. The fall color ranges from yellow to orange. The yellow-green and orange flowers are 3-4" across, tulip-shaped, and with six petals. The fruits are samaras that are produced in an upright cone-like structure, which persist after the leaves turn yellow in the fall. Squirrels and ten species of birds including quail eat the seeds. The flowers are a good source of pollen for bees. The leaves and twigs are eaten by deer and cotton-tail rabbits. Yellow Poplar is one of the most valuable timber producers of eastern North America. The wood is light, soft, even-grained, light yellow or brown with thin, creamy white sapwood, fairly stiff, moderately brittle, easily worked, and density = 28 lb per cu ft. The wood has been used for veneer, boxes, crates, plywood, pulp, hat blocks, furniture, cabinets, millwork, musical instruments, toys, novelties, barrel bungs, shingles, interior finish, boats, pulp, and fuel. Indians and pioneers made dugout canoes from the trunks. A tonic from the bark, mostly from the roots, has been reported to be used as a heart stimulant, hydrochlorate of tulipiferine, and to treat chronic rheumatism and dyspepsia. It was once believed to have antimalarial virtues. In well drained to mesic sites (FAC) mostly in the pine regions of the state. The current reports include 31 parishes.

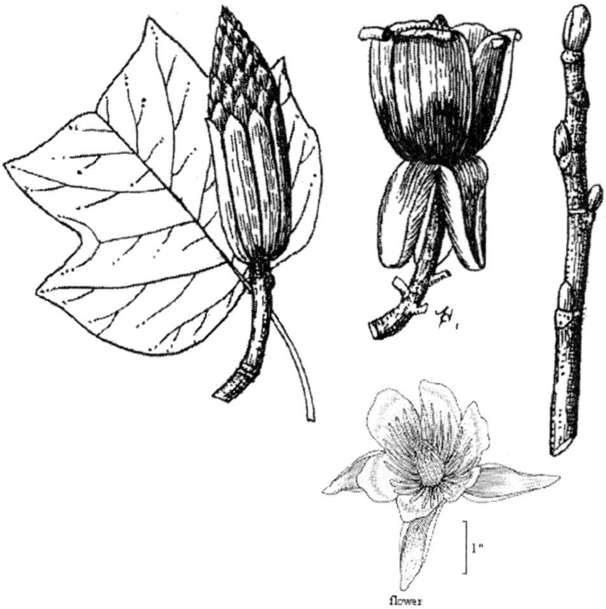

Liriodendron tulipifera (USDA-FS and USDA-NRCS)

HONEYSUCKLE - *Lonicera* - Caprifoliaceae

The genus name, "*Lonicera*", honors Adam Lonitzer (1528-1586), a German botanist. The genus includes vines or shrubs. The leaves are opposite, simple, and with pinnate major veins. The fruits are berries.

WINTER HONEYSUCKLE (***Lonicera fragrantissima)*** is also known as Standing Honeysuckle. It is a shrub to 15 ft with recurving branches. The twigs are slender, glabrous, and lustrous brown. The leaves are deciduous or partly evergreen and the blades are 1-3" long, 0.6-2" wide, elliptic to broadly ovate, apex acute, base broadly cuneate, upper surface dull blue green to dark green, lower surface yellow green midrib covered with bristles, and margins ciliate. The flowers are creamy white and very strongly lemon scented. The berry is red and about 0.25" diameter. It was first cultivated in 1845. Introduced from China and persisting at old home sites. It is more commonly escaped in the northern U.S. The current reports include 11 parishes. (Not Illustrated)

JAPANESE HONEYSUCKLE ***(Lonicera japonica)*** is also known as Chinese Honeysuckle, Southern Honeysuckle, and White Honeysuckle. The specific epithet, "*japonica*", denotes the plant's Asiatic origin. It is a twining or trailing vine with reddish brown stems that are pubescent when young but becoming glabrous with age. The bark on the stems also tends to become shreddy with age. The leaves are evergreen and the blades are ovate to oblong, 1.2-3.0" long, 0.6-1.2" wide, apex acute to obtuse, base cuneate, rounded or truncate, upper surface medium to dark dull green, lower surface pale green, and margins entire. The flowers are in pairs in leafy bracts, fragrant, and white but later turning yellow. The berries are subglobose black, 0.2-0.3" long. It provides excellent cover and food for several wildlife species, especially, birds, deer and rabbits. The flowers are adapted for moth pollination and are visited by hummingbirds. It is a very popular vine in the Southern states, where it is cultivated for its penetrating fragrance and showy flowers. It reproduces itself so abundantly, however, that it is now becoming a troublesome weed. Very common in disturbed areas but also spreading into natural forests (FAC). It is currently reported from all 64 parishes. (syn. *Nintooa japonica*)

CORAL HONEYSUCKLE (***Lonicera sempervirens***) is also known as Woodbine, Scarlet Trumpet, Trumpet Honeysuckle, Honeysuckle, Red Honeysuckle, and Red Woodbine. The specific epithet, "*sempervirens*", means "ever-living", because of its evergreen habit. It is a twining or trailing vine with glabrous stems that become covered with reddish-gray shredding bark. The leaves are evergreen and the blades are ovate to elliptic, 1.2-2.5" long, 0.8-1.6" wide, apex acute to acuminate, base attenuate to broadly cuneate, upper surface dark green, lower surface whitish, and margins entire. The upper one to several pairs of leaves on the tip of a stem are often perfoliate. The red, rarely yellow, flowers are in a pair of 3-flowered terminal sessile cymules. The berries are reddish-orange, translucent, obovoid, and 0.2-0.3" long. The fruit is eaten by a number of species of birds and hummingbirds like the flowers. In well drained habitats (FAC) but also widely cultivated with current reports, some possibly cultivated, from 43 parishes.

CAROLINA WOLF BERRY - *Lycium carolinianum* - Solanaceae

The genus name, "*Lycium*", is for the country of Lycia. The specific epithet, "*carolinianum*", is for the states of North or South Carolina. Other common names include Matrimony Vine, Box Thorn, and Christmas Berry. It is a weakly spiny shrub to 2 ft tall. The leaves are simple, alternate, deciduous, and with pinnate major veins. The blades are succulent, ligulate to clavate, about 1" long, and often have short, leafy branchlets in the axils. The flowers are usually

Lonicera japonica (USDA-NRCS)

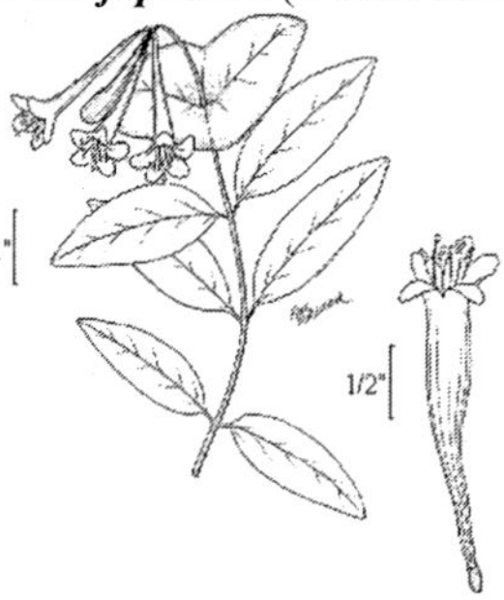

Lonicera sempervirens (USDA-NRCS)

Lycium carolinianum (USDA-NRCS)

solitary, corolla blue or lilac, or white. The red berry, which resembles a tomato, was formerly eaten raw or dried by the Indians. It also is consumed by wild fowl. In sandy soils (FACW) along the Gulf of Mexico. The current reports include eight parishes all adjacent to the Gulf.

FETTER BUSH, STAGGER BUSH - *Lyonia* - Ericaceae

The genus name, "*Lyonia*", honors John Lyon, American botanical explorer, who died in Asheville, North Carolina in 1818. The genus includes shrubs to small trees. The leaves are alternate, simple, and with pinnate major veins. The flowers are in clusters or racemes in the axils of the leaves. The fruit is a capsule with many very small seeds.

HE HUCKLE BERRY (***Lyonia ligustrina***) is also known as Male Berry, Pepper Bush, Seedy Buck Berry, White Wood, and White Alder. The specific epithet, "*ligustrina*", denotes its resemblance to Privet, or ***Ligustrum***. It is an erect, branching shrub up to 12 ft tall with terete twigs that are pubescent when young and glabrous with age. The leaves are deciduous and the blades are lanceolate to elliptic or obovate, 1.3-3.5" long, 0.3-2" wide, rounded or pointed at the tip, tapering to the base, and the margins are minutely toothed. The surface is usually black punctate dotted and hairy on the veins of both the upper and lower surface. The globose flowers are white, sometimes tinged with pink. The capsule is globose. It has been in cultivation since 1748. In wet to moist areas (FACW) in acid pH areas of the state with current reports from 14 parishes.

FETTER BUSH (***Lyonia lucida***) is also known as Stagger Bush, Hurrah Bush, Pipe Stem, and Shiny Lyonia. The specific epithet,"*lucida*", refers to its lustrous leaves. It is mostly shorter than 5-6 ft but may reach 12 ft. The twigs and bark are distinctly ridged with the young twigs green abut turning brown with age. The leaves are evergreen and the blades are leathery, dark shiny green, mostly broadly elliptic, and punctate with black dots. The margins are entire but closely paralleled by a distinctive and conspicuous vein and the margins are revolute. The cylindric flowers vary from pale to deep pinkish or almost red. The capsule is rounded to urn-shaped. It has been in cultivation since 1765. In wet to moist areas (FACW) in acid pH areas mostly in southern Louisiana with current reports from 8 parishes.

STAGGER BUSH (***Lyonia mariana***) is also known as Piedmont Stagger Bush, Sorrel Tree, and Wicks. The specific epithet, "*mariana*", means "of Maryland". It is a shrub to 5 ft with angled stems and black dots on the twigs. The leaves are tardily deciduous and the blades are elliptic to oblong, 0.8-3" long, glabrous above, lightly pubescent beneath with occasional black dots; narrowed or obtuse at base, obtuse to apiculate at apex, and entire margins. The cylindric flowers are white to pink. The capsule is ovoid-pyramidal and five angled. The plant was introduced into cultivation about 1736. The foliage is said to be poisonous to lambs and calves. Rare in mesic habitats (FAC) in extreme northwest Louisiana. It is currently reported only from Caddo and Claiborne parishes.

CAT'S CLAW - *Macfadyena unguis-cati* - Bignoniaceae

Cat's Claw is also called Yellow Trumpet Vine. The genus name "*Macfadyena*" honors James MacFadyen, a Scottish Botanist (1798-1850). The specific epithet "*unguis-cati*" means "cat's claw". It is a shrub to vine that climbs by tendrils. The stems can develop aerial roots. The leaves are opposite, evergreen, and one odd pinnate compound. The leaf is made of two leaflets plus three terminal tendrils. Each tendril ends in a stiff clawlike structure. The leaflets are ovate to lanceolate or elliptic, 2-3" long, 1-2" wide, apex acute to obtuse, base cuneate to obtuse and

Lyonia ligustrina (USDA-NRCS)

Lyonia lucida (USDA-NRCS)

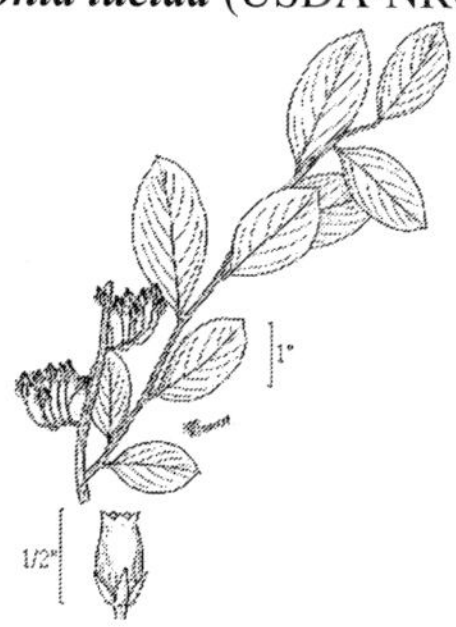

Lyonia mariana (USDA-NRCS)

Macfadyena unguis-cati (Godfrey 1988)

often slightly cordate, upper surface dark green, lower surface paler, and margins entire to toothed. The yellow flowers are large, trumpet shaped, to 3" long, to 4" wide, and in axillary clusters. The fruit is a flattened, linear, capsule to 20" long with oblong winged seeds. A native of Argentina that is cultivated, persists, and spreads mostly in south Louisiana. The current records include six parishes.

OSAGE ORANGE - *Maclura pomifera* - Moraceae

Osage Orange is also known as Hedge Apple, Bois d' Arc, Bodark, Bow Wood, Hedge, Naranjo Chino, Horse Apple, Mock Orange, and Yellow Wood. The genus name, "*Maclura*", honors William McClure, an early American geologist (1763-1840), and the specific epithet, "*pomifera*", means "fruit-bearing". The name Bois d' Arc was given to it by the French, meaning bow-wood with reference to the fact that the Osage Indians made bows from the wood. It is a small armed dioecious tree with milk sap that reaches a height of 60 ft and 20" diameter. The twigs are thick, brown, pubescent when young but becoming glabrous with age and usually with hard sharp spines to 1" long at leaf bases. The bark is gray to yellow-brown and divided into narrow forking ridges. The leaves are alternate, simple, deciduous, and with pinnate major veins. The blades are ovate, 2.5-5" long, 1.5-2.5" wide, usually acuminate, entire, glabrous, shiny dark green above, and pale green below. The flowers are very small and green. The fruit is a large (up to 5" diameter), yellow-green, roughened, spherical (ball) created by a fleshy calyx with the true fruit a nutlet embedded inside. The wood is bright orange or yellow, heavy, hard, durable, strong, and weighs 48 lb per cu ft. The wood is very resistant to decay and is used for fence posts. A yellow dye can be made from the stem or root bark. Also, the bark of the trunk was used for tanning leather. Squirrels and birds feed on the nutlets buried in the pulpy fruit, and deer and cattle browse the leaves and twigs. Before barbed wire came into general use, it was widely used as a spiny living fence to keep cattle out of crops. It has been cultivated since 1818. Widely distributed, mostly on floodplains of large streams (FACU). It is currently reported from 40 parishes.

MAGNOLIA - *Magnolia* - Magnoliaceae

The genus, "*Magnolia*", honors Pierre Magnol (1638-1715), French professor of botany. It includes shrubs to small and large trees. The twigs have a stipular ring scar at the base of each leaf. The leaves are alternate, simple, and with pinnate major veins. The blade margins are entire. The flowers are solitary, large, and with many free parts. The fruit is an aggregate (cone like) of follicles with red seeds.

CUCUMBER TREE (***Magnolia acuminata***) is also known as Cucumber Magnolia, Bitter Tree, Mountain Magnolia, Yellow Linn, and Black Linn. The specific epithet, "*acuminata*", is for the acuminate (long pointed) apex of the leaves. It is a medium-sized tree that reaches up to 80 ft. The bark is dark brown to gray and furrowed with narrow confluent ridges and on older trees, it breaks into thin scales. The twigs are thick, brown or gray, pubescent when young but becoming glabrous with age. The leaves are deciduous and the blades are elliptical to ovate, 5-10" long, 3-6" wide, apex acuminate, base rounded to broadly cuneate, rarely cordate, green and glabrous above, pale green and usually soft-hairy below. The flowers are bell-shaped, 2.4-3.5" wide, and with 6 greenish-yellow or yellow petals. The young fruits bear a striking resemblance to a cucumber and mature to a dark red and are 2.4-3" long. The wood weighs 34 lb per cu ft and is moderately hard, moderately strong, soft, light yellow brown with thin, lighter colored and often white sapwood, close-grained, fairly stiff, and moderately high in shock resistance. It is often marketed with Tulip Tree and mixed hardwoods and used for furniture, cabinets, flooring, interior finish, pump logs and troughs, Venetian blinds, boxes, crates, sidings,

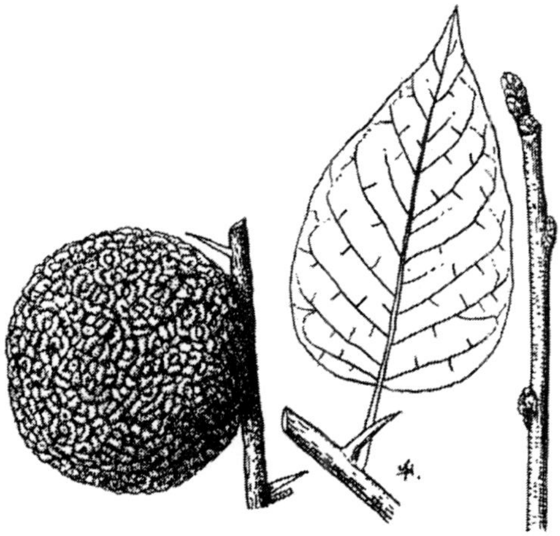

Maclura pomifera (USDA-FS)

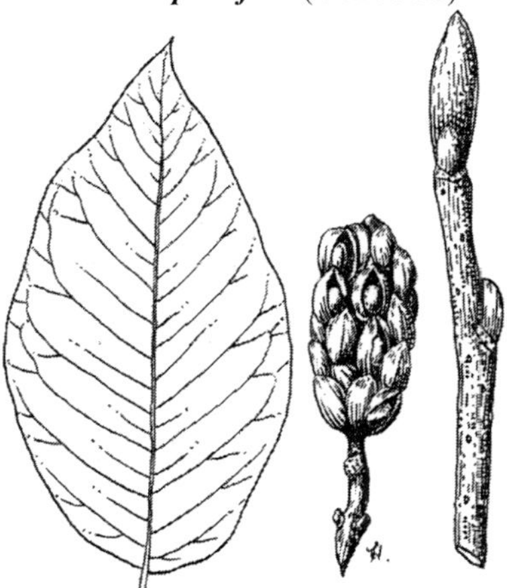

Magnolia acuminata (USDA-FS)

woodenware and general millwork. The fruits are infused in whiskey to give it pleasant bitter taste and to prevent intermittent fever. Herbal users have used the fruit to treat dyspepsia and general debility for many years. Herbalist used the bark and fruit prepared in the required form to give relief from pains of rheumatism. Midwives gave a tonic of Cucumber Tree for treatment in obstinate cases of suppressed menstruation. The twigs are browsed by deer. The seeds are eaten by several wildlife species. It has been in cultivation since 1736. In well drained hardwood slope forests throughout the pine regions of the state and absent from the large floodplains and the marsh. The current reports include 15 parishes. (syn. *Magnolia cordata*)

SOUTHERN MAGNOLIA (***Magnolia grandiflora***) is also called Evergreen Magnolia, Big Laurel, Bull Bay, Great Laurel Magnolia, and Loblolly Magnolia. The specific epithet, "*grandiflora*", means "grand flower" in reference to the large flowers. It is a large tree to 135 ft. The bark is grayish brown, smooth but breaking into small thin scales with age. The twigs are stout, green to olive, hairy when young but becoming glabrous with age. The leaves are evergreen and the blade leathery, elliptic to oval, 5-8" long, 2-3" wide, apex acute or obtuse, base cuneate, and the margins somewhat revolute. The upper surface is dark green, distinctly lustrous and glabrous while the lower surface is reddish brown pubescent. The twigs and leaves are aromatic when crushed. The flowers are large, white, aromatic, and 6-9" wide. The brown fruit is 2-4" long. It is the state flower of Louisiana and the state flower and tree of Mississippi. The creamy white wood is hard, fairly heavy, weak, is not durable, weighs 32 lb per cu ft, and turns brown upon exposure. It is used for fuel, baskets, crates, woodenware, furniture, shades, veneer, cabinets, and decorative trim. The seeds provide food to squirrels and five species of birds. The bark has been used medicinally as a stimulant, tonic, and diaphoretic. The fragrant seeds are sometimes used for flavoring salads. It has been in cultivation since 1734. It is found in moist to well drained soils (FAC+) along streams often associated with American Beech (***Fagus grandifolia***) and is more common in southern Louisiana. The current reports include 54 parishes; some reports could be from cultivated trees.

BIG LEAF MAGNOLIA (***Magnolia macrophylla***) is also known as Large Leaf Magnolia, Large Leaved Umbrella Tree, Big Bloom Cucumber Tree, Silver Leaf, and Elk Bark. The specific epithet, "*macrophylla*", means large leaved. It is a small tree that reaches a height of 50 ft. The bark is smooth, thin, and reddish brown to light gray. The twigs are stout, brittle, yellow green and pubescent when young but becoming glabrous and reddish brown to tan or gray with age. The leaves are deciduous and the blades are obovate to oblong, apex acute, rounded or obtuse, the base cordate with earlike lobes, 20-30" long, 9-10" wide, bright green and smooth above, and whitish pubescent below. The six petaled flowers are up to 1 foot across and creamy white with a rose center. The fruits are bright rose-color and downy. The wood is medium heavy (33 lb per cu ft), light brown, sapwood yellowish, satiny, close grained, weak, hard, and dry. It was first introduced into cultivation about 1800. Infrequent to rare on well drained hardwood forest slopes in southeastern and central Louisiana. The current reports include 13 parishes with some from cultivated trees.

PYRAMID MAGNOLIA (***Magnolia pyramidata***) is also known as Southern Cucumber Tree and Mountain Magnolia. The specific epithet, "*pyramidata*", refers to the pyramid fruit. It is a small tree that grows to a height of 30 ft. The bark is smooth and gray to reddish brown. The twigs are slender, green to reddish brown when young and turning gray with age. The leaves are deciduous and the blades are oblong to spatulate, apex obtuse or acute, base distinctly auriculate with large ear like lobes, 5.5-8.5" long, 3.5-4.5" wide, upper surface bright green, and lower surface pale, glabrous, and glaucous. The flowers are creamy white and 3-4" wide. The

Magnolia grandiflora (USDA-FS)

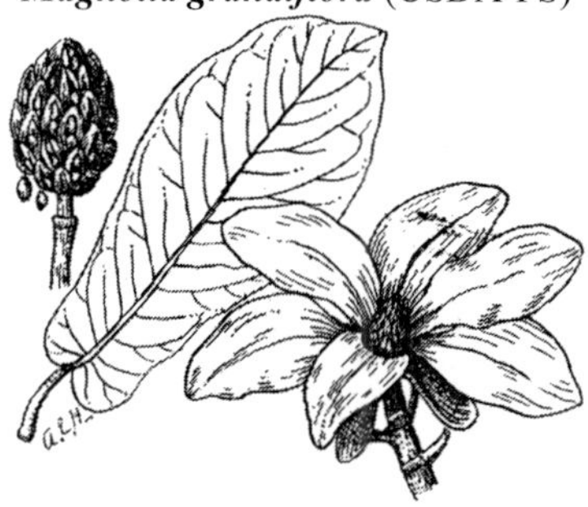

Magnolia macrophylla (USDA-FS)

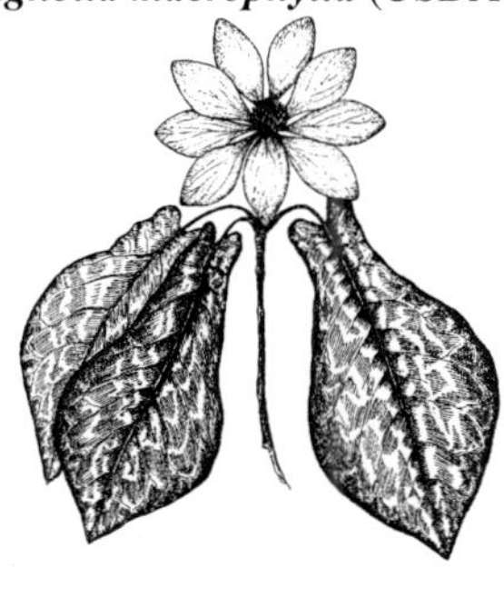

Magnolia pyramidata (Vines 1960)

2-2.5" long fruits are rose colored. It was first cultivated in 1825. In well drained to sandy soils sporadically across the state with non cultivated reports from Sabine, Washington, and West Feliciana parishes.

WHITE BAY (***Magnolia virginiana***) is also known as Sweet Bay, Swamp Bay, Southern Sweet Bay, Laurel Magnolia, Swamp Magnolia, Beaver Tree, Swamp Sassafras, Indian Bark, Sweet Magnolia, White Laurel, and Swamp Laurel. The specific epithet "*virginiana*" is for the state of Virginia. It varies in form from a shrub to a large tree to 30 ft. The bark is pale gray to brown and smooth. The twigs are slender, bright green, and glabrous to hairy. In Louisiana and other southern states it is tardily deciduous to evergreen but is deciduous in northern states. The blades are oblong to elliptic or narrowly oval, 3-6" long, 1-2.5" wide, apex obtuse, base broad cuneate, upper surface green, and lower surface distinctly white and glaucous. The creamy white flowers are 2-4" wide and very fragrant; it is reported that the flowers have been used in perfume manufacture. The red fruits are fleshy and 1-2" long. The wood is pale brown, sapwood white, soft, weighs 31 lb per cu ft and used for making light woodenware articles. The leaves are browsed by deer and the seeds are eaten by several songbirds. The dried leaves are used as a cooking spice. In bogs, baygalls, edges of streams (FACW+) in areas with acid pH throughout the state. The current reports include 32 parishes.

CRAB APPLE - *Malus* - Rosaceae

The genus name, "*Malus*", is the Latin name for apple. It includes shrubs to small trees. The leaves are alternate, simple, deciduous, and with pinnate major veins. The flowers are in simple corymbs and each has 5 petals, 5 sepals, and 5 styles. The fruit is a globose pome.

SOUTHERN CRAB APPLE (***Malus angustifolia***) is also called Narrow Leaf Crab Apple and Wild Crab Apple. The specific epithet, "*angustifolia*", refers to the narrow leaves. It is an armed, small tree to 35 ft. The bark is dark reddish brown to gray with deep fissures and narrow ridges and separating into platelike scales. The twigs are stout, light brown to reddish brown, pubescent when young and later glabrous. The twigs often end in a thorn. The blades are elliptic, oval, oblong, or obovate, 1-3" long, 0.5-2" wide, margins crenate to slightly lobed, apex acute to obtuse or rounded and often apiculate, base cuneate, dull green, pubescent when young but turning glabrous with age. The midrib usually retains the pubescence. The leaves often have the orange spots of cedar-apple rust. The flowers are at first pink and later fade to white. The wood is close grained, hard, heavy, weighs about 43 lb per cu ft, with a reddish brown heartwood and yellow sapwood. The wood is used for levers, tools, and small woodenware objects. The yellowish green pomes are 0.75-1" diameter and hang down. The pomes are made into jelly, preserves and cider. They are eaten by many species of birds and animals, including the bobwhite quail, blue jay, cardinal, ruffed grouse, prairie chicken, skunk, opossum, raccoon, cottontail, and red and gray fox. Throughout the state in many habitats and also planted in some locations. The current reports include 24 parishes. (syn. *Pyrus angustifolia*)

PRAIRIE CRAB APPLE (***Malus ioensis***) is also called Iowa Crab, Prairie Crab, and Western Crab Apple. The specific epithet, "*ioensis*", refers to the state of Iowa where it was first described. It is a small tree to 30 ft. The bark is red-brown to dark gray with small narrow scales. The twigs are red-brown to gray, densely tomentose when young but becoming less so with age. The twigs often end in a thorn. The blades are elliptic, oblong, or oblong-ovate, 1.5-5" long, 0.8-4" wide, dark green and lustrous above, glabrous to densely white tomentose beneath, apex acute, obtuse or rounded, base rounded or cuneate, and margins singly or doubly crenate-serrate and sometimes lobed. The yellowish green pomes are 0.8-1.5" diameter and hang down. The

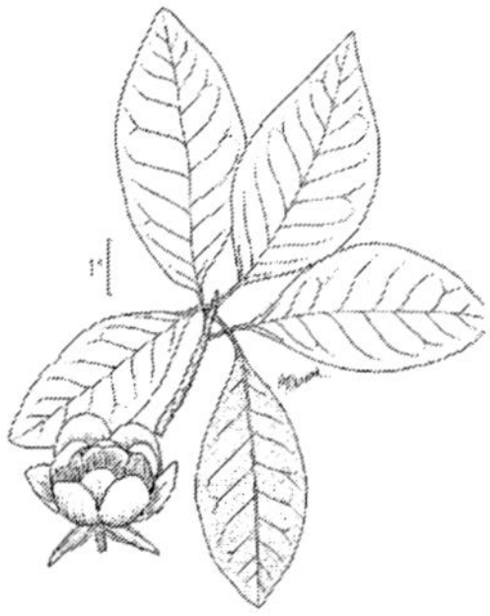

Magnolia virginiana (USDA-NRCS)

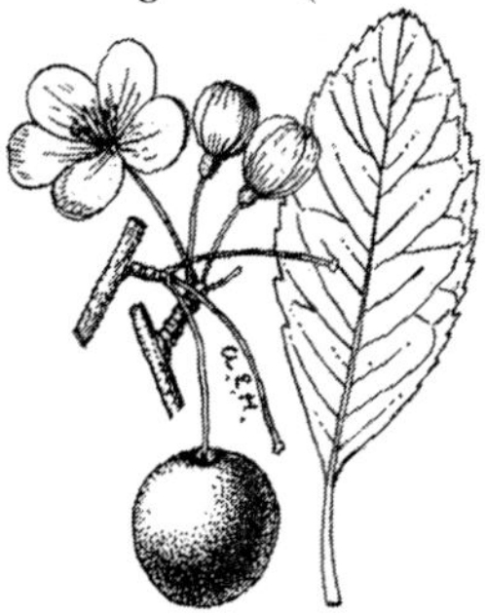

Malus angustifolia (USDA-FS)

Malus ioensis (Britton and Brown 1943)

pomes are made into jelly, preserves and cider. It has been cultivated since 1885. It is of considerable value as food for wildlife, the fruit known to be eaten by at least 20 species of birds and mammals, including bobwhite quail, ruffed grouse, ring-necked pheasant, gray and red fox, skunk, cottontail, woodchuck, red squirrel, and fox squirrel. Throughout the state in many habitats and also planted in some locations. The current reports include 14 parishes. (syn. *Pyrus ioensis)*

EUROPEAN CRAB APPLE (***Malus pumila***) is also called European Apple and Apple. The specific epithet "pumila", means "dwarf or small". It is a small, often armed tree to 40 ft. The twigs are stout, younger pubescent and older glabrous and lustrous reddish brown. The blades are oval, ovate, oblong-ovate to broad-ovate, 2-4" long, 0.75-3" wide, thick and veiny, mostly abruptly short-pointed, broad at base, unequally mostly rather bluntly serrate, becoming glabrous and more or less glossy above, pubescent beneath. The flowers are white, light pink, or rose. The pomes are variable but can be up to 3" diameter. Planted throughout the state and persisting or perhaps escaping. The current records include 21 parishes, mostly in north Louisiana. (syn. *Pyrus pumila, Malus sylvestris, Pyrus sylvestris)*

WAX MALLOW - *Malvaviscus arboreus* var. *drummondii* - Malvaceae

Wax Mallow is also known as Texas Mallow, Mexican Apple, Red Mallow, May Apple, Manzanilla, and Wild Turk's Cap. The genus name, "*Malvaviscus*", is from a Greek word meaning "sticky-mallow." The specific epithet, "*arboreus*", means woody and "*drummondii*" honors Thomas Drummond, who collected in Texas 1833-1834. It is a shrub to 10 ft that is herbaceous above and woody below. The twigs are green to brown and hairy. The leaves are alternate, simple, deciduous, and with palmate major veins. The blade is oval to ovate, 2-4" long and wide, apex acute or obtuse, base cordate, margins coarsely toothed or sometimes 3-lobed. The upper surface is dull green and glabrous or pubescent while the lower surface is velvety pubescent and with conspicuous veins. The red flowers are solitary, with five petals, and have the pistillate column long exerted beyond the petals. The fruit is a red fruit separating gradually into five one-seeded carpels. The leaves are used as an emollient, and the flowers are used in Mexico in a decoction for inflammation of the digestive tract and as an emmenagogue. The fruit is edible, either raw or cooked, and has a mealy taste. It is also eaten by a number of species of birds and mammals. Planted throughout the state and persisting or escaping. The current reports include 29 parishes.

CASSAVA - *Manihot grahamii* - Euphorbiaceae

Cassava is also called Cassada, Tapioca, and Manihot. The genus *Manihot* is the native Brazilian name. The specific epithet,"*grahamii*", honors Dr. Robert Graham (1786-1845), a Scottish Physician and Botanist. It is a shrub to small tree to 25 ft. The bark is smooth, light to dark brown, and marked with numerous horizontal lenticels. The young twigs are herbaceous, slender, smooth, glabrous, pale green but becoming woody and brown with age. The leaves are alternate, deciduous, with palmate major veins, and simple but deeply palmately divided into 6 to 9 segments so as to appear compound. The overall shape of the leaf is broadly ovate with each segment 3-6" long, oblanceolate, lanceolate, or narrowly elliptic, upper surface dark green, glabrous, semi lustrous, lower surface much paler with yellow veins, the margins entire or sinuate wavy, the apex acute or acuminate, and the base long attenuate. The flowers are axillary, solitary, and green. The fruit is a three-lobed, pale green, globose capsule. The plant is poisonous raw but can be made edible by processing. The seeds have emetic and purgative properties. The fleshy rhizomes are sometimes used as cassava, a starch derived after the

Malus pumila (Britton and Brown 1943)

Malvaviscus arboreus var. ***drummondii*** (Vines 1960)

Manihot grahamii (Nelson 1996)

poisonous acid is removed. Cultivated and persisting mostly in southern Louisiana with reports from 15 parishes.

CHINA BERRY - *Melia azedarach* - Meliaceae

China Berry is also called China Ball, Azedarach, Africa Lilac, Bead Tree, Pride of China, China Tree, Hag Bush, Hop Tree, Pride of India, Indian Lilac and Pride Tree. The genus, *Melia*, is a Greek name for ash, from the similar leaves. The specific epithet, "*azedarach*", is from a Persian word meaning "noble-tree". It is a moderate size tree to 40 ft. The bark is smooth, golden brown on young branches and darker brown and more or less scaly on older parts. The leaves are alternate, deciduous, and double even pinnately compound. The leaf is 8-20" long and wide, with thin green petiolules. The leaflets are numerous, paired except at end, 1-2" long, 0.4-0.8" wide, acuminate, usually serrate, dark green, with strong odor when crushed. The flowers are numerous, small, fragrant, pale purple, in panicles 4-10" long. The fruits are yellow drupes, 0.5-0.8" diameter, remaining attached in winter. All parts of the plant but especially the fruit is reported to be poisonous. It has been used medicinally as an anthelmintic, astringent, bitter tonic, emetic, emmenagogue, and purgative. The fruit is eaten by birds and swine, but if fermented it sometimes has a toxic effect. The fruit pulp is also used as an insect repellent and vermifuge. The wood is variegated, durable, somewhat brittle and has been used in cabinet work. The fruits are used as rosary beads in south Louisiana and as the ammo for the pop gun throughout the state. It is native to east Asia. Once a very popular ornamental throughout the state and now escaped and spreading in mesic to dry habitats. It is probably in all with current reports from 59 parishes.

WAX MYRTLE, BAY BERRY - *Morella* - Myricaceae

The genus includes dioecious shrubs to small trees. The leaves are alternate, simple, evergreen, and with pinnate major veins. The staminate catkins are oblong, cylindric, and reddish yellow while the pistillate catkins are short and ovoid. The fruit is a drupe.

BAYGALL WAX MYRTLE (***Morella caroliniensis***) is also known as Swamp Candle Berry and Evergreen Bay Berry. The specific epithet "caroliniensis" is for the state of North or South Carolina. It is a shrub to 9 ft. The smooth bark is grayish brown. The twigs are stout, gray, glabrous and somewhat glandular scaly. The leaves are tardily deciduous and the blades are oblong, oblanceolate, elliptic, or lanceolate, 2-4" long, 0.5-1.5" wide, apex acute to rounded, base attenuate, margins entire, slightly revolute and often with a few coarse teeth near the apex. The upper surface is dark green and semi lustrous while the lower surface pale dull green, with small brown peltate scales and small white hairs along the veins. The crushed blade has a strong odor. The drupe is white and about 0.25" diameter. In wet areas especially acid bogs and baygalls (FACW) throughout the state. The current records include 27 parishes. (syn. *Myrica heterophylla*)

SOUTHERN WAX MYRTLE (***Morella cerifera***) is also known as Wax Berry, Spice Bush, Candle Berry, Bay Berry, Sweet Oak, and Tallow Shrub. The specific epithet, "*cerifera*", refers to the waxy fruit. This species reaches a height of 40 ft. The bark is gray or grayish green and smooth. The twigs are reddish brown to gray with orange glands on the young twigs that disappear with age. The leaves are tardily deciduous and the blades 1.5-5" long, 0.25-0.75" wide, oblanceolate to elliptic, apex acute or rounded, cuneate or narrowed at base and decurrent on the short stout petiole, margin entire or coarsely serrate above the middle, shining above, and resinous with orange-colored glands beneath. The crushed blade has a strong aromatic odor. The drupe is about 1/8" diameter, light green to blue, covered with granules of bluish white wax. The

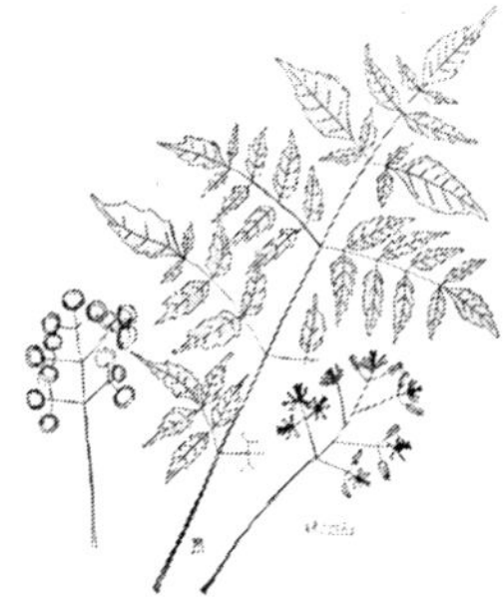

Melia azedarach (Moore 1960)

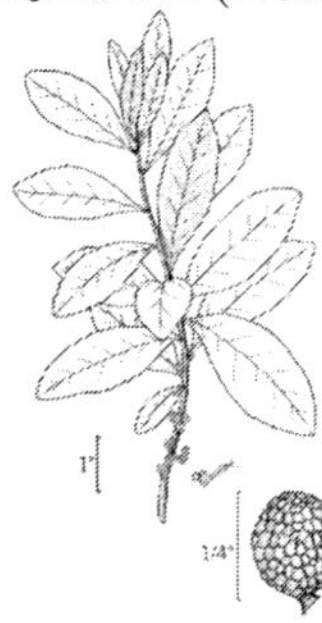

Morella caroliniensis (USDA-NRCS)

Morella cerifera (USDA-NRCS)

fruit is eaten by about 40 species of birds, especially bobwhite quail and turkey. It was first cultivated in 1699. Candles were formerly made by boiling the waxy drupes. The bark and leaves are reputed to have medicinal properties. The dark brown wood is light, brittle, soft, finegrained, with lighter sapwood, weighing 35 lbs per cu ft. One of the most common and widely distributed shrubs in the state with a wide range of habitats from the marsh to dry sandy soil (FAC+). It probably occurs in all with current reports from 55 parishes. (syn. *Myrica cerifera, Myrica pusilla)*

ODORLESS WAX MYRTLE (***Morella inodora)*** is also known as Wax Berry, Candle Berry, and Wax Tree. The specific epithet, "*inodora",* refers to the odorless leaves. The plant was named by William Bartram, who found it near Taensa, Alabama, in 1778. This species reaches a height of 20 ft. The bark is grayish brown to gray. The young twigs are slender, reddish brown, pubescent while the older twigs are gray and glabrous or pubescent. The leaves are evergreen and the blades 2-4" long, 0.75-1.5" wide, oblong-obovate to ovate, apex obtuse to rounded, often apicuate, base cuneate and decurrent on the petiole, margin entire or rarely with a few teeth near the apex and slightly revolute, thick and coriaceous, veins rather obscure, upper surface lustrous dark green, and lower surface paler green. The crushed blade has no odor. The drupe is about 3/16" diameter, black, and papillose. Rare along streams (OBL) in southeastern Louisiana with reports only from St. Tammany and Washington parishes. (syn. *Myrica inodora)*

MULBERRY - *Morus* - Moraceae

The genus name, "*Morus*", is the Latin name of the mulberry. It includes small dioecious trees with milky sap. The leaves are alternate, simple, deciduous, and with three palmate major veins. The flowers are green and in drooping catkins.The fruit is a multiple of achenes emersed in a fleshy calyx with the entire structure usually called a berry.

WHITE MULBERRY (***Morus alba***) is also known as Silkworm Mulberry, Russian Mulberry, Morera, and Morea. The specific epithet, "*alba",* refers to the white fruit. It is a small tree that reaches a height of 60 ft. The bark is dark brown to gray tinged with red and divided into irregular elongated plates. The young twigs are thin, reddish brown, glabrous to slightly pubescent but becoming glabrous and gray with age. The blades are ovate to oval or heart-shaped, often 3 or 5 lobed, coarsely serrate, 2.5-6" long, 2-4" wide, shiny green and glabrous above, pale green and more or less hairy below, with large teeth, glossy green above, with pointed tips, and nearly hairless. The fruit is white or occasionally reddish black. The fruit is not as juicy as the native Red Mulberry, and is somewhat smaller. The fruit has some wildlife value, being eaten readily by a number of species of birds, opossum, and raccoon, as well as by poultry and hogs. The wood is hard, durable, and is used for furniture, utensils, and boat building. It has been said that White Mulberry is cultivated in more countries than any other tree, and that more has been written about it than any other tree. The reason is that it is the favorite food of the silkworm caterpillar, and silkmaking was very important. It is native to Europe or China. Escaping and spreading across many habitats in the state especially in disturbed areas (FACU). The current records include 44 parishes.

BLACK MULBERRY (***Morus nigra***) is a native of western Asia. The specific epithet, "*nigra*", refers to the black color of the fruit. It reaches a height of 30 ft. The twigs of young ones are green to brown and pubescent and the older ones are darker brown to gray or glabrous. The blades are ovate to oval, blades 1.5-6" long, apex acute or short-acuminate, margin coarsely toothed or sometimes with one or more lobes, base rounded, cordate or semitruncate, upper surface dull dark green, usually rough and becoming glabrous, lower surface paler and sparingly

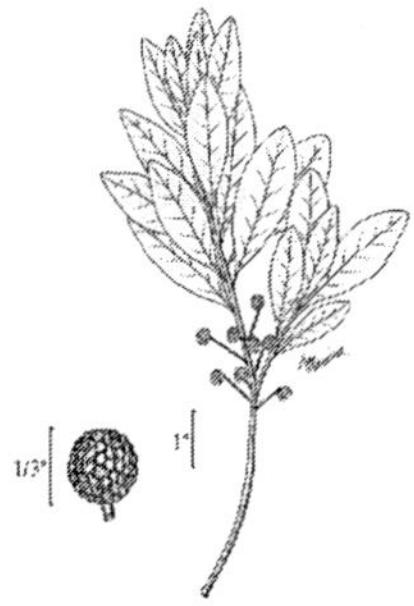

Morella inodora (USDA-NRCS)

Morus alba (Britton and Brown 1943)

Morus nigra (USDA-FS)

pubescent on the veins to glabrous. The fruits are dark red or black. It is difficult to separate from White Mulberry even in fruit as there is a black fruited form of White Mulberry. The current records include 10 parishes mostly in south Louisiana.

RED MULBERRY (***Morus rubra***) is also known as Moral and Common Mulberry. The specific epithet, "*rubra*", a Latin word for "red", has reference to the red, immature fruit. It is a medium-sized tree that reaches a height of 70 ft. The bark is dark brown to gray with irregular elongate plates separated into flat, compressed scales. The twigs are thin, light brown, and usually glabrous. The blades are ovate, variable in degree of lobing, sometimes unlobed but often 3-lobed, coarsely serrate, 4-7" long, 2.5-5" wide, oval in outline, dark green and rough above, pale green and usually hairy below. The twigs contain a milky juice. The berry is red to purplish black. The yellow wood is durable, soft, close-grained, light, weak, and weighs about 45 lb per cu ft. It has been used for boats, fencing, cooperage, cross-ties and fence posts. Indians used the fibrous bark for clothing and the milky sap from the twigs to cure ringworms. The bark of the Red Mulberry root is a traditional European remedy for tapeworms. Mulberry bark has also long been used as a laxative preparation for home use. The bark was used by Desoto's men to make ropes for the boats to float down the Mississippi River. The fruits are sweet and can be eaten raw or used for pie or jelly. Also, birds and squirrels eat the fruits. The young shoots and leaves of the Red Mulberry can be cooked like greens. Deer browse the leaves and beaver eats the bark. It has been in cultivation since 1629. Very common in many habitats (FAC) throughout the state. It probably occurs in all with current reports from 59 parishes.

NANDINA - *Nandina domestica* - Berberidaceae

Nandina is also known as Heavenly Bamboo. The genus, "*Nandina*" is from the Japanese name nanten. The prefix "nan" means "small or dwarf bush". The specific epithet, "*domestica*" means "cultivated". It is a sparsely branched, upright rhizomatous shrub to 6 ft. The leaves are usually borne near the ends of branches and stems. The large leaves are evergreen, to 20" long and 30" wide, alternate, bi- to tri-pinnately compound, and with a purplish rachis. The numerous leaflets are ovate to lanceolate, entire, 1-2" long, 0.5-1" wide, dark shiny green above when mature, much paler and soft when young. The lateral veins of the leaflets are depressed on the upper surfaces, at least near the central vein. The white flowers are small, borne in terminal panicles. The fruit is a bright red, globose berry. The berry ir reported to be poisonous. It is native to China and Japan. A very popular cultivated plant with most yards in the state containing one or more of the plants. Persisting and recently beginning to spread with reports from 24 parishes.

OLEANDER - *Nerium oleander* - Apocynaceae

Oleander is a native of Asia. The genus name, "*Nerium*", is from the Greek "*neros*" ("moist"), referring to the places the wild plants grow. The specific epithet, "*oleander*", is from the Latin, meaning "olive-like", referring to the leaves. It is a shrub that reaches 20 ft. The twigs are erect or arching, young ones green, older ones light brown to gray; lenticels numerous, oval. The leaves are evergreen, simple, whorled or opposite, and with pinnate major veins. The blades are narrowly oblong-lanceolate, 4-12" long, 0.5-1.5" wide, apex acuminate, tapering at the base into a short petiole, dark green above, and paler beneath with a prominent yellow midrib. The blades are glabrous and the lateral veins are distinctly parallel to each other. The flowers are white to red or purple, 1-3" across, odorless, in compound, terminal cymes, and often double. The fruit is 2 ovaries forming follicles, erect or nearly so, and 4-8" long. The seeds are twisted. The plant, including the flowers, are poisonous; some reports of people getting sick from using the twigs to roast hog dogs. The leaves have been known to kill cattle and are used as a rat poison in Europe.

Morus rubra (Britton and Brown 1943)

Nandina domestica (Nelson 1996)

Nerium oleander (USDA-FS)

It persists at old house sites or from roadside plantings. It is currently reported from 17 parishes.

BLACK GUM, TUPELO GUM - ***Nyssa*** - Nyssaceae

The genus name, "*Nyssa*", is from Nysa, a water nymph. It includes dioecious trees that are mostly aquatic. The twigs have chambered pith. The leaves are deciduous, alternate, simple, and with pinnate major veins. The flowers are green with the male in spherical clusters, the female solitary on long stalks. The fruit is a drupe and the seeds are flattened and longitudinally ridged.

TUPELO GUM (***Nyssa aquatica***) is also known as Water Tupelo, Cotton Gum, Bay Poplar, Swamp Tupelo, Black Gum, Sour Gum, and Hornbeam. The specific epithet, "*aquatica*", refers to the tree's aquatic habitat. It reaches up to 100 ft with a much swollen fluted base. The bark is light or dark gray, breaking up into thin scales. The blades are ovate to oblong, pointed at the tip, rounded or slightly heart-shaped at the base, with an occasional coarse tooth or toothless. It is smooth or somewhat hairy on the upper surface, paler and soft hairy on the lower surface, up to 18" long and up to 4" wide, on hairy stalks up to 3" long. The drupes are oblong, dark purple with pale speckles, about 1" long. The wood is light brown to white, soft, light, tough, close grained, weighes 32 lb per cu ft. It is used for rough lumber, veneer core stock, boxes, fruit crates, woodenware, and paper pulp. The wood from the swollen base is light weight and used for duck decoys, corks, and net floats. The fruit is eaten by at least 10 species of birds, and the wood is made into boxes, woodenware, and fruit crates. The wood of the light spongy roots is made into net floats and corks. The flowers have some value as a bee food. In swamps, edges of lakes, ponds and other water bodies (OBL) with current reports from 46 parishes.

SWAMP BLACK GUM (***Nyssa biflora***) is very similar to Black Gum and is often considered a variety of Black Gum. The specific epithet "*biflora*" refers to the paired drupes which is one of the characters used to separate the two species. The leaves of Swamp Black Gum tend to be a little narrower and the seeds have more prominent ridges. Another good character is the habitat; Swamp Black Gum is always in wetter sites (OBL) than Black Gum. The trunks of Swamp Black Gum in the wetter habitat tend to be swollen and fluted at base. The current records include 22 parishes. (syn. *Nyssa sylvatica* var. *biflora*)

BLACK GUM (***Nyssa sylvatica***) is also known as Black Tupelo, Sour Gum, Tupelo, Pepperidge, Gum Tree, Swamp Hornbeam, Yellow Gum, Snag Tree, Beetlebung, Hornbeam, Hornpipe, Hornpine, Hornbine, and Bee Gum. The specific epithet, "*sylvatica*", refers to the wooded habitat. It is a large tree reaching 100 ft with distinct horizontal branches. The bark is gray to brown or black, deeply fissured, and broken into small irregular shaped blocks giving it an "alligator-skin" appearance. The twigs are thin, red-brown, glabrous to slightly hairy. The blades are very variable, elliptic to oblong, 2-5" long, 1-3" wide, entire or rarely with remote teeth, shiny glabrous green above, paler and usually sparsely hairy below, turning bright red or sometimes purple in late Summer or early Fall. The apex is acute to acuminate and the base cuneate or rounded. The ovoid drupes are about 0.5" long and bluish-black. Robins, in particular, gather to feast on these berries. The wood is tough, heavy (39.5 lb per cu ft), hard, not durable, warps easily, light brown, and closed grained. The wood is used for veneer, plywood, cross-ties, boxes, cooperage, pulp, woodenware, hubs, wharf piles, handles, and planing mill products, rollers in glass factories, ox-yokes, wooden pipes, fruit boxes, soles of shoes and gun stocks. The fruits are used by 32 species of birds, foxes, and squirrels and the foliage is used by beaver, black bear and white-tailed deer. Blackgum is an excellent "den tree" because it usually has numerous hollows for squirrels, rabbits, otters, opossum, raccoons, and owls. The flowers are a good nectar source. Take a twig, fuzz it by twisting, and you have

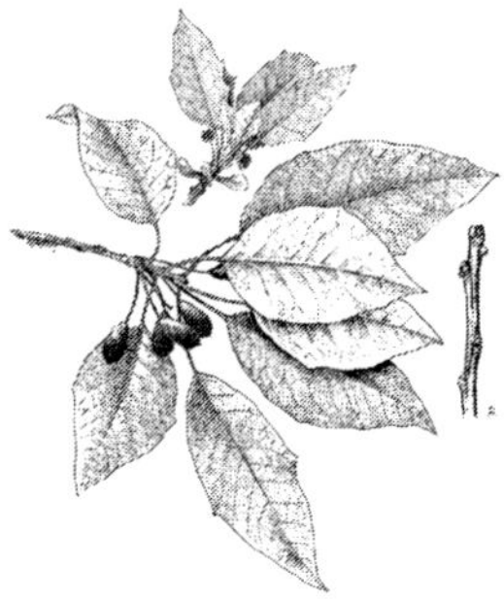

Nyssa aquatica (USDA-FS)

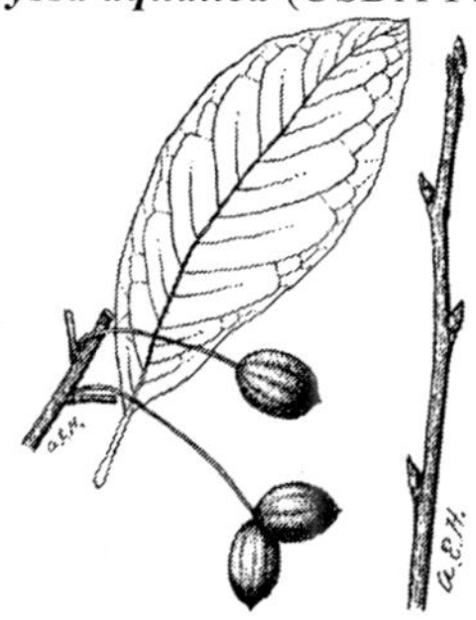

Nyssa biflora (USDA-FS)

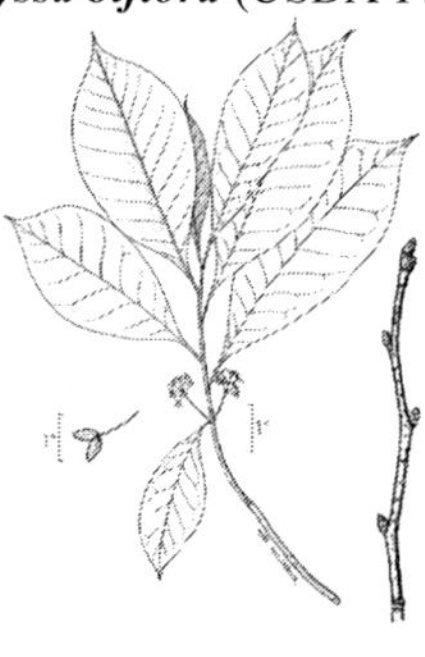

Nyssa sylvatica (USDA-FS and USDA-NRCS)

Nature's toothbrush. Some early settlers would dip this "toothbrush" in a jar of snuff and place in their jaw. This was their way of dipping snuff. A very common tree with a wide range of habitats in the state from somewhat wet to dry sandy soils (FAC). The current records include 46 parishes.

PRICKLY PEAR CACTUS - *Opuntia* - Cactaceae

The genus *Opuntia* is a Greek name given by Theophrastus to a plant that grew near Opus, a town in ancient Greece. The genus includes succulents with armed green stems. The leaves are quickly deciduous leaving the paired stipular spines on either side of the leaf scar. Most species are low growing, spreading plants that can produce large clumps. Most are shorter than 3 ft. There are several species reported for the state with some probably native to the sandy soils of extreme northwest and southwest Louisiana. There are other records scattered across the state for plants that are persisting or spreading from old home and/or dump sites.

DEVIL WOOD - *Osmanthus americana* - Oleaceae

Devil Wood is also known as American Wild Olive. The genus name, *"Osmanthus"*, is from the Greek word *"osme"* (odor) and *"anthus"* (flowers), referring to the fragrant flowers. The specific epithet, *"americana"*, refers to its North American habitat. It is a shrub or small tree to 50 ft. The bark is gray to reddish brown and breaking into small scales with age. The twigs are slender, reddish to brown to gray, pubescent to glabrous, and slightly angled or terete. The pith is white and homogenous. The leaves are opposite, simple, evergreen, and with pinnate major veins. The blades are leathery, thick, elliptic, oblong-elliptic or lanceolate, 2-6" long, 1-2.5" wide, upper surface bright green, glabrous and shiny, and paler below. The flowers are small, greenish to creamy white in axillary panicles on twigs of the previous season. The fruits are dark blue drupes 2.5-4" long, 0.25-1" diameter, with a solitary seed. The wood is dark brown to reddish brown, sapwood lighter, close-grained, hard, strong, specific gravity 0.81, and difficult to work. It has been cultivated since 1758. In southeastern Louisiana along the edges of streams (FAC) with current non-cultivated reports from five parishes.

HOP HORNBEAM - *Ostrya virginiana* - Betulaceae

Hop Hornbeam is also known as Iron Wood, Eastern Iron Wood, Wooly Hop Hornbeam, Rough Barked Iron Wood, Lever Wood, American Hop Hornbeam, Deer Wood, Hardhock, Ostryer de Virginie, Bois de Fer, and Indian Cedar. The genus name, "*Ostrya*", is from the ancient Greek word "Ostrys" = shell for the shell like bracts. The specific epithet, "*virginiana*", refers to the state of Virginia. It is a small tree that reaches up to 60 ft. The light reddish brown bark is broken into small, narrow, shreddy scales. The leaves are simple, alternate, deciduous, and with pinnate major veins. The blades are ovate or oblong to lanceolate, 2-5" long, 1.2-2" wide, base rounded, heart-shaped, asymmetrical, and yellowish-green. The blade apex is acuminate, the base is rounded, hear shaped, or cuneate and often asymmetrical. The upper surface is yellowish green, the lower surface is pubescent, and the margins are doubly serrate. The parallel lateral veins are forked near the end to form a "y" which helps to distinguish it from Iron Wood. The flowers are catkins and produced from March to June before the leaves. The male flowers are in drooping catkins that form the previous summer and the female flowers are in erect catkins on newly formed spring twigs. The fruits are nutlets in drooping cone-like paper sac structures that look like hops, an ingredient found in beer. The hard, tough, strong, close-grained, light brown to white in color with pale sapwood, durable wood weighs about 51 lb per cu ft. The wood is used for posts, golf clubs, tool handles, mallets and woodenware. An infusion was made from the chips of the heart of the tree. It was used on intermittent fevers, neuralgic affections, dyspepsia, and scrofula with a dose reported to be one teaspoon three times a day before meals.

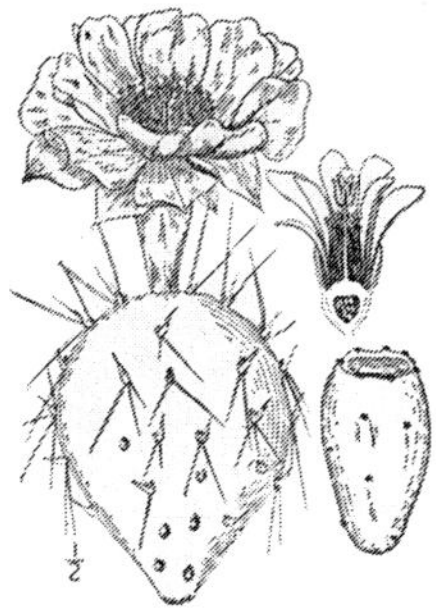

Opuntia humifusa (Britton and Brown 1943)

Osmanthus americanus (Main and Mills 1981)

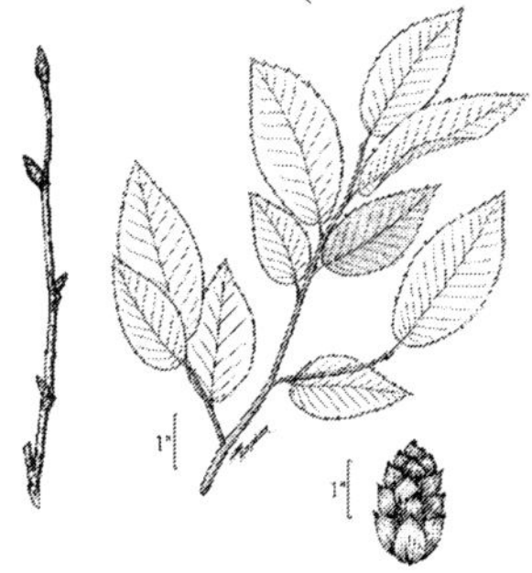

Ostyra virginiana (USDA-FS and USDA-NRCS)

Native Americans are also reported to have used Hop Hornbeam for flux, consumption, cancer of the rectum, female weakness, kidney trouble, coughs, and toothaches. Five species of birds and some mammals use the fruit. White-tail deer browse on the twigs and foliage. Squirrels eat the seeds and buds. It has been in cultivation since 1690. In Louisiana, Hop Hornbeam has a very similar distribution pattern as Iron Wood; it is absent from the marsh and also from the Mississippi and Red River floodplains but is usually at a slightly higher and drier site (FACU-) than Iron Wood. It is currently reported from 39 parishes.

SOUR WOOD - *Oxydendrum arboreum* - Ericaceae

Sour Wood is also known as Sorrel Tree, Sour Gum, Arrow Wood, Lily of the Valley Tree, Elk Tree, and Titi. The genus, "*Oxydendrum*", is from the Greek oxys (acid) and dendron (a tree) referring to the acid tasting leaves. The specific epithet, "*arboreum*", means tree-like. It is a small to medium-sized tree that reaches heights of 70 ft. The bark is gray to reddish brown and smooth with short deep fissures with broad ridges as it ages. The twigs are slender, smooth, glabrous, terete, green to brown or reddish brown when young and turning gray with age. The leaves are simple, alternate, deciduous, and with pinnate major veins. The blades are elliptic to oblong, 4-7" long, 1-3" wide, apex acute to acuminate, and base cuneate. The margins are serrate, the upper surface is dark shiny green, and the lower surface is paler with distinct white hairs along the midrib. The leaves turn a bright red (steak red) in the fall and are among the first to do so. The white, fragrant flowers are in drooping racemes. The fruits are a persistent capsule containing numerous small seeds. The wood is light, red brown, sapwood yellowish brown or pinkish brown, hard, moderately strong in shock resistance and endwise compression, shrinks considerably, takes a fine polish, hard, very close-grained, strong in bending, strong in shock resistance, and weighs 46 lb per cu ft. It is used for fuel, tool handles, machinery bearings, and sled runners. The acid leaves are used to relieve thirst, as a tonic, refrigerant, diuretic and to treat cardiac ailments. Some eclectic physicians say it can be used to treat dropsy and fever and is administered as a semisolid or fluid extract. The leaves are browsed by white-tailed deer. Honeybees seek out the flowers and Sour Wood honey can be found in some areas. In well drained soils in southeastern and central Louisiana. The current records include 15 parishes.

ALLGEHENY SPURGE - *Pachysandra procumbens* - Buxaceae

The genus *Pachysandra* is from the Greek words pachys meaning "thick" and "andros" meaning male in reference to the thick stamens. The specific epithet "*procumbens*" is from the word "procumbo" which means "to fall forwards", "to become prostrate" in reference to the stems. It is a shrub/vine with stolons that reaches a height of 2 ft but mostly shorter. The lower stems have scales with regular leaves concentrated near the tips. The leaves are alternate, simple, evergreen, and with palmate major veins. The blades are ovate to widely elliptic, 2-5" long, 1-2" wide, apex acute, base rounded to widely cuneate, margins coarsely serrate, and dark green mottled with lighter green. The blades often take on purple coloration, especially in the winter and early spring. The white flowers are in dense spikes. The fruit is a 3-locular capsule with two seeds per locule. In well drained hardwood slope forests in West Feliciana Parish only.

Oxydendrum arboreum (USDA-FS)

Pachysandra procumbens (Britton and Brown 1943)

JERUSALEM THORN - *Parkinsonia aculeata* - Fabaceae

Jerusalem thorn is also known as Horse Bean, Paloverde, Cloth of Gold, Crown of Thorns, Barbados Fence Flowers, and Mexican Paloverde. The genus name, "*Parkinsonia*", honors John Parkinson, an English botanical author. The specific epithet, "*aculeata*", refers to the spines. It is an armed shrub or small tree that reaches a height of 40 ft. The bark is thin, smooth, green, and later brown to reddish with small scales on old trunks. The twigs are green to yellowish when young and gray to brown or orange with age. The spines are green brown or black and up to 1" long. The leaves are alternate, deciduous and bipinnate or rarely pinnate compound and with a spinescent green rachis. The leaflets are numerous, small, mostly shorter than 0.5", linear to oblanceolate, inequilateral, and early deciduous leaving the rachis on the tree. The flowers are fragrant, yellow, and in distinct racemes. The fruit is a linear, brown to orange or reddish legume. The wood is light brown to yellow, hard, heavy, close-grained, and with a specific gravity of 0.61. The wood is occasionally used as fuel and was formerly used for papermaking. An infusion of leaves is used in tropical America as a febrifuge, for diabetes, for epilepsy, and as a sudorific and abortifacient. The leaves and pods are eaten by horses, cattle, and deer, particularly in times of stress. The Indians formerly pounded the seeds into flour to make bread. The flowers are sometimes important as bee food. Planted throughout the state and persisting or perhaps spreading (FAC). The current records include 17 parishes.

VIRGINIA CREEPER - *Parthenocissus quinquefolia* - Vitaceae

Virginia creeper is also known as Woodbine, False Grapes, Five Leaves, Wild Woodbine, Wild Wood Vine, Woody Climber, Five Finger Ivy, Five Finger Creeper, Five Leaf Ivy, American Woodbine, and American Ivy. The genus name, "*Parthenocissus*", is from the Greek "parthenos" (a virgin) and "kissos" (ivy) referring to the common name Virginia Creeper. The specific epithet, "*quinquefolia*", refers to the five leaflets. It is a high climbing or trailing woody vine that climbs by branched tendrils that are often tipped with adhesive disks. The stems are glabrous, flattened, somewhat grooved, green to red or brown when young and brown with age. The leaves are alternate, deciduous, and palmately compound with 5 leaflets (rarely 7). The leaflets are elliptic to oblanceolate or lanceolate with serrated margins, 2-6" long , 1-3" wide, elliptic to obovate, apex acuminate, base cuneate, margins coarsely serrate above the middle of the blade, glabrous and dull green above, paler and sometimes pubescent below. The small green flowers are borne in branched racemes. The fruits are globose black or dark blue drupes that are 0.2-0.4" diameter with 1-3 seeds. The Indians used the fruits to dye skin and feathers pink. The bark was used as a tonic, alterative, astringent, expectorant, and used for dropsy. The roots are used for diarrhea and the bark and twigs are made into a syrup for use in coughs and colds. Birds greedily eat the fruits. In a wide range of habitats from dry to mesic and slightly moist (FAC) throughout the state. It is probably in all with current records from 56 parishes. (syn. *Ampelopsis quinquefolia)*

PRINCESS TREE - *Paulownia tomentosa* - Vitaceae

Princess Tree is also known as Empress Tree and Royal Paulownia. The genus name, *Paulownia*, honors Anna Paulownia (1795- 1865), Princess of the Netherlands. The specific epithet, "*tomentosa*", refers to the densely hairy underside of the leaves. It is an introduced species from China that is a fast growing tree reaching a height of 60 ft. The bark is scaly, dark brown to gray. The twigs are smooth, brown, with prominent white lenticels. The leaves are opposite, deciduous, simple, and with palmate major veins. The blades are broad ovate to ovate or suborbicular, 5-12" long and wide, apex acuminate, base cordate, margins entire or 3-lobed, dark green above, and densely tomentose below. The vanilla scented flowers are about 2" long, violet or pinkish, and in panicles. The fruit is an oval, 2-part capsule with very small, winged

Parkinsonia aculeata (USDA-FS)

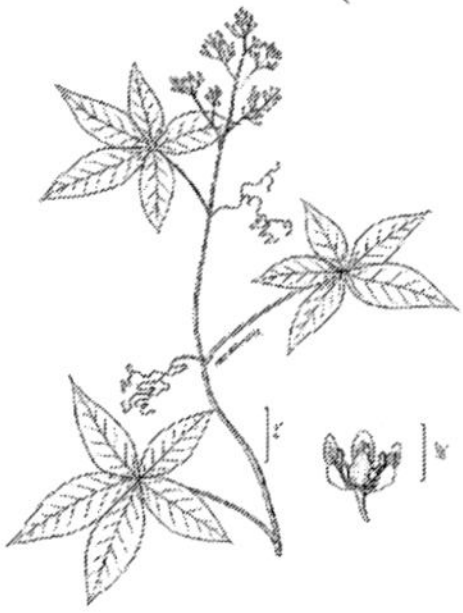

Parthenocissus quinquefolia (USDA-NRCS)

Paulownia tomentosa (USDA-FS)

seeds. Planted throughout the state (FACU) as an ornamental with some recent interest in its wood. The current records include 14 parishes.

RED BAY - *Persea* - Lauraceae

The genus, "*Persea*", is the ancient Greek name of an unidentified Egyptian tree. It includes shrubs to small trees to 70 ft but usually much shorter. The bark is reddish brown to grayish brown with deep fissures and broad and scaly ridges. The twigs are slender, angled, dark green when young but turning light brown with age, and with gray or brown hairs. The leaves are alternate, simple, evergreen, and with pinnate major veins. The blades are elliptic-oblong, apex acute or acuminate, base cuneate, 3-4" long, 1-2" wide, margins entire, upper surface bright lustrous green, and lower surface glaucous and hairy. The crushed leaves and stems have an aromatic odor. Many plants have leaves with colorful galls along the margins of the blade. The flowers are pale yellow to green, small, and borne in axillary panicles. The fruits are obovoid to globose, lustrous dark blue to black drupes. The wood is reddish brown, sapwood lighter, close-grained, fairly hard and strong, and weighs about 40 lb per cu ft. The wood is used for furniture, boat building, interior finish, and cabinetmaking. The fruit is eaten by at least two species of birds. Soups and meats are often flavored with the dried leaves. Two species are usually listed for Louisiana with both receiving various combinations of the following common names: Red Bay, Swamp Bay, Swamp Sweet Bay, Sweet Bay, Florida Mahogany, Tiss Wood, Isabella Wood, Silk Bay, Red Bay Persea, and Laurel Tree.

RED BAY (***Persea borbonia***) is the less common species in Louisiana with reports only from coastal areas in Calcasieu and LaFourche parishes. The specific epithet, "*borbonia*", is the old generic name of Persea. It is different from the more common ***Persea palustris*** by having the underside of the blade appearing glabrous with the naked eye except for a few hairs along the midrib. If the underside of the blade is examined with magnification, it has a sparse amount of short appressed pubescence. In sandy soils near the coast (FACW).

SWAMP RED BAY (***Persea palustris***) is the more common and widespread species with reports from 23 parishes. The specific epithet,"*palustris*", is the Latin word for marsh or swamp, which refers to the common habitat for these plants. The underside of the blade is densely covered with brown shaggy hairs. In baygalls and other moist areas in the pine region and also in wet areas in the marsh regions (OBL) of the state. (syn. *Persea pubescens)*

MOCK ORANGE - *Philadelphus* - Hydrangeaceae

The genus name, "*Philadelphus*", honors Ptolemy Philadelphus (283-247 B.C.), King of Egypt or from Greek name meaning brotherly love from "Phileo" = love and "adelphus" = brother. The genus includes shrubs to 12 ft with arching branches. The leaves are opposite, simple, deciduous, and with pinnate major veins. The white flowers have four petals and resemble the flowering dogwood. The fruit is a capsule.

SCENTLESS MOCK ORANGE (***Philadelphus inodorus***) is also called Summer Dogwood, English Dogwood, and Syringa. The specific epithet, "*inodorus*", refers to the plant being unscented. The bark peels off in papery strips on last year's growth and is gray to brown. The blades are ovate to elliptic, entire, 1.5-3" long, 1-1.5" wide, apex acuminate, base truncate to rounded, and usually glabrous on both surfaces. The flowers are in terminal cymules of 3 or axillary. Planted throughout the state and persisting with current records from 17 parishes.

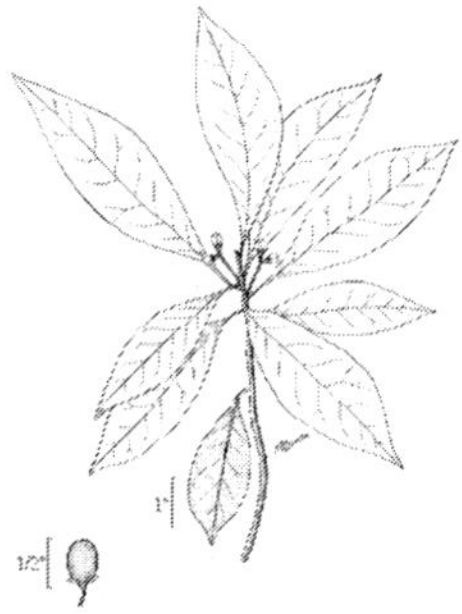

Persea borbonia (USDA-NRCS)

Persea palustris (USDA-FS)

Philadelphis inodorus (Britton and Brown 1943)

HOARY MOCK ORANGE (***Philadelphus pubescens***) is also called Summer Dogwood, English Dogwood, and Syringa. The specific epithet, "*pubescens*", is for the short hairs on the foliage. The twigs are glabrous and gray to brown. The bark peels off in papery strips on older stems. The blades are ovate to ovate-elliptic, 2-3" long, 1-2" wide, apex acuminate, base truncate, margins usually dentate, dark green glabrous above, and gray-pubescent below. The flowers are in a raceme with 5-9 flowers. It was introduced into cultivation in 1800. Cultivated and persisting with records from only Jackson, Morehouse, Natchitoches, Ouachita, and Union parishes.

MISTLETOE - *Phoradendron leucarpum* - Viscaceae

Mistletoe is also known as Christmas Mistletoe and American Mistletoe. The genus name, "*Phoradendron*", is Greek for a tree thief and aptly describes its semi-parasitic nature on hardwood trees. The species epithet, "*leucarpum*", refers to the pale white fruits. It is a dioecious semi-parasite on mostly deciduous trees. The plants are attached to and penetrate the branches of host trees. The stems and leaves form bushy growths up to 3 ft in diameter. The stems are brittle, much-branched, and usually stellate-pubescent. The leaves are opposite, entire, succulent, yellow-green, usually obovate, 0.8-2.4" long, 0.6-0.8" wide. Staminate inflorescence 0.8-1.4" long, divided into 3-6 segments, with numerous very small yellowish flowers. Pistillate inflorescence 0.6-1.4" long, divided into 3-5 segments each with 4-11 very small yellowish flowers. The fruits are globose, white, 0.15-0.25" diameter drupes with sticky pulp. Birds and mammals eat the fruits. Mistletoe is the state flower of Oklahoma. It is used for Christmas decorations. All parts of Mistletoe are poisonous. It has been used medicinally; it is oxytoxic and used to arrest post-partum hemorrhages, to treat people with fits (epilepsy), and dogs with fits. During the Middle Ages, branches of this plant were hung from the ceilings of homes to keep away any evil spirits. In southeastern Louisiana, the preferred host seems to be Sweet Gum while most of the rest of the state, the preferred host seems to be Water Oak. Throughout the state and parasitic only on hardwood woody species. It probably occurs in all with current records from 48 parishes. (syn. *Phoradendron flavescens*) Note: there could be two species in Louisiana since ***Phoradendron tomentosum*** is often listed for the state.

PHOTINIA, RED CHOKE BERRY - *Photinia*- Rosaceae

The genus name, *"Photinia",* is from Greek from Photos for light referring to the shiny leaves of some species. The genus includes shrubs to small trees. The leaves are alternate, simple, and with pinnate major veins. The flowers are white and the fruit is a small tough pome.

JAPANESE PHOTINIA (***Photinia glabra***) is also known as Chinese Photinia, Red Tipped Photinia, and Red Tip. The specific epithet, "*glabra*", means "hairless, smooth, or bald". It is mostly a shrub to 12 ft. The leaves are evergreen; the blades are narrowly oval to slightly obovate, apex acuminate, base cuneate to attenuate, shallowly serrate, bronzy red when emerging, finally glossy dark green, and petiole always glabrous. The flowers are in clusters 2-4" wide and petals with tufts of pubescence. The pomes are red when young but turning black at maturity. It is a native of Japan that is widely cultivated, especially as a hedge, and persists in a few locations. The current records, mostly from cultivated plants, include Lafayette, Lincoln, Ouachita, Tangipahoa, and Winn parishes. (Not Illustrated)

Philadelphis pubescens (Vines 1960)

Phoradendron leucarpum (Vines 1960 as ***Phoradendron flavescens***)

RED CHOKE BERRY (***Photinia pyrifolia***) is also known as Red Choke Cherry, Choke Pear and Dog Berry. The specific epithet, *"pyrifolia"*, refers to the *Arbutus*-like leaves. This is a shrub to 12 ft tall. The twigs are brown to gray and pubescent. The leaves are deciduous; the blades are elliptic to oblong or slightly obovate, 1-3" long, 0.5-1.5" wide, apex acute to short acuminate and apiculate, base cuneate to attenuate, upper surface dark green and glabrous except for glands along the midrib, lower surface densely gray tomentose, and margins crenate with the teeth incurved and glandular. The petioles are glabrous to pubescent. The leaves turn red in the autumn. The flowers are in clusters 0.5-2.5" wide and the petals are glabrous. The pome is bright red at maturity. The fruit is a valuable wildlife food in fall and winter, being eaten by at least 13 species of birds, including bobwhite quail, ruffed grouse, ring-necked pheasant, and cedar waxwing. It has been cultivated since 1700. This species is native and restricted to the acid pH bogs, baygalls, and pine flatwoods (FACW) of the pine regions of the state. The current records include 27 parishes. (syn. *Pyrus arbutifolia, Sorbus arbutifolia, Aronia arbutifolia)*

TAIWANESE PHOTINIA (***Photinia serratifolia***) is also called Chinese Photinia, Red Tipped Photinia, and Red Tip. The specific epithet, "*serratifolia*", refers to the serrate margins of the leaves. It is a shrub to small tree to 40 ft. The leaves are evergreen; the blades are oblong, 4-8" long, 2-3.5" wide, red when young, older dark green above and pale to yellow green below, and margins distinctly serrate. The petioles are pubescent. The flowers are in clusters 4-8" wide and petals glabrous. The pomes are red when young and turning dark blue to black at maturity. It is a native of China that is widely cultivated, especially as a hedge, and persists in a few locations. The current records, mostly from cultivated plants, include seven parishes. (syn. *Photinia serrulata*)

YELLOW BAMBOO – *Phyllostachys aurea* - Poaceae

Yellow Bamboo is also known as Golden Bamboo. The genus "*Phyllostachys*" means "leafy spikes". The specific epithet "*aurea*", refers to the golden color. This is a perennial, rhizomatous, woody shrub that reaches 15 feet tall and is introduced from China or Japan. The stems are yellow and often brilliant. The internodes at the base of the plant are very short. The leaves are evergreen, up to 6" long, 0.5-0.75" broad, narrowed to a long acuminate point, minutely serrulate on one edge, finely tessellate, light green and mostly glaucous beneath, and dark green above. The sheath is with few small purplish spots and one tuft of cilia-like bristles on each side of the petiole. The inflorescence is reported to be axillary and a solitary spikelet. Yellow Bamboo is cultivated throughout the state, persists, and often spreads from old home sites. It is currently reported from 11 parishes. Note: a second species, ***Phyllostachys bambusoides*** (Japanese Timber Bamboo) was recently discovered in West Feliciana Parish.

PINE - *Pinus* - Pinaceae

The genus name, *"Pinus"*, is the Old Latin name for pines. Pines are evergreen trees with needles; in Louisiana all the species have needles in clusters of twos or threes. Pines are monoecious and produce their pollen and ovules (seeds) in cones. The pollen cones are small and produced from late February through early April. After pollen production, these cones are shed from the tree. The seed or ovule cones are much larger than the pollen cones and persist on the tree for several years. The new seed cones are produced yearly beginning in late February and these cones increase in size over the current year and to the fall of the following year. Seed production in pines requires two years. The bark on pine is rough and brownish to reddish. The seeds have long wings, which accounts for their wide distribution. A number of species of birds and rodents use the seeds, bark, and wood as a food source. Pine needles are used as a nesting source by several birds.

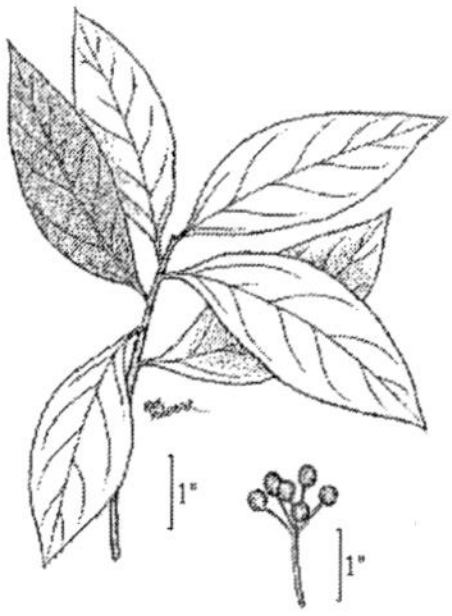

Photinia pyrifolia (USDA-NRCS)

Photinia serratifolia (Loudon 1838)

Phyllostachys aurea (Pohl 1980)

SHORT LEAF PINE (***Pinus echinata***) is also known as North Carolina Pine, Yellow Rosemary, Southern Yellow Pine, Arkansas Soft Pine, Arkansas Pine, Forest Pine, Old Field Pine, Shortleaf Yellow Pine, Longtag Pine, Short Straw Pine, Bull Pine, Pitch Pine, Slash Pine, Carolina Pine, and Yellow Pine. The specific epithet, *"echinata"*, refers to the hedge-like, or echinate, bristly needles. It is a tree that reaches up to 120 ft and matures in about 170 years, but very old trees may reach the four-century mark. The bark of young trees is black and later becomes thick, brownish red, and broken into large angled scaly plates. The needles are in 2's (sometimes 3's) and 3-6" long, nearly sessile, slender, often twisted, and persistent for several years. The terminal bud is brown and small. This pine may be distinguished from Loblolly Pine at quite a distance by the yellow green foliage, which is not nearly so luxuriant looking as that of Loblolly Pine. The seed cones are sessile or short-stalked, reddish brown, up to 2.5" long, ovoid to oblong, with thin scales, and often clinging to the twigs for 3 or 4 years. The wood is softer and less resinous than any other southern pine. The heartwood is orange brown and the sapwood yellowish. It weighs 38 lb per cu ft and is medium hard. The wood is used for general construction, exterior and interior finish, planning mill products, veneer, cooperage, excelsior, boxes, crates, agricultural implements, low-grade furniture, posts, poles, woodenware, and toys. Earlier it was used much on ships. It was, and is, used for flooring in houses because of the close growth rings. When used as edge-grain flooring, it is almost wear resistant. Shortleaf pine is not tapped for resin. It is the state tree of Arkansas. Shortleaf Pine is usually found growing naturally on drier, poorer soils than the other species. The distribution in the state is throughout the non marsh and non bottomland hardwood regions. The current reports include 40 parishes.

SLASH PINE (***Pinus elliottii***) is also known as Swamp Pine and Southern Pine. The specific epithet, *"elliottii"*, honors Stephen Elliot (1771-1830), a botanist who studied plants in the southeastern United States, particularly South Carolina and Georgia. It is a fast growing tree that reaches up to 120 ft tall. The bark is medium-thick, forming large, dark colored scales. The bark of old trees forms heavy ridges of large thin scales. The leaves are in twos and threes, stout, 5-10" long and dark green. The seed cones are ovoid, shiny, rich brown, 2-3.5" wide and 2-6" long. The scales are thin and flexible bearing very small curved prickles and turn a rich dark "chocolate" brown. The seeds are 1/4" long with a wing 0.75-1" long. Old seed cones of Slash Pine characteristically bend backward along their twigs. The wood is heavy (48 lb per cu ft), very hard, strong, durable, coarse-grained, and dark brown with much lighter sapwood. The wood is used for general construction, railroad ties, and one of the chief sources of naval stores. The wood is considered to be the hardest, strongest and heaviest of all conifers in the United States. Slash Pine is often planted as an ornamental along highways in the South. The natural range of Slash Pine in Louisiana was the southeastern part, east of the Amite River in the pine flatwoods (FACW). The species has been widely planted throughout the state in many habitats and the current records include 40 parishes. (syn. *Pinus caribea*)

SPRUCE PINE (***Pinus glabra***) is also known as White Pine, Cedar Pine, and Walter's Pine. It was given the name Spruce Pine by John Bartram. He saw it for the first time on 25 September 1765, near Savannah, Georgia. The specific epithet, "*glabra*", refers to the smoothish bark. The cones resemble those of Shortleaf Pine somewhat, but are more globose, often smoother, and the prickles are minute and weak. It grows rapidly reaches a height of 120 ft. The bark is much smoother than that of most pines and may even resemble that of an oak. It is pale gray on young trees to almost black on the older specimens. The dark brownish bark of large trees is divided into narrow ridges and shallow furrows. The branches are smooth. The needles occur in clusters of 2 and twisted, dark bluish green, and 1.5-4" long. The staminate cones are 0.25" long, clustered, and yellow. The pistillate cones are raised on slightly ascending

Pinus echinata (USDA-FS)

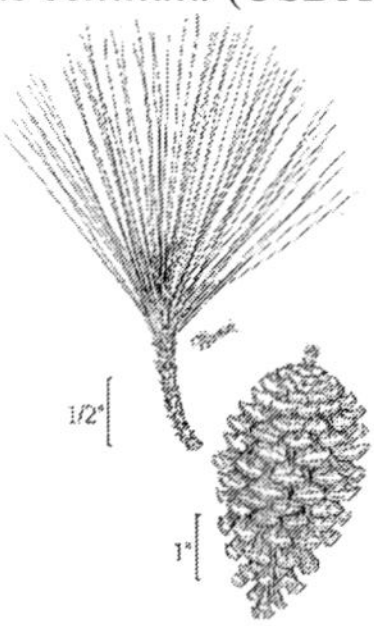

Pinus elliottii (USDA-NRCS)

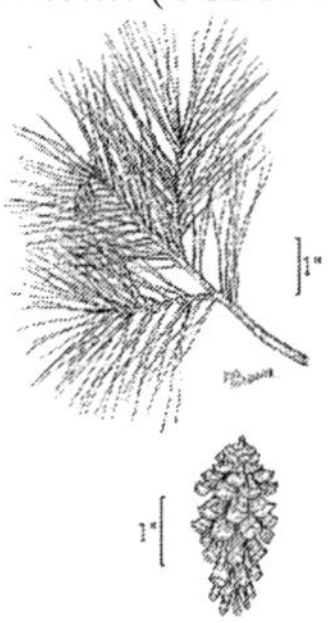

Pinus glabra (USDA-NRCS)

peduncles. The shiny brown, seed cones vary from egg shaped to oblong, in clusters of 2-3, and measure 1-2.5" long and 2" wide. They remain on the tree for many years and turn gray with age. The wood is light brown with thick pale sapwood, close-grained, not strong, soft, and brittle. The original distribution was along streams (FACW) in southeastern Louisiana from the Mississippi River eastward. It has been planted in many locations and today the distribution includes 23 parishes.

LONG LEAF PINE (***Pinus palustris***) is also known as Southern Pine, Long Leaf Yellow Pine, Georgia Pine, White Rosin Tree, Pine Broom, Yellow Pine, Texas Yellow Pine, Fat Pine, Turpentine Pine, Florida Pine, Southern Yellow Pine, Long Straw Pine, Hill Pine, Pitch Pine, Hard Pine, and Heart Pine. The specific epithet, "*palustris*", refers to a marshy or swampy habitat. It is a tall and slender tree with few branches that reaches 125 ft and is longer lived than most of the other pines, but grows slowly. Seedlings grow very little for the first three to five years. They are attractive and almost grass-like as a young tree. During this period, they produce a dense tuft of needles. This stage can last for 3 to 25 years. The bark is thin, light orange brown and forming large, close, papery scales. The buds at the tips of the branches are silvery-white and to 0.75" diameter. The leaves are in threes, three-sided, 8-18" long, slender, and drooping. The seed cones are slender conic, slightly curved, and usually 6-12" long. The scales are dull brown with small, reflexed prickles. The wood is moderately heavy (44 lb per cu ft), very hard and durable, light red or orange with almost white sapwood. The wood is very valuable for general construction, masts, spars, bridges, flooring, and interior finishing. It is a principal source of naval stores. Longleaf Pine was originally the most important timber tree of the pinelands of southeastern, southwestern and central Louisiana. Longleaf Pine is a slow-growing tree that is gradually coming back where protected from livestock and fires. The date of its earliest cultivation is 1727. It thrives in deep sandy soil (FACU+) and usually does poorly in heavy soils. The current reports include 29 parishes.

LOBLOLLY PINE (***Pinus taeda***) is also known as Rosemary Pine, Old Field Bull Pine, Old Field Pine, Yellow Pine, Southern Yellow Pine, Black Pine, Arkansas Soft Pine, Longleaf Pine, Indian Pine, Long Straw Pine, Frankincense Pine, Lowland Short Leaf Pine, Sap Pine, Swamp Pine, Bastard Pine, Long Shucks Pine, Foxtail Pine, Torch Pine, Slack Pine, and Short Leaf Pine. Loblolly Pine's common name is derived from a word (old English in origin) meaning a wet or muddy place. It is thought, the name Loblolly was given by the early colonists. It seems they associated the moist depressions and swamps that these trees grew near with a lumpy gruel called "loblolly" that was served for seasickness on their voyage. The specific epithet, "*taeda*", is for the resinous wood. Loblolly Pine is a large tree reaching heights of 170 ft and an age of 300 years under favorable conditions. The bark is dark, reddish brown, thick, deeply furrowed, and scaly plates on older trees. The needles are in three's (occasionally 2's), three sided, slightly twisted, rich dark green, and 5-10" long. The tips of the branches turn upward. The small terminal bud is reddish brown. The seed cones are up to 6" long with reddish brown, but not shiny scales that are spiny. Loblolly Pine is not used for turpentine because the gum does not flow readily. However, the wood is used more than all other southern pines put together in Louisiana. The resinous, soft, brittle wood weighs about 34 lb per cu ft. It is used for lumber, cooperage, poles, plywood, pulp, boxes, cross-ties, posts, and fuel. The heartwood is orange brown and the sapwood is light yellow. The date of the earliest cultivation of Loblolly Pine is 1713. Loblolly Pine originally occurred on the slopes of the small streams in the state, and was absent from the Mississippi River floodplain, the prairie section, and the clay terrace soils (FAC). Due to planting, it is much more widespread and abundant than when the white men first started to clear the forests. The current records include 53 parishes.

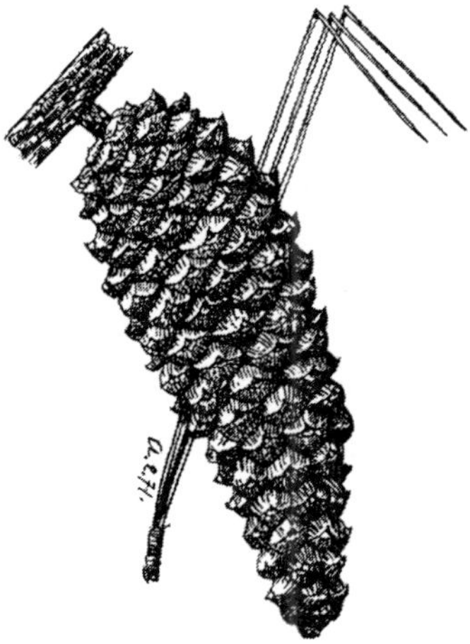

Pinus palustris (USDA-FS)

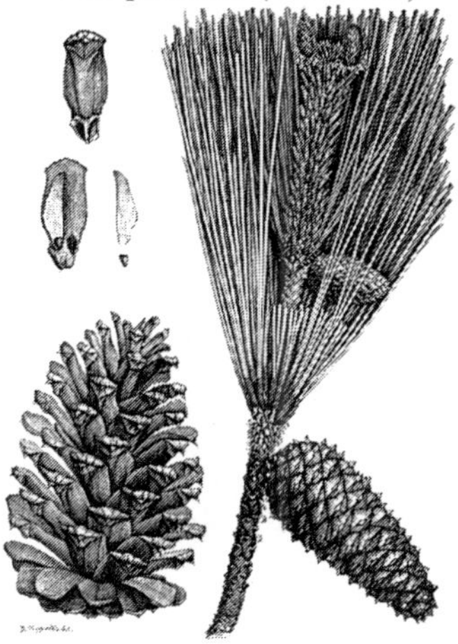

Pinus taeda (USDA-FS)

WATER ELM - ***Planera aquatica*** - Ulmaceae

Water Elm is also known as Planer Tree. The genus name, "*Planera*", honors the German botanist Johann Jakob Planer (1743-1789), a professor at the University of Erfurt and the specific epithet, "*aquatica*", refers to the swampy habitat of the tree. It is a monoecious shrub to small tree, often distorted, that reaches a height of 40 ft. The bark is distinctive, light to dark brown, rather thin, falling away in long strips or scales, giving the trunk a ragged appearance and disclosing the red-brown inner bark. The twigs are thin, brown to gray, and hairy when young. The leaves are alternate, simple, deciduous, and with pinnate major veins. The blades are ovate or oblong lanceolate, 2-4" long, 0.8-1" wide, apex is acute to obtuse, base unequal and oblique or cordate, margins coarsely crenulate-serrate with gland-tipped teeth, upper surface rough and dull green, and lower surface paler and glabrous. The tiny flowers are small, green or greenish yellow, and appearing with the leaves. The fruit is a drupe that is covered with warty projections. The wood is light (33 lb per cu ft), coarse-grained, soft, and light brown with a broad zone of nearly white sapwood. The wood has no known commercial value except for fuel. Water Elm is a favorite beaver food. The fruit is considered to be an important duck food in swampland areas. Squirrels also eat the fruit. Restricted to edges of swamps, lakes and other wet areas (OBL) throughout the state. The current reports include 49 parishes.

SYCAMORE - ***Platanus occidentalis*** - Platanaceae

Sycamore is also known as Cotton Wood, Button Wood, Button Ball Tree, Plane, Plane Tree, American Plane Tree, White Wood, Water Beech, and Virginia Maple. The genus name, "*Platanus*", is the classical name of the Plane-tree, and the specific epithet, "*occidentalis*", means "western". It is the most massive hardwood tree of Eastern United States and reaches heights of 170 ft. The bark is mottled reddish brown and scales off in thin plates, yearly, like jigsaw puzzle pieces to expose the yellowish/whitish bark underneath. The twigs are slender, orange brown to gray, and hairy when young but becoming glabrous with age. The buds are small, orange, covered by one shiny scale, and hidden by the enlarged base of the petiole. The leaves are alternate, simple, deciduous, and with palmate major veins. The blades are 5-8" long, 4-10" wide, broadly ovate, margins with three to five large acuminate lobes and teeth between, base truncate to cordate, upper surface bright green, and lower surface paler and densely pubescent along veins. The green stipules are large, toothed, and sheathing. Sycamore produces the largest single-bladed leaf native to our American forest. The green flowers are in globose clusters. The light brown globe-shaped fruit is composed of many hairy, one-seeded nutlets. The fruit hang on the bare tree all winter, breaking up in early spring, when the seeds are carried far by the wind. The rather weak, hard, difficult to split, tan or light brown and not durable on exposure wood weighs about 35 lb per cu ft. It is used for crates, interior finish, furniture, cooperage, ox-yokes, rollers, lumber, pulpwood, butcher blocks, and tobacco boxes. The Indians used Sycamore to make dugout canoes; one canoe was reported to be 65 feet long. Also, American Indians used this tree for sugar and syrup production. The seeds are used by a number of species of birds and muskrat for food. Along the edges of streams (FACW-) on the natural levee throughout the state but especially along the larger streams, *i.e.* Mississippi and Red rivers. It is also widely planted as an ornamental. It probably occurs in all with current records from 57 parishes.

Planera aquatica (USDA-FS and USDA-NRCS)

Platanus occidentalis (USDA-FS)

JOINT WEED - *Polygonella* - Polygonaceae

The genus name, "*Polygonella*", is a diminutive of polygonum, which is a word from the Greek meaning "many -knees" and refers to the swollen joints of some species. The genus includes small shrubs that are woody at base and herbaceous above. The brown twigs are jointed with ocrea at the nodes and ridged along the internodes. The ocreae are brown, nerved, and apex scarious. The leaves are alternate, simple, deciduous, with pinnate major veins, and jointed at base. The white to pink flowers are in axillary racemes. The fruits are 3-angled brown achenes.

SOUTHERN JOINT WOOD (***Polygonella americana***) is also called Small Joint Weed. The meaning of the specific epithet, "*americana*", is for North America. It has perfect flowers and reaches a height of 4 ft. The ocrea ends in an acute to obtuse apex. The leaves are often fascicled; the blades are linear to spatulate, up to 1.5" long, and the apex is acute, scarious, and somewhat revolute. In very sandy soils in Bienville, Caddo, Natchitoches, Sabine, and Winn parishes.

OCTOBER FLOWER (***Polygonella polygama***) is also called Joint Weed. The specific epithet, "*polygama*", means with both single and dual sex flowers on one or different plants. It is dioecious and reaches a height of 3 ft. The ocrea ends in an acuminate to aristate apex. The blades are linear-spatulate to narrowly obovate, three nerved, up to 0.5" long, and the apex is rounded to obtuse. It is currently found in very sandy soils in Natchitoches Parish.

TRIFOLIATE ORANGE - *Poncirus trifoliata* - Rutaceae

The genus name, "*Poncirus*", is from the French word "poncire", a kind of citrus and the specific epithet, "*trifoliata*", refers to the three leaflets. Other common names include Mock Orange, Bitter Orange, Hardy Orange, and Limoncito. It is an armed shrub to small tree to 30 ft. The bark is green with brown streaks. The flattened twigs are dark green and with flattened 0.5-3" long thorns. The leaves are alternate, deciduous, and palmately compound with three leaflets. The leaflets are elliptic, oblong to obovate, 0.5-2.5" long, glabrous, olive green above, paler beneath, apex obtuse or rounded, base cuneate or rounded, and margins crenate. The leaflets are sessile or with very short petiolules and the rachis is broadly winged. The fascicles of white flowers are axillary on the previous year's growth. The fruit is a globose yellow to yellow-orange 1.5-2" diameter hespiridium with many seeds. The crushed stems, leaves, and fruits are aromatic. This species was introduced from China and Korea as an ornamental, to create fences, and as the rootstock for grafted citrus plants. It persists in dense thickets near old home sites and has escaped and spread in many locations throughout the state. The current records include 49 parishes. (syn. *Citrus trifoliata)*

COTTON WOOD, POPLAR - *Populus* - Saliacaceae

The genus name, "*Populus*", is the ancient Latin name given by Pliny. It includes small to large trees. The leaves are alternate, simple, deciduous, and with palmate major veins. The plants are dioecious with the naked flowers in drooping catkins. The fruit is a catkin with many seeds each with tufts of cottony hairs.

WHITE POPLAR (***Populus alba***) is also known as Abele. The specific epithet, "*alba*", refers to the white undersurface of the leaves. It is a tree to 100 ft. The bark is greenish gray to white, usually smooth on branches or young trunks, toward the base of old trunks roughened into firm dark ridges. The blades are 2.5-5" long, base rounded to subcordate, upper surface dark green, and lower surface distinctly white tomentose. The margins are palmately 3-5 lobed and

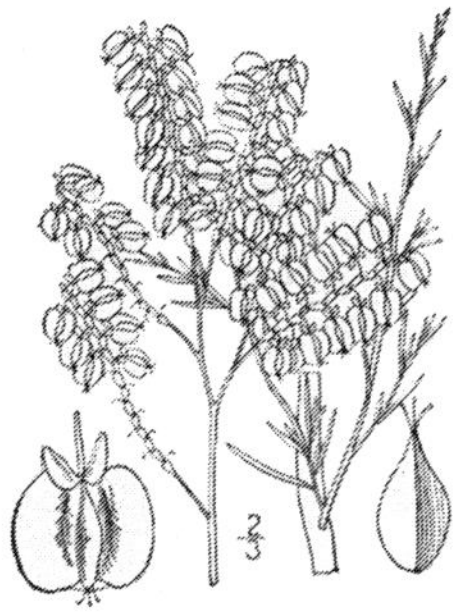

Polygonella americana (Britton and Brown 1943)

Polygonella polygama (Nelson 1996)

Poncirus trifoliata (Vines 1960)

Populus alba (USDA-FS)

the lobes also have large teeth or small lobes. The petiole is terete and also densely white tomentose. The wood is reddish to yellowish, sapwood nearly white, and is tough but light and soft. Widely planted as an ornamental and a native of Europe and Asia. Persisting at old homes sites and spreading throughout the state. The current records include 20 parishes.

EASTERN COTTON WOOD (***Populus deltoides***) is also known as Carolina Poplar, Necklace Poplar, Water Poplar, Southern Cotton Wood, Yellow Cotton Wood, and Alamo. Alamo is the Spanish name for the tree, and was also the name given the famous Texas fort which was surrounded by the trees. The specific epithet, "*deltoides*", refers to the triangular shape of the leaf. It is a tree to 80 ft. The bark is yellowish and smooth on young branches, becoming pale gray, rough, thick, and deeply furrowed on old branches and trunks. The twigs are thick, green to yellow-brown, glabrous, with large triangular leaf scars. The blades are triangular, 3-7" long, 2.5-5" wide, apex acuminate, base truncate (nearly straight across), glabrous, shiny green on both surfaces, and margins crenate-serrate. The petiole is flattened and glabrous. The wood is dark brown and sapwood is white. The wood is weak, soft, moderately weak in bending, weak in endwise compression, and low in shock resistance. It is also moderately easy to work with tools, takes paint well, easy to glue, warps and shrinks considerably, low in durability, and weighs about 24 lb per cu ft. The wood is used for paper pulp, cases and crates, tubs and pails, excelsior, veneer for plywood, musical instruments, dairy and poultry supplies, laundry appliances, and fuel. The foliage is known to be browsed by cattle, black-tailed deer, and cottontail. The seeds are eaten by rose-breasted grosbeak and evening grosbeak. In a wide range of habitats (FAC+) but especially on the natural levee of many streams throughout the state. It is very common in the Mississippi River floodplain. The current records include 42 parishes.

SWAMP COTTON WOOD (***Populus heterophylla***) is also called Black Cotton Wood, Black Poplar, Swamp Poplar, and Downy Poplar. The specific epithet, "*heterophylla*", refers to the variously shaped leaves. It is a tree to 90 ft. The bark is brown to reddish and has broad and flattened ridges that separate into long, loose flakes, and shallow fissures. The twigs are thick, and varying shades of gray, red, or brown, tomentose when young but glabrous with age, and with large triangular leaf scars. The blades are ovate to oval, 2.5-7" long, 1.5-5" wide, apex obtuse or rounded, base broadly cuneate, rounded, or cordate, upper surface dark green and glabrous, lower surface often white tomentose, and margins crenate-serrate. The petiole is terete and glabrous to pubescent. The wood is dull brown, close-grained, moderately soft when dry, low in durability, weak in bending, and low in shock resistance. It also warps somewhat, is resistant to splitting, shrinks considerably, does not hold nails well, moderately hard to work, and has lighter sapwood. The wood is used for excelsior, veneer, musical instruments, boxes and crates, poultry coops and brooders, ironing and washboards, tubs and pails, and concealed parts of furniture. Along the edges of lakes, bayous, and other water bodies (OBL) in northeastern and extreme southeastern Louisiana (St. Tammany and Washington parishes). The current records include 10 parishes.

HONEY MESQUITE - *Prosopis glandulosa* var. *glandulosa* - Fabaceae

The genus, "P*rosopis*", is from a Greek name for a kind of prickly fruit or for burdock which has a prickly fruit. The specific epithet, "*glandulosa*", means "full of glands". Other common names include Mesquite, Glandular Mesquite, and Algaroba. It is an armed small tree or shrub that reaches a height of 30 ft. The bark is rough, reddish brown, and with shallow fissures and thick scales. The twigs are glabrous, brown, with usually paired (stipular) sharp spines at the nodes. The leaves are alternate, deciduous, drooping, and twice pinnate compound. The 7-17 pairs of

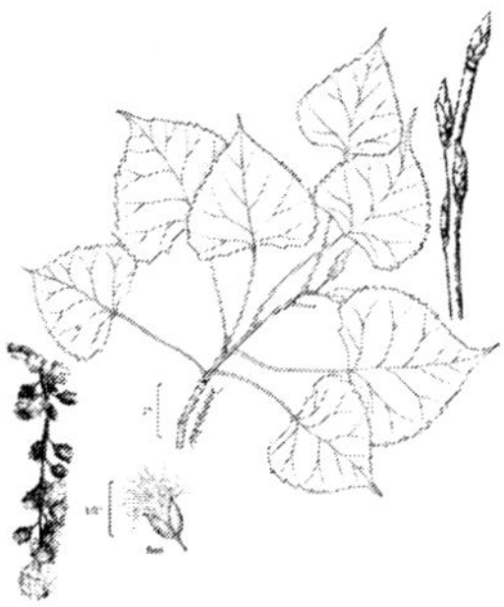

Populus deltoides (USDA-FS and USDA-NRCS)

Populus heterophylla (USDA-FS)

Prosopsis glandulosa var. ***glandulosa*** (USDA-FS)

leaflets are linear to narrowly oblong, 0.4-1.2" long, glabrous, yellowish-green, apex acute, base rounded, and margins entire. The yellowish green flowers are born in cylindric, axillary, pedunculate spikelike racemes. The yellow to brown fruit is a flattened, glabrous, linear, straight or curved legume that ends in a point. The seeds are oblong, compressed, shiny, brown, and set in spongy tissue. The wood is reddish brown, heavy, hard, durable, close-grained, and the sapwood yellow. The wood is used for charcoal, fuel, furniture, building blocks, crossties, and posts. The foliage and pods are eaten by livestock, and the seeds pass through the digestive tracts and grow where they fall. The seeds are also considered an important wildlife food, being eaten by several birds and mammals. The beans were part of the Southwestern United States Indians diet. It was ground into a meal and was used for bread and an intoxicating substance. The gum in the branches and trunk is used to make candy, to produce a black dye and cement for mending pottery. It is a wonderful honey tree. In extreme northwestern Louisiana (Bossier and Caddo parishes) with one report from East Carroll Parish. (syn. *Prosopis chilensis* var. *glandulosa*, *Prosopis juliflora* var. *glandulosa*)

CHERRY, PLUM - *Prunus* - Rosaceae

The genus names "*Prunus*", is the classical Latin name for a plum of Europe. The genus includes large shrubs to small trees and large trees. The leaves are alternate, simple, and with pinnate major veins. The white or pink flowers are in racemes or umbels and the fruit is a drupe.

AMERICAN PLUM (***Prunus americana***) is also called Horse Plum, Hog Plum, Goose Plum, Native Plum, River Plum, Wild Yellow Plum, Red Plum, and Thorn Plum. The specific epithet, "*americana*", is for North America. It is a shrub to small tree to 35 ft with somewhat thorny stems. The bark is dark brown to red-brown and platey. The twigs are green to red-brown with minute lenticels, glabrous, and the tips often end in a thorn. The leaves are deciduous and the blades are ovate, obovate, or oblong, 2.5-4" long, 1.5-2" wide, glabrous to slightly pubescent along veins of lower surface, dark green above, paler beneath, apex acuminate, base rounded or cuneate, and margins singly or doubly serrate. The petioles are 0.5-0.75", long, slender, glabrous or somewhat pubescent, and often with a dark gland at the apex or on the base of the blade. The white flowers are in umbels, with 2-5 flowers on slender glabrous pedicels. The red to dark blue drupes are subglobose, and 0.75-1" diameter. The fruit is too tart to be eaten raw but makes excellent jellies and preserves. The fruit is also eaten by many species of birds. The wood is dark brown, with lighter colored sapwood, close-grained, strong, hard, and weighs 45 lb per cu ft. In well drained soils of hardwood slopes and also pine forests (FACU-) throughout the pine regions of the state. The current records include 31 parishes. Note: it is very difficult to separate this species from ***Prunus mexicana*** so the distribution of both species should be considered.

CHICKASAW PLUM (***Prunus angustifolia***) is also called Sand Plum and Mountain Cherry. The specific epithet, "*angustifolia*", refers to the narrow foliage. It is a shrub or small tree to 10 ft with thorny stems and forms dense thickets. The bark is red-brown to dark gray with thin scales and prominent lenticels. The twigs are lustrous red-brown and pubescent. The leaves are deciduous and the blades are lanceolate or oblong-lanceolate, 1-2" long, 0.3-0.6" wide, glabrous or slightly pubescent beneath, lustrous green above, paler beneath, apex acuminate to acute, base rounded and broadly cuneate, and margins sharply serrate with glandular teeth. The petioles are 0.25-0.5" long, slender, glabrous to pubescent, and usually have 2 red glands near the apex. The white flowers are umbellate and appear before the leaves. The red or yellow drupes are globose, and 0.5-0.75" diameter. The wood is reddish brown, sapwood lighter, rather soft, not strong, fairly heavy, and weighs 43 lb per cu ft. The drupe is red or yellow. The drupes make excellent jelly and preserves and can be eaten raw. The drupes

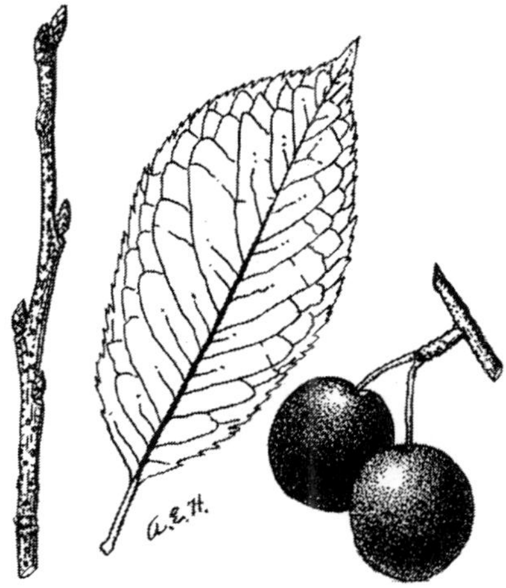

Prunus americana (USDA-FS)

Prunus angustifolia (USDA-FS)

are also eaten by several species of birds and mammals. This species was widely planted near homes and dense thickets and can often be found still persisting today. Very few truly wild populations are found and thus the question of whether this species was originally native to Louisiana is difficult to ascertain. The current records include 44 parishes.

CAROLINA CHERRY LAUREL (***Prunus caroliniana***) is also called Wild Peach, Cherry Laurel, Carolina Cherry, and Mock Orange. The specific epithet, *"caroliniana",* refers to North or South Carolina. It is shrub or small tree to 40 ft. The bark is gray, smooth at first but becoming blotchy with age. The twigs are green to red or grayish brown, slender, and glabrous. The leaves are evergreen and the blades are elliptic to oblanceolate, 2-4.5" long, 0.5-1.5" wide, coriaceous, apex acute or acuminate and mucronate, base cuneate, upper surface lustrous and dark green, lower surface paler, and margins entire or with a few remote spine-tipped teeth. The crushed blades have a "Dr. Pepper" or almond extract odor. The petiole is stout, 0.25-0.5" long, and glabrous. The white flowers are in short axillary racemes. The drupe is dry, dark red to dull black, ellipsoidal, and shorter than 0.3" and narrower than 0.25". The wood is reddish brown, hard, heavy, strong, and close-grained. A number of birds feed on the drupes. The leaves contain prussic acid which is poisonous to livestock. Widely planted as an ornamental but the original habitat seems to be well drained soils along slopes and natural levees. The current records include 44 parishes.

OKLAHOMA PLUM (***Prunus gracilis***) is also called Hog Plum and Sour Plum. The specific epithet, *"gracilis",* refers to the slender branches. It is a shrub to 10 ft and often forms dense thickets. The bark is red-brown to dark gray with thin scales and prominent lenticels. The twigs are red-brown and pubescent when young but turning gray and glabrous with age. The leaves are deciduous and the blades are elliptic, oval or ovate, 1-2" long, apex acute to obtuse, base cuneate, upper surface glabrous to slightly pubescent, lower surface densely pubescent and with distinct reticulate veins, and margins sharply serrate with appressed teeth. The petiole is short and pubescent. The white flowers are in umbels. The red drupes are subglobose and 0.5-0.8" diameter. The drupes are often prepared as jelly or preserves and are eaten by birds and mammals. The plant was introduced into cultivation in 1916. In loose sandy soils only in Caddo Parish.

MEXICAN PLUM (***Prunus mexicana***) is also called Wild Plum, Fall Plum, and Big Tree Plum. The specific epithet, *"mexicana",* refers to this species' Southwestern distribution near Mexico. It is a shrub or small tree to 30 ft. The bark is dark to black, roughly furrowed and peeling. The twigs are metallic gray, pubescent when young, becoming glabrous with age, and end in a thorn like structure. The leaves are deciduous and the blades are ovate, elliptic, or obovate, 1.8-3.5" long, 1-2" wide, apex acuminate, base cuneate or rounded, both surface light to yellow green and with distinct reticulate veins, glabrous above, hairy beneath, and margins singly or doubly serrate. The petioles are 0.4-0.8" long, stout, pubescent, and sometimes with one or two reddish glands near apex. The white flowers are in umbels. The drupes are subglobose to oblong, dark purple-red, glaucous, and 1-1.5" diameter. The fruit is too tart to be eaten raw but makes excellent jellies and preserves. The fruit is also eaten by many species of birds and mammals. In well drained soils of hardwood slopes and also pine forests throughout the pine regions of the state. The current records include 41 parishes. Note: it is very difficult to separate this species from ***Prunus americana*** so the distribution of both species should be considered.

Prunus caroliniana (USDA-FS)

Prunus gracilis (Britton and Brown 1943)

Prunus mexicana (USDA-FS)

PEACH (***Prunus persica***) is commonly cultivated especially in northern Louisiana. The specific epithet, *"persica"*, means "Persian," and is also an old generic name for peach. It is a shrub to small tree to 25 ft. The bark is dark reddish brown with blotches. The twigs are dark gray to brown with lines. The leaves are deciduous often fascicled and the blades are elliptic-lanceolate to oblong lanceolate, 3-6" long, apex long acuminate, base acute to attenuate, both surfaces lustrous bright green and glabrous, and margins serrate. The petiole is glabrous, with two glands near apex, and the blade extends down the top as a narrow pair of wings. The pink flowers are solitary or in pairs. The yellow to red drupes are subglobose, grooved on one side, and 2-3.5" diameter. Widely planted and persisting throughout the state with current records from 36 parishes.

BLACK CHERRY (***Prunus serotina***) is also known as Mountain Black Cherry, Black Choke, Choke Cherry, Rum Cherry, and Wild Cherry. The specific epithet "*serotina*" means "late-flowering". It is a tree to 100 ft. The bark is red-brown, gray or white, smooth when young but broken into plates with age. The twigs are red-brown to gray, and when crushed with a strong unpleasant odor. The leaves are deciduous and the blades are oval to oblong-lanceolate, 2-6" long, 1-2" wide, glabrous or with hairs on midrib beneath, apex acuminate, base cuneate, upper surface lustrous dark green, lower surface dull light green, and margins serrate. The petiole is glabrous with one or more glands on the upper side near the apex. The white flowers are in racemes. The drupes are ovoid, 0.25-0.5" diameter, black, thin skinned, and with a bittersweet taste. It was once famous for the making of "cherry bounce" in the South. The wood is reddish-brown with thin yellow sapwood, light (35 lb per cu ft), hard, close-grained and takes a beautiful finish. It is much sought for use in interior finish and the making of furniture, cabinetmaking, printer's blocks, veneer, patterns, panels, handles, woodenware, toys, scientific instruments, and handles for tools. The leaves and twigs contain cyanide (hydrocyanic acid) and are poisonous to livestock but not to deer. The bark is used medicinally as a cough remedy. American Indians had many uses for Black Cherry: One tribe used a tea made from the inner bark to ease pain during labor; others used a tea of the bark for diarrhea and lung problems; a decoction of the inner bark served one tribe as an enema for hemorrhoids; and another tribe cured dysentery by drinking the juice of the ripe cherries, which had been allowed to ferment for a year. The fruit is eaten fresh, used as a flavoring extract, and popular for wines and jellies. Black Cherry is also a favorite of many birds and wildlife; however the berries ferment and cause the birds that eat them to become intoxicated. The flowers produce considerable nectar for bees and possibly some pollen. One of the first New World trees to be introduced to English gardens, this tree was first recorded in 1629. Very widely distributed from mesic to very dry sites (FACU) throughout the state. It probably occurs in all with current records for 58 parishes.

FLATWOODS PLUM (***Prunus umbellata***) is also called Hog Plum, Sloe, and Black Sloe. The specific epithet, *"umbellata"*, refers to the umbellate flower clusters. It is a large shrub to small tree to 20 ft and sometimes forms thickets. The bark is dark gray with small scale. The twigs are slender, spurlike, metallic gray to dark reddish brown, ridged, and end in a thorn like structure. The leaves deciduous and the blades are oblong, lanceolate, or oval, 1.5-2.5" long, 0.5-1.5" wide, apex acute to acuminate, base rounded, cordate, or broadly cuneate, upper surface medium green and glabrous, lower surface pale green and pubescent along midrib, and margins glandular serrate. The white flowers are in umbels. The drupe is globose, dark purple to red, glaucous, and 0.3-0.5" diameter. The fruit is used in considerable quantities for jellies and preserves. The wood is reddish brown with lighter sapwood, close-grained, heavy, and hard. Throughout the state usually in slightly moist sites with current records from 34 parishes.

Prunus persica (Vines 1960)

Prunus serotina (USDA-FS)

Prunus umbellata (USDA-FS)

COMMON HOP TREE - ***Ptelea trifoliata*** - Rutaceae

The genus name, *"Ptelea",* is the classical name for the elm. The specific epithet, *"trifoliata",* refers to the three leaflets. Other common names include Three Leaf Hop Tree, Shrubby Trefoil, Swamp Dogwood, Wafer Ash, Nickles Dimes and Quarters, Skunk Bush, Potatochip Tree, Quinine Tree, Ague Bark, Pickaway Anise, Prairie Grub, Cola de Zorillo, and Wing Weed. It is a shrub or small tree to 25 ft. The bark is smooth and light to dark gray. The twigs are green to red-brown, slender and pubescent. The crushed twigs and leaves have an unpleasant odor. The leaves are alternate or opposite, deciduous and palmately compound with three leaflets. The leaflets are ovate-oblong, 4-6" long, 2-4" wide, apex acute or acuminate, base cuneate and often oblique, dark green and lustrous above, pubescent and glandular below, and margins entire or finely serrate. The green white small flowers are in terminal compound corymbs. The fruit is a samara, that is 0.8-1" long, thin and wafer-like, suborbicular to obovate with 2-3 reddish brown seeds. The wood is yellowish brown, close-grained, hard, heavy, and weighs about 51 lb per cu ft. The fruit was once used as a substitute for hops in beer brewing. A root tincture is used to treat dyspepsia and as a mild tonic; it contains the alkaloid berberine. It has been cultivated since 1724. In well drained soils often hardwood forests or along streams (FAC). The current reports include 28 parishes.

CHINESE WING NUT - ***Pterocarya stenoptera*** - Juglandaceae

The genus, "P*terocarya*", is from the Greek word *"pteron"* (a wing) and "karyon" (a nut) referring to the winged fruit. The specific epithet, "*stenoptera*", comes from the Greek word "*stenos"* which means "narrow", and *"ptera"* is from "pteron" which is "winged". It is a tree to 50 ft tall. The twigs are stout, light to dark gray, ridged, and with distinct lenticels. The leaves are alternate, deciduous and odd pinnately compound. The 11-23 leaflets are oval-oblong, 1-4" long, 0.5-1.5" wide, apex acute to obtuse, base cuneate to rounded and often oblique, upper surface medium dark green and glabrous, lower surface pale green and pubescent along veins, and margins serrate. The plants are monoecious with the male and female flowers in separate catkins. The fruit is a black nut with a distinct pair of brown wings. Cultivated and persisting with records from East Baton Rouge and Washington parishes. (Not Illustrated)

POMEGRANATE - ***Punica granatum*** - Punicaceae

The genus name, *"Punica",* is from the Latin words for "Punic apple." The specific epithet, "*granatum",* means "many-seeded apple". It is an armed shrub or tree to 20 ft. The bark on older limbs and trunks is gray to brown, smooth at first but eventually breaking into small, thin scales, shallowly reticulate. The twigs are slender with the younger ones green to reddish-brown, somewhat striate or angular and the older ones gray and more terete. The leaves are simple, mostly opposite, deciduous, and with pinnate major veins. The blades are oblong or oval-lanceolate, 0.75-3.5" long, apex obtuse or acute, base attenuate, upper surface glabrous and shining, lower surface pale green with a distinct midrib, and margins entire. The flowers are orange-red, 1-1.5" across and in groups of 1-5 on the tips of axillary shoots. The fruit is a globose 2-4" diameter brownish yellow to red berry with a fleshy aril around the seeds. The wood is white to yellowish, hard, close-grained, and with a specific gravity of about 0.93. The wood is sometimes used by engravers as a substitute for Boxwood (***Buxus***). From the days of Solomon, pomegranate has been used for making cooling drinks and sherbets; it is also eaten in its natural state. The astringent rind yields a black or red dye and has been used for tanning morocco leather and for ink. The fruit is sweet in the better varieties, and a spiced wine is made from the juice. The soft seeds are also eaten, sprinkled with sugar, or, when dried, as a confection. It has been used medicinally as an astringent in dysentery and diarrhea and as a

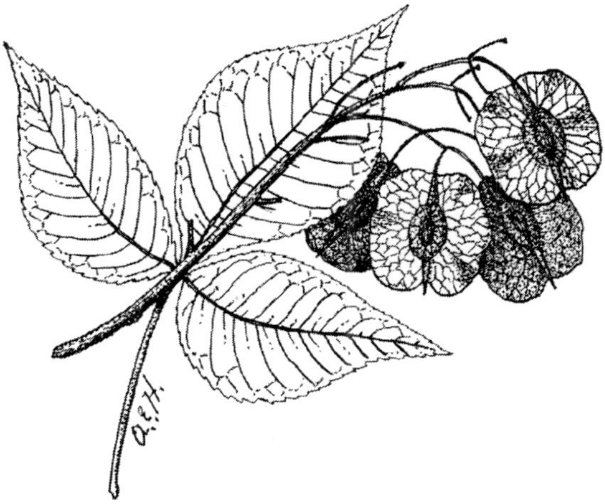

Ptelea trifoliata (USDA-FS)

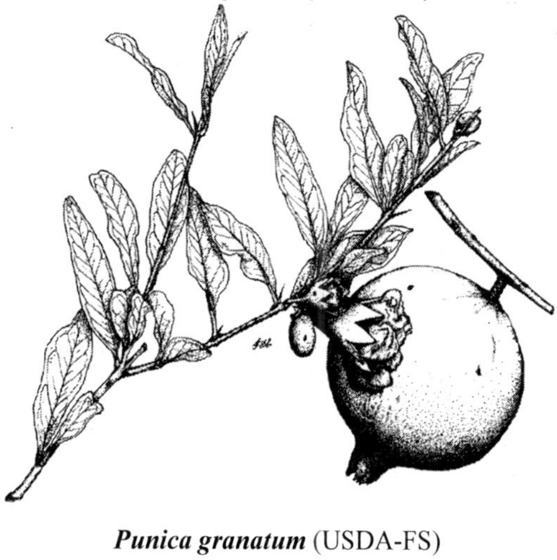

Punica granatum (USDA-FS)

gargle. The bark was used to treat tapeworms. A lot of symbolism revolves around the fruit, in particularly, fertility due to the many seeds inside. Grenadine syrup is made from the seeds and the name is derived from the fruit looking like a grenade. Cultivated and persisting in a few locations with reports from 12 parishes.

SCARLET FIRETHORN - *Pyracantha coccinea* - Rosaceae

The genus, "P*yracantha*", comes from the Greek word "*pyr*" (fire) and "*akantha*" (a thorn) referring to the red berries and spiny shoots. The specific epithet, "*coccinea*", means "berry of the scarlet oak", used as a red dye. Another common name is Pyracantha. It is an armed shrub to small tree to 20 ft tall. The twigs end in thorns and are grayish pubescent when young and gray to brown and glabrous with age. The leaves are evergreen to semi-evergreen, alternate, simple, and with pinnate major veins. The blades are elliptic to oblong or oblanceolate, 1-2.5" long, 0.3-0.76" wide, apex acute to obtuse, often emarginate, upper surface lustrous dark green,lower surface pale green, and margins entire. The white flowers are in corymbs. The fruit is an orange to red fleshy globose pome about 0.25" diameter. Many birds, including robins and cedar waxwings, eat the pomes. It is widely cultivated throughout the state and persists and perhaps spreads. The current records include 29 parishes. (Not Illustrated)

PEAR - *Pyrus* - Rosaceae

The genus name, *"Pyrus",* is the classical name for the Pear-tree. The genus includes small to medium trees. The leaves are alternate, usually on short lateral spurs, simple, deciduous, and with pinnate major veins. The flowers are white, rarely pink, and in cymes. The fruit is a pear shaped pome.

BRADFORD PEAR (***Pyrus calleryana***) is also known as Callery Pear. The specific epithet, "*calleryana*", honors J. Callery, a French missionary who collected the type of specimen. It reaches a height of 40 ft. The bark is lightly furrowed and grayish brown at maturity. The twigs are somewhat pubescent at first, glabrous later, reddish brown to gray or black. The blades are ovate to elliptic, 1-4" long, apex acute to acuminate, base rounded to truncate, upper surface olive green, lower surface paler, glabrous, and margin finely serrate to entire. The pomes are dark purple, russet dotted, ovoid, and 0.5-2" long. It is a native of Europe and Asia and is widely planted. It persists near old homes sites and along streets and highways. It is currently reported from 19 parishes but is cultivated in many more, if not all 64 parishes.

COMMON PEAR (***Pyrus communis***) is also known as Domestic Pear. The specific epithet, "*communis*", means common. It reaches a height of 75 ft but is usually much shorter. The twigs are somewhat pubescent at first, glabrous later, reddish brown to gray or black. The blades are ovate to elliptic or obovate, 1.5-4" long, apex acute to acuminate, base usually rounded, upper surface olive green, lower surface paler, glabrous, and margin finely serrate to entire. The pomes are yellow to red, and 2-5" long. The wood is reddish brown, hard, fine-grained, and weighs 51 lb per cu ft. It is a native of Europe and Asia and is widely planted. It persists near old homes sites and has been reported from 36 parishes.

OAK - *Quercus* - Fagaceae

The genus, "*Quercus*", is for the ancient classical name of oak. It includes trees with alternate, simple leaves with pinnate major veins. The margins vary from entire through various shapes and sizes of teeth to lobes and the lobes also vary from shallow to deep. Most species have deciduous leaves but some are evergreen as in ***Quercus virginiana*** (Live Oak) and others are tardily deciduous as in ***Quercus laurifolia*** (Laurel Oak). The buds are clustered near the tips of

Pyrus calleryana (Southwestern Forestry College)

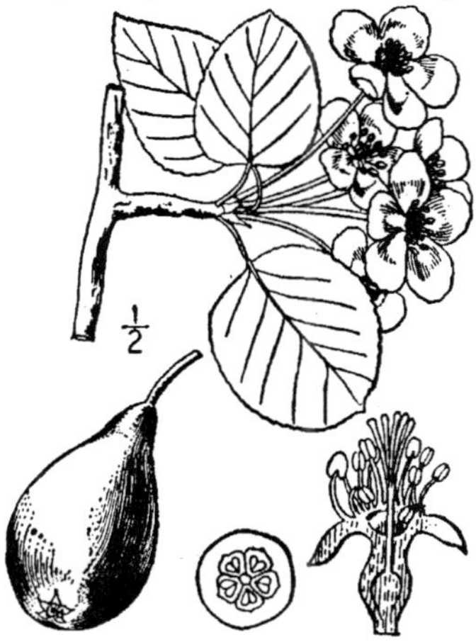

Pyrus communis (Britton and Brown 1943)

the branches and the twigs are often five angled. The plants are monoecious and the flowers are naked with the staminate in catkins and the pistillate single or few in a cluster. The fruit is nut that is usually termed the acorn; the acorn sits above a cup like structure. Oaks are divided into two major groups; the White Oak Group and the Red Oak Group. Plants in the White Oak Group have leaves without bristles, cup scales that are woody, a smooth inner surface of the acorn shell, and the acorns mature in one year. Plants in the Red Oak Group have leaves with bristles, cup scales that are thin, a hairy inner surface of the acorn shell, and the acorns require two years to mature. A good way to check for the one year or two year maturation of acorns is to examine several acorn bearing twigs from a tree. If all the acorns are the same size and are on this year's growth, the species has the one year or annual maturing acorns. If there are different sized acorns on the twigs and some are on last year's growth, the species requires two years for acorn maturing. Acorns from the white oak group contain less tannins than those from the red oak group and thus are more sought after by wildlife and were more often the acorn of choice of native Americans. The wood of White Oaks has blocked pores so liquid cannot pass through producing tight cooperage while the wood of Red Oaks are porous and thus termed loose cooperage. Oaks are very variable especially in leaf form and identification can be difficult, if not impossible, especially on young plants. And, hybridization is common between many species of oaks. Since the shape of the leaf is variable, the trichomes (hairs) on the leaves are very important in identification. To quote the North American Flora Volume 3 page 445; "Among the most important diagnostic characters within *Quercus* and particularly the White Oak Group are features of the foliar trichomes. Often these can be seen with a 10X or 15X lens. The large number of misidentified specimens in herbaria that can be easily identified properly with the use of trichome characters illustrates this point."

SAWTOOTH OAK (***Quercus acutissima***) is in the Red Oak Group and is usually a small tree but can reach 60 ft. The specific epithet "*acutissima*" means very sharp pointed for the leaf margins. The bark is ash brown with lighter vertical fissures on young trunks and a corky appearance to old trunks. The twigs are gray to gray brown and glabrous. The leaves are deciduous and the blades are 3.5-7" long, 1-2.25" wide, oblong to obovate, apex acuminate, base rounded, upper surface lustrous dark green and glabrous, lighter green and glabrous except for axillary tufts of hairs. The acorns are 0.75-1" long, dark brown and the cup covers about 2/3's of acorn. The cup scales are recurved. This species is reported to have acorns maturing in one year unlike other members of this group. The acorns are produced by very young trees making this popular for wildlife plantings. Widely planted and persisting and spreading throughout the state. The current reports include 16 parishes.

WHITE OAK (***Quercus alba)*** is also called Fork Leaf Oak, Ridge White Oak, and Stave Oak. The specific epithet "*alba*" refers to the white bark or the underside of the blades. It is in the White Oak Group and reaches a height of 150 ft. The bark is thin, light gray to white or reddish brown and slightly furrowed to scaly. The twigs are slender and gray. The leaves are deciduous and the blades are elliptic to oblong, 4-10" long, 5-9 lobed with lobes "finger" shaped, apex of lobes rounded, base cuneate, upper surface dark green, lower surface glaucous white, and turning red to brown in the fall. The acorns are ellipsoid-ovoid to oblong, light brown, 0.5-1.5" long, cup bowl shaped, warty and encloses about 1/4 of the acorn. The wood is strong, tough, close-grained, light brown, hard, heavy, nonporous, durable, and weighs about 48 lb per cu ft. It has been used for fuel, ties, baskets, cabinets, barrels, tools, furniture, construction work, agricultural implements, shipbuilding, wagons, crossties, fence posts, interior finish, and tight cooperage. Acorns provide food for turkey, squirrels, hogs, white-tailed deer, turkey, and quail. White Oak is usually considered the most important of the many species of oak and is considered to be the most important hardwood tree in America. There are reports of parents

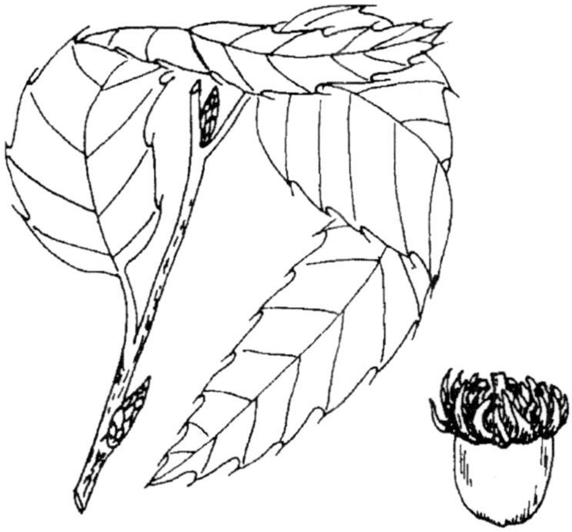

Quercus acutissima (Miller 1978)

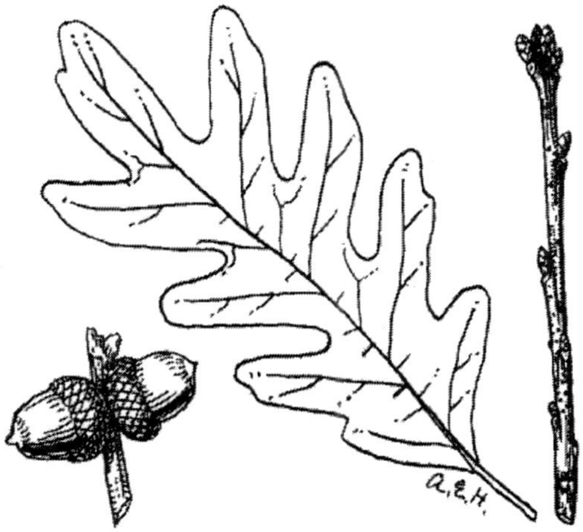

Quercus alba (USDA-FS)

planting a white oak when a child was born so the child would grow up strong like the White Oak. The early colonists and Indians ground the acorns into meal and poured water through it to remove the tannic acid before baking into bread. Also, Indians used the bark for medicinal teas; the bark is reported to be astringent and tonic and used to treat hemorrhoids. Early European settlers are also reported to have made coffee from the roasted acorns. White Oak is the state tree of Connecticut and Maryland. In mesic to well drained soils especially hardwood slope forests (FACU) in the pine regions of the state. The current reports include 39 parishes. It hybridizes with ***Quercus stellata* = *Quercus x fernowii*.**

ARKANSAS OAK (***Quercus arkansana***) is also known as Water Oak and Arkansas Water Oak. The specific epithet, "*arkansana*", refers to the state of Arkansas, where it occurs in most abundance. It is in the White Oak Group and reaches a height of 70 ft. The bark is dark to grayish brown or black, with deep fissures, long and narrow ridges, and thick and tight scales. The twigs are slender, brownish gray, pubescent when young but becoming glabrous with age. The leaves are deciduous and the blades are broadly obovate, 2-23" long and wide, apex rounded toacute and often mucronate, rarely shallowly 3-lobed, base broadly cuneate, upper surface yellowish green and glabrous, and lower surface paler and glabrous or pubescent in the axils. The acorns are ovoid, apex rounded, light brown, slightly striate, 0.25-0.3" long, 0.5-.75" wide, with the cup covering 1/4 to 1/3 of acorn. In mesic to well drained sites mostly in northern Louisiana with reports from 10 parishes. It hybridizes with ***Quercus marilandica***.

SCARLET OAK (***Quercus coccinea)*** is also known as Spanish Oak and Red Oak. The specific epithet, "*coccinea*", refers to the scarlet leaves in autumn. It is a member of the Red Oak Group that reaches a height of 100 ft. The bark is brownish, with rather fine fissures, and often flaky on the upper branches. The leaves are deciduous and the blades are obovate to oval, 3-7" long, 2-5" wide, upper surface bright green, lower surface paler often with axillary tufts of rusty hairs, base truncate to broadly cuneate, and margins deeply 5-9 lobed. The lobes extend halfway to midrib, are divergent or ascending, oblong to obovate and sometimes falcate, apex repand with bristle tips, and sinuses wide and rounded. The acorn is ovoid o ovate, apex rounded, brown, often with striations, 0.4-1" long, and cup covering 1/3 to 1/2 of acorn. The wood is light to reddish brown, coarse-grained, heavy, hard, strong, and weighs 47 lb per cu ft. The foliage is occasionally browsed by deer and the acorn is eaten by red and gray squirrel, blue jay, and sharp tailed grouse. It was introduced into cultivation in 1691. The only report is from sandy soil in Washington parish.

SOUTHERN RED OAK (***Quercus falcata***) is also known as Swamp Red Oak, Swamp Spanish Oak, Bottomland Red Oak, Three Lobe Red Oak, Finger Oak, Spanish Oak and Turkey Oak. The specific epithet, "*falcata*", refers to the scythe-shaped leaves. It is a tree that reaches up to 80 ft and is in the Red Oak Group. The bark is dark gray, rough, furrowed into broad plates and ridges on trunks and large branches. The twigs are thick, rusty hairy in spring, becoming shiny brown later, end in a cluster of small pointed reddish brown buds. The leaves are deciduous and the blades are elliptical, 4-5" long, base cuneate or vase shaped often asymmetrical, shiny green above, and gray or rusty soft hairy below. The margins are deeply 3-5 lobed, with each ending in a long narrow end lobe and several usually curved side lobes, each tip with 1-3 bristle tipped teeth, and most usually falcate. The sinuses are rounded to pointed and not at right angles to the midrib. The acorns are about 0.5" long, orange-brown, globular or hemispheric, often striated, pubescent and with the cup enclosing about 1/3 of acorn. The durable, red with thick, lighter colored sapwood, heavy (43 lb per cu ft), hard, strong, close-grained, wood is used for general purpose, rough lumber, furniture (chairs and tables), and

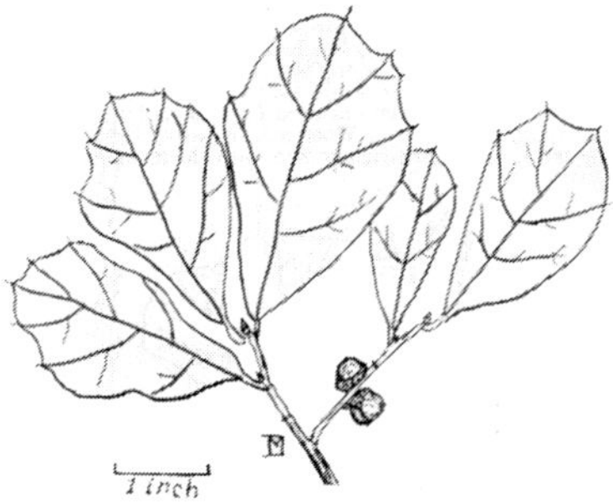

Quercus arkansana (Moore 1960)

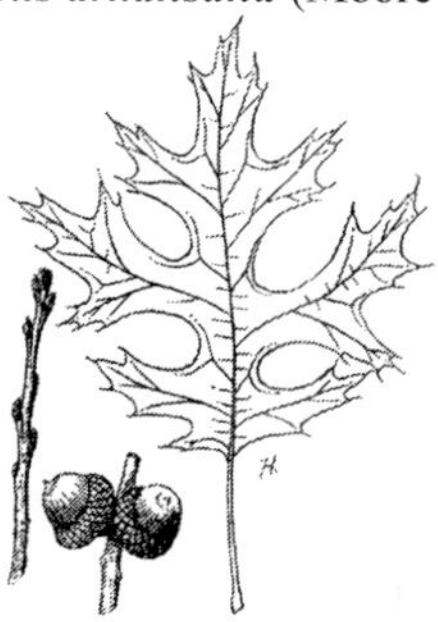

Quercus coccinea (USDA-FS)

Quercus falcata (USDA-FS)

interior finish. The bark is used for tanning and astringent. The acorns are eaten by wildlife. In well drained, often sandy habitats (FACU-), throughout the state. The current records include 45 parishes. It hybridizes with ***Quercus nigra* = *Quercus x garlandensis***, with ***Quercus shumardii* = *Quercus x. joorii***, with ***Quercus phellos* = *Quercus x subfalcata***, with ***Quercus incana* = *Quercus x subintegra***, and with ***Quercus velutina* = *Quercus x wildenowiana***.

SAND LIVE OAK (***Quercus geminata***) is also called Bay Live Oak. The specific epithet "*geminata*" means paired, probably in reference to the paired acorns. It is similar to Live Oak (***Quercus virginiana***) and is in the White Oak Group. It is always a small tree that is restricted to the coast where it is subjected to strong winds and often has an irregular, dwarfed, and contorted shape. The trunk is often leaning, with light to dark gray bark which is shallowly and irregularly fissured. The leaves are evergreen and the blades are often curved and boat-shaped in appearance, 0.5-3" long, 0.3-1" wide, oblong to elliptic or somewhat narrowly elliptic, apex acute, base narrowed gradually or rounded, upper surface dark green and glabrous, lower surface pubescent, and margins conspicuously rolled back and thickened. Reported for Louisiana and should be looked for along the coast, in particular the eastern portion. (syn. *Quercus virginiana* var. *geminata, Quercus virginiana* var. *maritima*)

UPLAND LAUREL OAK (***Quercus hemisphaerica***) is also called Darlington Oak. The specific epithet "*hemisphaerica*" refers to the shape of the fruit and comes from "semi" = "half" and "sphaerica" = "globe or sphere". It is in the Red Oak Group and reaches a height of 100 ft. The bark is dark brown to black with flat ridges and deep furrows. The twigs are glabrous and light to dark red brown. The leaves are tardily deciduous and are positioned on the twigs at an angle. The petioles are very short and the blades are narrowly ovate or elliptic or narrowly oblanceolate, 1-5" long, 0.5-1.5" wide, upper surface medium dark green, lower surface, lower surface slightly paler, base broadly cuneate, and apex bristle tipped and acute to acuminate or obtuse. The margins are entire or with teeth or lobes. The acorns are hemispheric, dark brown, shorter than 0.5", and the cup covers 1/4 to 1/3 of acorn. In dry sandy soils to somewhat mesic sites throughout the pine region mostly. The current records include 28 parishes.

SHINGLE OAK (***Quercus imbricaria)*** is also known as Lea Oak, Jack Oak, and Northern Laurel Oak. The specific epithet, "*imbricaria*", means "overlapping", as with shingles. Michaux gave the name to the plant when he found early settlers making handhewn shingles from the tree. It is a member of the Red Oak Group that reaches a height of 65 feet. The bark is thick, gray-brown to brown, and shallowly furrowed and ridged. The twigs are slender, lustrous brown, glabrous, and end in a cluster of small shiny-scaled buds. The leaves are deciduous and the blades are elliptical to oblong-obovate, 3-6" long, 0.8-2" wide, apex acute to rounded and usually apiculate, base cuneate to rounded, dark green and shiny above, and paler and sparsely white-hairy below. The margins are entire, rarely shallowly 3-lobed, undulate, and often revolute. The acorns are ovoid to subglobose, light to dark chestnut brown, .3-.8" long, often striated, and cup covering 1/3 to 1/2 of acorn. The wood is brownish red, sapwood lighter, coarse-grained, hard, heavy, strong, and weighs 47 lb per cu ft. It is used for construction and fuel. The acorns are eaten by birds and mammals. It has been in cultivation since 1724. Rare in mesic to dry habitats (FAC-) in north Louisiana with uncultivated records only from Bossier, Morehouse, and Ouachita parishes.

Quercus geminata (Vines 1960 as ***Quercus virginiana*** var. ***maritima***)

Quercus hemisphaerica (Godfrey 1988)

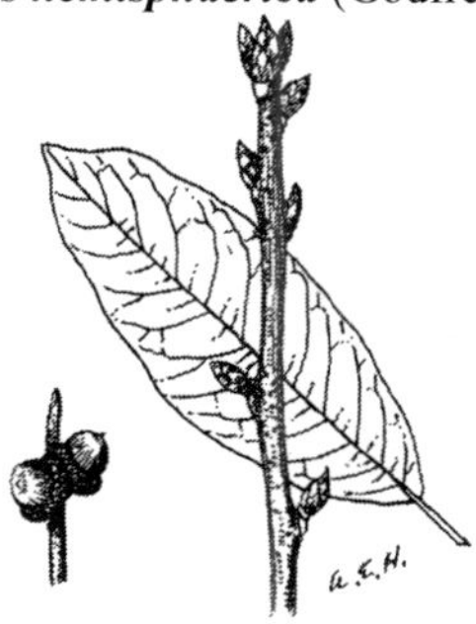

Quercus imbricaria (USDA-FS)

SANDJACK OAK (***Quercus incana***) is also known as Upland Willow Oak, High Ground Willow Oak, Bluejack Oak, Cinnamon Oak, Shin Oak, Sand Oak, and Turkey Oak. The specific epithet, "*incana*", refers to the grayish green lower surface of the leaves. It is a member of the Red Oak Group and is shrub or a small tree that reaches a height of 35 ft. The bark is dark gray to black and deeply furrowed into nearly square plates. The twigs are slender, dark brown, hairy when young, and end in a cluster of glabrous short-pointed brown buds. The leaves are deciduous and the blades are elliptic to lanceolate, 2-4" long, 0.5-1" wide, apex apiculate and acute to rounded, base cuneate to rounded, upper surfaces bluish to ashy green, and lower surface white to gray pubescent. The margins are entire or rarely very shallowly 3-lobed. The acorn is globose to ovoid, brown with grayish pubescence, often striated, about 0.5" long and with the cup covering 1/4 to 1/2 of the acorn. The wood is reddish brown, close-grained, hard, and strong. It is used for posts and fuel. Its acorns provide food for wildlife living in its bleak environment. It is in very sandy soils of the pine regions of the state usually in the same area as Blackjack Oak. The current records include 22 parishes. (syn. *Quercus cinera)* It hybridizes with ***Quercus laevis*** = ***Quercus x ashei*** , with ***Quercus nigra*** = ***Quercus x caduca***, and with ***Quercus marilandica*** = ***Quercus x cravensis***.

TURKEY OAK (***Quercus laevis***) is also known as Catesby Oak and Scrub Oak. The specific epithet, "*laevis*", refers to the smooth leaves. It is a member of the Red Oak Group that is a small tree that reaches a height of 30 feet. The bark is rather thick, nearly black, and deeply furrowed with rough, scaly ridges. The leaves are deciduous and the blades are generally shaped like a turkey foot, deltoid to obovoid, 2.25-12" long, 1-10" wide, base cuneate, upper surface shiny yellowish green, and lower surface glabrous with axillary tufts of reddish hairs. The margins are deeply 3-5, rarely 7, lobed, terminal lobe elongate and entire or dentate, and lateral lobes spreading, usually curved, and 3-lobed or toothed at the apex. The acorns are ovoid to oval, 0.75-1" long, apex rounded, light brown, pubescent apically, and cup covering about 1/3 of acorn. The wood is reddish brown, sapwood paler, close-grained, hard, heavy, and strong. The wood is sometimes used for fuel. The acorns are eaten by wildlife. In very sandy soils in southeastern Louisiana with reports only from St. Tammany and Washington parishes.

LAUREL OAK (***Quercus laurifolia)*** is also known as Water Oak, Willow Oak, and Live Oak. The specific epithet, "*laurifolia*", refers to the laurel-like foliage. It is a member of the Red Oak Group and is a tree that reaches up to 100 feet. The bark on young trees is nearly smooth and dark brown tinged with red while on mature trees it is dark gray to black with furrows and ridges. The twigs are reddish brown to gray, glabrous, and slender. The leaves are tardily deciduous to evergreen and the blades are elliptic to oblong to obovate, sometimes falcate, 3-6" long, 0.5-1" wide, base cuneate, glossy dark green above, and lighter and duller beneath. The margins are entire and the apex is acute and apiculate. The acorns are ovoid to hemispheric, about 0.5" long, dark brown, and the cup covers about 1/4 of the acorn. The wood is reddish brown, coarse-grained, hard, strong, and used probably for the same purpose as that of Water Oak. It is used mostly for fuel and charcoal. In wet sites (FACW) in pine flatwoods and along streams throughout the state. The current records include 57 parishes. (syn. *Quercus obtusa*) It hybridizes with ***Quercus velutina*** = ***Quecus x cocksii***, with ***Quercus nigra***, and with ***Quercus phellos***.

Quercus incana (USDA-FS)

Quercus laevis (Vines 1960)

Quercus laurifolia (USDA-FS)

OVERCUP OAK (***Quercus lyrata***) is also known as Water White Oak, Swamp White Oak, Swamp Post Oak, and White Oak. The specific epithet, "*lyrata*", refers to the lyrate-pinnatifid leaves. It is a member of the White Oak Group and may attain a height of 100 ft. The gray to white bark is rough, breaking into small flakes or scales. The twigs are slender, gray, becoming glabrous, and end in a cluster of small blunt light-brown buds. The leaves are deciduous and the blades are obovate or oblong, 5-8" long, 1.5-4" wide, shiny dark green above, usually gray-green below, long tapering at base. The margins are 5 to 9 lobed, with lobe apex acute to acuminate or rounded, middle lobes usually the broadest, and sinuses deep and broadly rounded. The acorns are ovoid, chestnut brown, 0.5-1" long, and the cup covers almost all of the acorn. The wood is heavy (52 lb per cu ft), hard, rich dark brown with thick, lighter colored sapwood, strong, tough, and durable. Commercially it is classed with white oak and used for the same purposes. Ducks, squirrels and deer eat the acorns. Deer browses young trees. It was first introduced into cultivation in 1786. In wet (OBL) bottomland habitats throughout the state. The current records include 47 parishes. It hybridizes with ***Quercus virginiana = Quercus x comptoniae.***

BUR OAK (***Quercus macrocarpa)*** is also known as Mossy Cup Oak, Mossy Over Cup Oak, and Prairie Oak. The specific epithet, "*macrocarpa*", refers to the large acorn. It is a member of the White Oak Group that reaches a height of 150 feet. The bark is thick, gray-brown, deeply furrowed, and ridged. The twigs are thick, sometimes winged, gray-hairy, and end in a cluster of eggshaped gray hairy buds. The leaves are deciduous and the blades are obovate, base cuneate, 4-12" long, 3-6" wide, upper surface dark green and glabrous, and lower surface gray-green and pubescent. The margins are 5-9 lobed with rounded apices and terminal lobe the largest and with smaller lobes or coarse teeth. The sinuses are very broad. The acorns are ellipsoid to ovoid, 0.75-2" long, apex pubescent, and cup covering 1/2 to 3/4 of acorn. The cup scales end in a bristle and creates a "mossy" look to the acorn. The wood is dark or light brown, close-grained, heavy, hard, strong, tough, durable, and weighs about 46 lb per cu ft. It is used for baskets, lumber, ties, fences, cabinets, ships, and fuel. The acorns are eaten by deer and squirrels. The young plants are browsed by livestock. It has been in cultivation since 1811. In mesic to well drained soils (FAC) and cultivated in some areas with current records from six parishes mostly in north Louisiana.

RUNNER OAK (***Quercus margarettiae)*** is also known as Sand Post Oak, Dwarf Post Oak, Post Oak, and Scrubby Post Oak. The specific epithet "*margarettiae*" was named by W. W. Ashe to honor Margaret Wilcox, who later became Mrs. Ashe. It is a member of the White Oak Group and is a small tree that reaches a height of 20 feet. The bark is light gray, fissured, and scaly. The twigs are pubescent with stellate hairs. The leaves are deciduous and the blades are broadly obovate, 1.5-4" long, base cuneate or rounded, leathery dark green and glabrous above, and stellate pubescent below. The margins are 5-lobed with deep rounded sinuses, apex rounded, and middle lobes almost square and opposite creating a cross. The acorns are ovoid, striated, dark shiny brown, and the cup covers 1/3 to 1/2 of acorn. In very sandy soil throughout the pine regions of the state. The current records include 26 parishes. (syn. *Quercus drummondii*, *Quercus stellata* var. *margarettiae*)

BLACKJACK OAK (***Quercus marilandica***) is also known as Iron Oak, Black Oak, Ridge Oak, Jack Oak, Barren Oak, and Scrub Oak. The specific epithet, "*marilandica*", refers to the state of Maryland. It is a member of the Red Oak Group that is a small tree to 60 ft. The bark is dark brown and broken into rough, blocklike plates. The twigs are grayish brown, stout, densely pubescent when young but becoming glabrous with age. The leaves are deciduous and

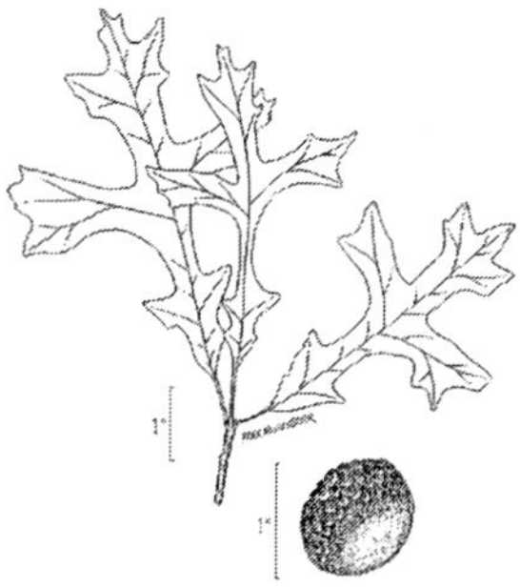

Quercus lyrata (USDA-NRCS)

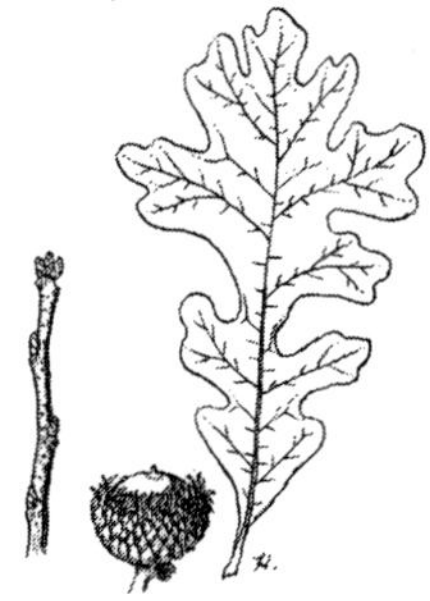

Quercus macrocarpa (USDA-FS)

Quercus margarettiae (Vines 1960)

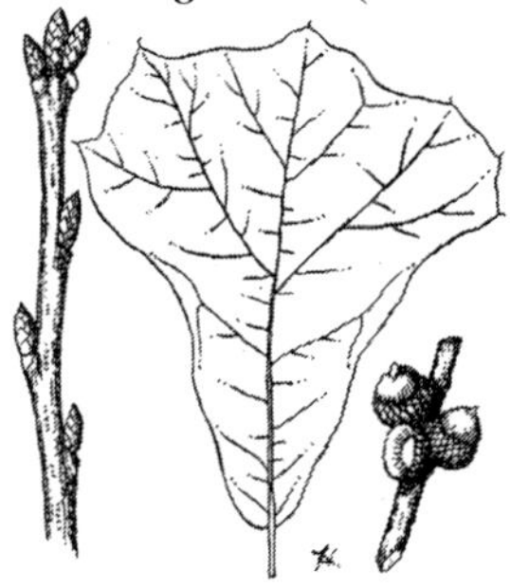

Quercus marilandica (USDA-FS)

the blades are broadly obovate to clavate, base rounded, cordate or cuneate, 3-7" long, 2-5" wide, upper surface dark glossy green, lower surface paler, usually pubescent and with conspicuous veins. The margins are 3-lobed at apex or entire, revolute, and bristle tipped. The acorns are ovoid-oblong to subglobose, 0.5-0.8" long, often striated, light brown, pubescent, and the cup covering about ½ of acorn. The wood is dark brown, sapwood lighter, heavy, hard, strong, and weighs 46 lb per cu ft. The wood is used mostly for posts, fuel, and charcoal, and the acorns are sought by wild turkey and white-tailed deer. In sandy soil to mesic or wet pine flatwoods throughout the pine regions of the state. The current records include 35 parishes. It hybridizes with ***Quercus velutina* = *Quercus x bushii*** and with ***Quercus phellos* = *Quercus x rudkinii***.

COW OAK (***Quercus michauxii***) is also known as Basket Oak, Swamp Chestnut Oak, Swamp White Oak, Michaux Oak, and Swamp Oak. The specific epithet "*michauxii*" honors Francois Andre Michaux, 1770-1855, a French botanist who originally described it as a new species. It is a member of the White Oak Group and is a large tree that reaches 100 feet. The bark is pale gray and broken into long, shaggy strips. The twigs are gray or brown, hairy when young, and end in a cluster of small pointed reddish brown finely hairy buds. The leaves are deciduous and the blades are obovate-oblong, 5-8" long, 3-4" wide, base cuneate to truncate often cordate, pale green and lustrous above, silvery white and pubescent below. The margins are crenate with 10-14 teeth on each side, and teeth apex rounded. The acorns are shiny brown, ovoid, 0.75-1.2" long, and cup covers about 1/3 of the acorn. The heavy (51 lb per cu ft), hard, very strong, tough, close-grained wood is used for posts, tools, baskets, splints, cooperage, wheels, boards, veneer, and fuel. The wood is light brown with thin, darker sapwood. The bark is rich in tannin and used for tanning leather. The acorns are eaten by cows, mourning doves, ducks, quails, squirrels, turkey, and deer in the fall of the year. As with other White Oaks, these acorns can be eaten by humans and were very popular with the Indians. The name "Cow Oak" was derived from the consumption of the acorns by cows. The other common name Basket Oak comes from the use of the wood to make baskets. In mesic habitats along streams, especially the natural levees, throughout the state. The current records include 50 parishes.

DWARF LIVE OAK (***Quercus minima***) is also called Minimal Oak. The specific epithet "*minima*" means small or little. It is a member of the White Oak Group that is a rhizomatous clone forming shrub to 8 ft. The bark is light gray or brown and smooth. The twigs are light gray and glabrous to pubescent. The leaves are evergreen and the blades are obovate to oblanceolate and often asymmetrical, 1.5-5" long, 0.5-2" wide, dark green above, paler green and pubescent below, apex acute to rounded, and base cuneate. The margins are revolute, and entire or with 1-5 teeth and teeth mucronate. The acorns are narrowly ovoid or subcylindric, dark brown, 0.5-1" long, and cup covers 1/2 to 3/4 of acorn. In very sandy soil along the Mississippi River in Tensas Parish and along the coast in St. Tammany Parish.

CHINQUAPIN OAK (***Quercus muehlenbergii***) is also known as Pin Oak, Shrub Oak, Scrub Oak, Yellow Oak, Chestnut Oak, Rock Oak, and Chinkapin Oak. The specific epithet, "*muehlenbergii*", honors G.H.E. Muhlenberg (1753-1815), botanist and minister in Pennsylvania. It is a member of the White Oak Group and is a medium tree to 80 ft tall. The bark is light grayish brown, broken into narrow, loose plates. The twigs are reddish brown to gray, slender, hairy to glabrous, and the ovoid terminal bud is orange to reddish brown. The leaves are deciduous and the blades are oblong to lanceolate or obovate, 4-6" long, 1-3.5" wide, apex acute to acuminate, base cuneate to rounded or cordate, upper surface lustrous dark green, and lower surface pale gray, pubescent, and with conspicuous veins. The margins are toothed and the teeth are coarse, acute, mucronate, and recurved. The acorns are broadly ovoid, dark

Quercus michauxii (USDA-FS)

Quercus minima (Nelson 1996)

Quercus muehlenbergii (USDA-FS)

shiny brown, 0.5-0.75" long, and about half covered with a thin, sometimes hairy cup. The wood is reddish brown, sapwood lighter, close-grained, durable, hard, heavy, strong, and weighing 53 lb per cu ft. It is used for posts, ties, cooperage, furniture, and farm implements. In well drained hardwood and Inland Prairie Forests in the pine regions of the state. The current records include 13 parishes. (syn. *Quercus prinoides* var. *acuminata)*

WATER OAK (***Quercus nigra***) is also known as Pin Oak, Bluejack Oak, Duck Oak, Spotted Oak, Barren Oak, Punk Oak, and Possum Oak. The specific epithet, "*nigra*", refers to its black bark. It is a member of the Red Oak group and is a tree that reaches up to 100 feet. The smooth tight bark is grayish black to light brown becoming covered with smooth, close scales, as the tree grows older. The twigs are slender, gray-brown, glabrous, and in a cluster of small pointed dark red finely hairy buds. The leaves are deciduous to tardily deciduous and the blades are very variable obovate to spatulate, 2-4" long, 1-2" wide, upper surface glabrous dark green, lower surface slightly lighter and pubescent along vein axils, and base acute to attenuate. The margins vary from entire to lobed with lobes shallow to deep and bristle tipped. The apex varies from rounded and truncate to acute. The acorns are globose-ovoid, light yellowish brown, often striated, 0.3-0.6" long, and the cup covers about 1/3 to 1/2 of acorn. The wood is heavy (45 lb per cu ft), light brown, with lighter colored sapwood, hard, and strong. It is used for piling, crossties, and fuel. It is one of the better wildlife food trees for deer, squirrel, turkey, and ducks. It is often attacked by mistletoe. In mesic to well drained soils (FAC) and planted in many locations. It probably occurs in all with current records from 61 parishes. It hybridizes with ***Quercus phellos = Q x capesii***, with ***Quercus rubra = Quercus x heterophylla***, with ***Quercus velutina = Quercus x demareei***, with ***Quercus shumardii = Quercus x neopalmeri***, and with ***Quercus texana***.

OGLETHORPE OAK (***Quercus oglethorpensis***) is a member of the White Oak Group that reaches a height of 50 ft. The specific epithet *"oglethorpensis"* is for Oglethorpe County, Georgia where the species was first discovered in 1940. The bark is light gray to whitish and scaly. The twigs are brownish red and sparsely pubescent. The leaves are deciduous and the blades are narrowly elliptic to oblanceolate, 2-6" long, 1-2" wide, apex acute, rounded, or obtuse, base cuneate to cordate, upper surface dark green, and lower surface yellowish green and covered with branched hairs. The margins are entire to sinuate. The acorns are ovoid, gray brown, pubescent, 0.3-0.5" long, and the cup covering about 1/3 of acorn. In Inland Prairie Forest (FAC+) only from Copenhagen in Caldwell Parish.

CHERRY BARK OAK (***Quercus pagoda)*** is also known as Swamp Red Oak, Spanish Oak, Southern Red Oak and Swamp Spanish Oak. The specific epithet "*pagoda*" is for the pagoda shaped leaf lobes. It is a member of the Red Oak Group that reaches a height of 120 feet. The bark is nearly black on older trunks but lighter on younger, and broken into thin plates similar to the bark of old Black Cherry trees. The twigs are yellowish brown and pubescent. The leaves are deciduous and the blades are ovate to broadly elliptic or obovate, 3.5-12" long, 3.5-6.5" wide, upper surface dark glossy green, lower surface paler and hairy, and base truncate to rounded, often asymmetrical. The margins are 7-11 lobed, lobes pagoda shaped and distinctly positioned at right angles to midrib, sometimes falcate, and apex acute or acuminate and bristle tipped. The acorns are subglobose, orange brown, often striate, 0.3-0.7" long, and the cup covers 1/4 to 1/3 of the acorn. The fruit is used by birds, squirrels, and deer. In mesic to well drained soils (FAC+), often along streams, usually at slighter lower and moister sites that ***Quercus falcata***. The current reports include 54 parishes. (syn. *Quercus falcata* var. *pagodifolia*, *Quercus*

Quercus nigra (USDA-FS)

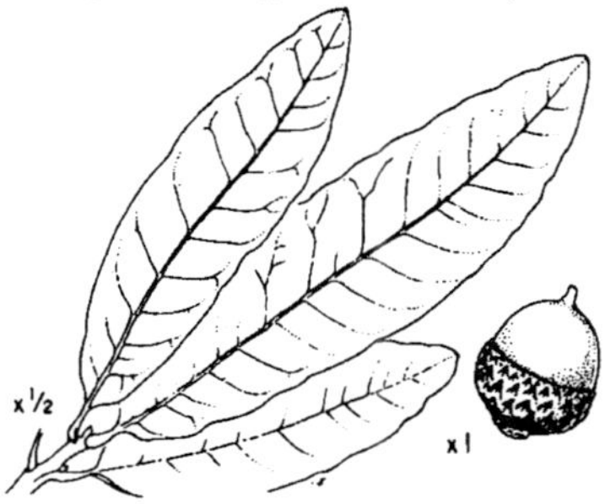

Quercus oglethorpensis (Radford, Ahles, and Bell 1968)

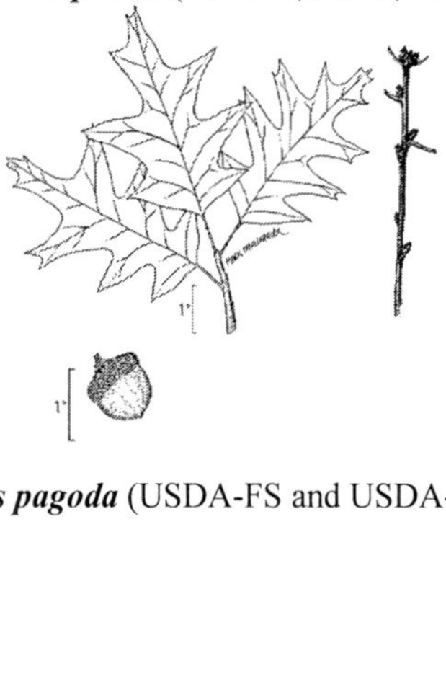

Quercus pagoda (USDA-FS and USDA-NRCS)

falcata var. *leucophylla* and *Quercus pagodifolia*) It hybridizes with ***Quercus phellos* = *Quercus x ludoviciana***.

WILLOW OAK (***Quercus phellos***) is also known as Pin Oak, Water Oak, Peach Oak, Sandjack Oak, Red Oak, Swamp Oak, and Swamp Willow Oak. The specific epithet, "*phellos*", means "cork" and is the ancient Greek name of Cork Oak (***Quercus suber***). Willow Oak is often called Pin Oak in many sections of the South, but that name should apply to the true Pin Oak, ***Quercus palustris***. Willow Oak is a member of the Red Oak Group that may reach a height of 130 ft. The bark on young trees is smooth and light gray; on older trees, it is brown, roughened and divided into narrow ridges. The twigs are reddish brown and pubescent when young but turning gray and glabrous later. The leaves are tardily deciduous and the blades are linear-lanceolate to elliptic, 2-4" long, 0.25-0.75" wide, upper surface light to dark green, lower surface paler and pubescent along midrib, base cuneate to narrowly rounded, and apex acute and bristle tipped. The margins are entire. The acorns are subglobose to ovoid, yellowish to dull brown, sometimes striated, 0.4-0.6" long, and the cup covering 1/4 to 1/3 of acorn. The strong, heavy (49 lb per cu ft), coarse-grained, light brown tinged with red, sapwood thin and lighter colored, hard wood is used for crossties and occasionally for construction. Acorns provide wildlife food and are edible to humans after boiling to remove bitter tannin. It has been in cultivation since 1723. In moist to somewhat wet sites (FACW-) throughout and often growing in monocultures creating the "Willow Oak Flat". The current records include 50 parishes. It hybridizes with ***Quercus shumardii* = *Quercus x molutonensis*** and with ***Quercus velutina* = *Quercus x filialis***.

DWARF CHINQUAPIN OAK (***Quercus prinoides***) is also known as Scrub Chestnut Oak, Dwarf Chestnut Oak, Running White Oak, and Chinquapin Oak. The specific epithet, "*prinoides*", refers to its resemblance, especially in the leaves, to ***Quercus prinus***. It is a member of the White Oak Group and is a small tree or shrub that reaches a height of 15 feet. The bark is gray and flaky to papery. The twigs are reddish brown and pubescent when young and turning gray and glabrous. The leaves are deciduous and the blades are obovate to oblanceolate or elliptic, 2-6" long, 2-3" wide, base cuneate, apex acute to acuminate, upper surface olive to bright green, lower surface paler, pubescent, and with distinct lateral veins. The margins are serrate with 3-7 pairs of teeth and the teeth are acute or obtuse. The acorns are ovoid-oblong, chestnut brown, 0.4-1" long, and the cup covers ½ or more of the acorn. It was first introduced into cultivation in 1730. The acorn is known to be eaten by a number of species of birds and mammals, including the gray squirrel and ruffed grouse, and the foliage is browsed by cottontail. In Inland Prairie Forests (UPL) only in Caldwell and Natchitoches parishes.

SWAMP CHESTNUT OAK (***Quercus prinus***) has been reported from Louisiana but the most recent comprehensive work (Flora of North America Volume 3, page 476) report that this name has been misapplied to ***Quercus montana***. And, ***Quercus montana*** has not been reported for Louisiana. (Not Illustrated)

NORTHERN RED OAK (***Quercus rubra***) is also known as Leopard Oak, Gray Oak, Mountain Red Oak, Champion Oak, Black Oak, and Spanish Oak. The specific epithet, "*rubra*", refers to its reddish wood. It is a member of the Red Oak Group that reaches 150 ft. The bark on young trunks is smooth and greenish brown while on older trees it is dark gray or black, furrowed into broad hard scaly ridges, and with the inner bark reddish to pink. The twigs are slender, dark reddish brown, and end in a cluster of pointed dark red glabrous buds. The leaves are deciduous and the blades are oblong to obovate or oval, 5-9" long, 4-6" wide, base broadly

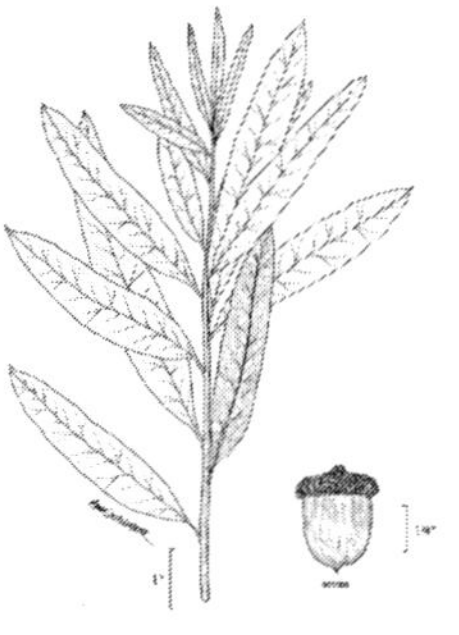

Quercus phellos (USDA-NRCS)

Quercus prinoides (Britton and Brown 1943)

Quercus rubra (USDA-FS)

cuneate to truncate or rounded, often asymmetrical, upper surface dull dark green, and lower surface paler and with occasional tufts of axillary hairs. The margins are 7-11 lobed with lobes extending less than halfway to midrib, apex acute to acuminate, bristled tipped, and 1-3 toothed. The sinuses are oblique and rounded at base. The acorns are ovoid, pale brown, lustrous, 1-1.25" long, and the cup covers about 1/4 of the acorn. The wood is heavy, strong, hard, close-grained, and weighs 41 lb per cu ft. It is used for furniture, interior finishing, crossties, posts, mine timbers, and general construction. The acorns are eaten by bobwhite quail and ruffed grouse, and the foliage browsed by white-tailed deer and cottontail. It was first introduced into cultivation about 1880. In well drained soils (FACU) of hardwood slope forests. The current reports include nine parishes.

SHUMARD RED OAK (***Quercus shumardii***) is also known as Texas Oak, Swamp Red Oak, Spotted Oak, Leopard Oak, and Spanish Oak. The specific epithet, "*shumardii*", honors Benjamin Franklin Shumard (1820-1869), state geologist of Texas. It is a member of the Red Oak Group that reaches up to 150 feet. The bark is gray to reddish brown and smooth or broken into small ridges. The twigs are grayish brown, glabrous, and smooth. The leaves are deciduous, obovate or elliptic-oblong, 3-7" long, 2-4.5" wide, base truncate, upper surface medium dark green, and lower surface slightly paler and with conspicuous tufts of hairs in the leaf axils. The margins are 7 to 9 lobed with lobes symmetrical, deeply cut, and apex lobes bristle tipped and sometimes also lobed. The sinuses are elliptic at base and closed; an imaginary line drawn from the ends of each sinus would meet. After the leaves fall to the ground, many black spots develop on the leaves hence the name Spotted or Leopard Oak. The acorns are ovoid to oblong-ovoid, light brown, sometimes with narrow stripes, 0.75-1" long and the cup covers about 1/4 of the acorn. If the cup covers more than 1/3 or more of the acorn, those plants are often separated as var. ***schneckii***. The hard, strong, heavy, close-grained wood weighs 57 lb per cu ft and is used for veneer, cabinets, furniture, flooring, interior trim, and lumber. Birds, squirrels, and deer use the fruit. In mesic to well drained sites (FACW-) throughout the state. The current reports include 38 parishes.

DELTA POST OAK (***Quercus similis***) is also called Mississippi Valley Post Oak and Yellow Oak. The specific epithet "*similis*" means similar or alike and could be in reference to the similarity of this species to ***Quercus stellata***. It is a member of the White Oak Group that reaches 65 ft. The bark is gray to reddish brown and divided into irregular fissures with platelike scales. The twigs are brown, stout, and pubescent but becoming glabrous with age. The leaves are deciduous, broadly obovate, 3-5" long, 2.5-4" wide, base cuneate to rounded, pale to dark leathery green above, paler and pubescent with stellate hairs below. The margins are 5-lobed with deep rounded sinuses, apex rounded, and middle lobes almost square and opposite creating a cross. The acorns are ovoid to ovoid-oblong, striated, dark shiny brown, and the cup covers 1/3 to 1/2 of acorn. In bottomland hardwood forests (FACW) especially in northeast Louisiana. The current records include 39 parishes. (syn. *Quercus stellata* var. *mississippiensis*). It hybridizes with ***Quercus margarettiae*** and with ***Quercus stellata***.

BASTARD OAK (***Quercus sinuata***) is also called Bastard White Oak, Durand Oak, Shortlobe Oak, Bluff Oak, Pin Oak, White Oak, and Basket Oak. The specific epithet "*sinuata*" refers to the wavy margins of the leaves. It is a member of the White Oak Group that reaches a height of 50 ft. The bark is gray to light brown, flaky to papery and exfoliating. The twigs are pale gray to brown and stellate pubescent when young but becoming glabrous with age. The leaves are deciduous and the blades are variable, obovate or elliptic, 2-7" long, 0.5-3.5" wide, base cuneate, apex rounded or lobed, upper surface dark lustrous green, and lower surface white-

Quercus shumardii (USDA-FS and USDA-NRCS)

Quercus similis (Vines 1960 as ***Quercus stellata*** var. ***mississippiensis***)

Quercus sinuata (Vines 1960 as ***Quercus durandii***)

silvery stellate tomentose. The margins are entire to irregularly lobed. The acorns are ovoid to ellipsoid, chestnut brown, 0.5-0.75" long, and cup covers 1/4 or less of the acorn. In Inland Prairie Forest at Copenhagen in Caldwell Parish. (syn. *Quercus durandii*)

POST OAK (***Quercus stellata***) is also known as Iron Oak, Cross Oak, Branch Oak, Rough Oak, and Box Brash Oak. The specific epithet, "*stellata*", refers to the stellate hairs of the leaves and petioles. It is a member of the White Oak Group that reaches a height of 70 ft. The bark is gray to reddish brown and divided into irregular fissures with platelike scales. The twigs are brown, stout, and pubescent but becoming glabrous with age. The leaves are deciduous, oblong-obovate, 4-7" long, 3-4" wide, base cuneate to rounded, pale to dark leathery green above, paler and pubescent with stellate hairs below. The margins are 5-lobed with deep rounded sinuses, apex rounded, and middle lobes almost square and opposite creating a cross. The acorns are ovoid to ovoid-oblong, striated, dark shiny brown, and the cup covers 1/3 to 1/2 of acorn. The durable, light or dark brown, heavy, hard, close-grained, and rather durable wood weighs 52 lb per cu ft and is used for cross-ties, fuel, fenceposts, furniture, lumber, and cooperage. A coffee like drink is made from the acorns. The Oklahoma Indians used the leaves as cigarette wrappers. Deer and turkey eat the acorns. In well drained to mesic sites (FACU) throughout the state. The current records include 44 parishes. Note: this is very similar to ***Quercus margarettiae*** and to ***Quercus similis*** with both often considered as varieties of this species. It hybridizes with ***Quercus margarettiae***.

NUTTALL OAK (***Quercus texana)*** is also known as Red Oak, Texas Red Oak, Rock Oak, Hill Oak, Spotted Oak, Smooth Bark Red Oak, Tight Bark Red Oak, Yellow Butt Oak, Spanish Oak, Red River Oak, Pin Oak, and Striped Oak. The specific epithet, "*texana*", refers to the state of Texas, its native habitat. The common name Nuttall Oak is attributed to the specific epithet "*nuttallii*" that was applied to this species for many years. It is a member of the Red Oak Group that reaches 120 ft. The bark on young trees is greenish brown, smooth, and shiny but on older trees is light to dark gray or brownish to black, slightly fissured, and often with warts or burls. The young twigs are olive green to reddish brown and older twigs are brown to gray. The branches are easily broken. The leaves are deciduous and the blades oval to very broadly elliptic to oblong, 4-8" long, 2-5" wide, upper surface dull dark green, lower surface paler and often with tufts of pale hairs in the axils of the veins, and base cuneate to truncate. The margins are 5 to 7 lobed with lobes at nearly right angles to the mid-vein, and with the middle or upper pair the longest. Each lobe slightly tapers to the end and terminates in bristle-tipped teeth. The sinuses are rounded at base and open at the tip; an imaginary line drawn from the ends of each sinus would not meet. The acorns are oblong-obovoid, light to dark reddish brown, distinctly striated, 0.75-1.5" long, and the cup covers 1/3 to 1/2 of acorn. The wood is reddish brown, sapwood yellowish brown, hard, and heavy. The wood is used locally for fuel and posts and the acorns are used by wildlife. In wet to moist habitats (OBL) along streams and lakes throughout the state. The current records include 38 parishes. (syn. *Quercus nuttallii*)

BLACK OAK (***Quercus velutina)*** is also known as Yellow Oak, Dyer's Oak, Tan Bark, Smooth Bark, Yellow Bark Oak, and Quercitron Oak. The specific epithet "*velutina*", is for the velvety buds and/or the lower leaf surface. It is a member of the Red Oak Group and may attain a height of 100 feet. On young branches and very young trunks, the bark is smooth and light; on older trunks, it becomes almost black, extremely rough and very deeply fissured. The bright orange-yellow inner bark is quite distinctive. The leaves are deciduous and the blades are oval-obovate, 3-10" long, 3-8" wide, base cuneate or truncate, upper surface glossy green, and lower surface yellow green, glabrous to pubescent, and with conspicuous, rusty hairs in the forks of the

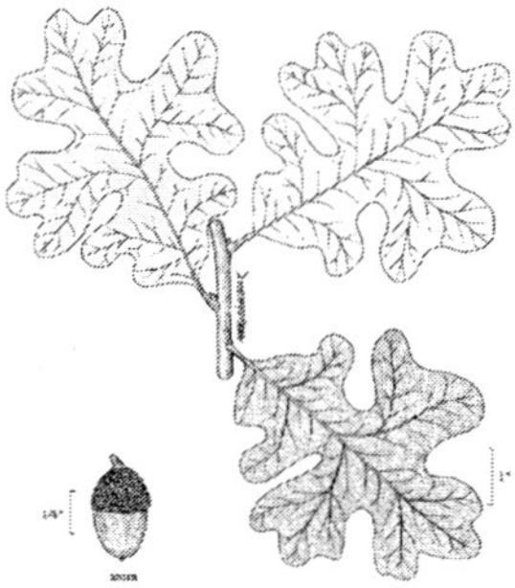

Quercus stellata (USDA-NRCS)

Quercus texana (USDA-FS)

Quercus velutina (USDA-FS)

veins. The margins are 7-lobed with lobes oblique, apex acute to acuminate and bristle tipped, and middle lobes usually the longest. The acorn is hemispherical, 0.5-1" long, brown, often striate, pubescent, and enclosed for at least half its length in a thin, scaly brown cup. The wood is hard, strong, heavy (43 lb per cu ft), bright red brown with a thin outer edge of paler sapwood, and coarse-grained. It is used for cross-ties and sometimes sawed into lumber; but it is used chiefly for fuel. The bark is used in tanning as a yellow dye and in medicine. The fruit is used by birds, squirrels, and deer. It has been in cultivation since 1802. In well drained soils of hardwood slope forests throughout the pine regions of the state. The current reports include 29 parishes.

LIVE OAK (***Quercus virginiana***) is a member of the White Oak Group with wide spreading, often drooping branches. The specific epithet, "*virginiana*", refers to the state of Virginia. The bark is dark brown to black with narrow interlacing furrows, and small, closely appressed scales. The twigs are grayish brown, slender, rigid, and glabrous. The leaves are evergreen and the blades are oblong to elliptic or obovate, 2-5" long, 0.5-2.5" wide, apex rounded to acute, base cuneate, rounded or cordate, coriaceous, upper surface dark shiny green, and lower surface paler and glabrous to pubescent. The margins are entire to dentate apically, and usually revolute. The acorns are ellipsoid obovoid, shiny brownish black, 0.3-0.5" long, and the cup covers about ½ of acorn. The wood is light brown, close-grained, tough, hard, strong, heavy, and weighs about 59 lb per cu ft. The wood is used for hubs, cogs, shipbuilding, or for any other purpose where a hard and strong wood is required. The bark was formerly much used in production of tannin, and acorn oil was used in cooking by the Indians. In well drained (FACU+) soils with the original distribution reported to be in southern Louisiana from the coast north to about U.S. Hwy 190. It is now widely planted and probably occurs in all with current reports from 58 parishes.

LANCE LEAF BUCKTHORN - *Rhamnus lanceolata* - Rhamnaceae

The genus, *Rhamnus*, is the Greek name of a shrub. The specific epithet, "*lanceolata*", refers to the lance-shaped leaves. Another common name is Ozark Buckthorn. It is a shrub to 9 ft. The twigs are slender, reddish brown and pubescent when young and becoming gray and glabrous with age. The leaves are alternate, simple, deciduous, and with pinnate major veins. The blades are elliptic to lanceolate or oblong, 1-3.5" long, 0.5-1" wide, apex acute to acuminate, base cuneate to rounded, upper surface glabrous and dark green, lower surface pubescent and paler, and margins finely serrulate with incurved glandular teeth. The flowers are greenish, axillary, and solitary or in groups of 2-4. The fruit is an obovoid to globose black two-seeded drupe that is 0.3-0.5" long. The drupes are eaten by at least 5 species of birds. Rare in Louisiana with the only non-cultivated report from Caldwell Parish.

AZALEA - *Rhododendron* - Ericaceae

The genus "Rhododendron" is Greek and means "rose tree". It includes shrubs to small trees. The leaves are alternate but tend to be clustered apically, simple, deciduous, and with pinnate major veins. The flowers are large and the fruit is a capsule. At times these plants produce large, gall-like, green, translucent structures on the twigs. This structure is edible and was used by early settlers for pickling. The structure is probably caused by bacterial action on the twig.

Quercus virginiana (USDA-FS)

Rhamnus lanceolata (Britton and Brown 1943)

SWEET AZALEA (***Rhododendron canescens***) is also known as Hoary Azalea, Pinxter Bloom, Honeysuckle Azalea, Wild Honeysuckle, Pinxter Flower, Early Azalea, Pinxter Azalea, Piedmont Azalea, Bush Honeysuckle, and Wild Azalea. The specific epithet, "*canescens*", refers to the hairs on the branches and leaves. This plant reaches a height of 12 ft. The bark is smooth and brown to mottled gray. The young twigs are brown and densely pubescent with stalked glands and the older twigs are glabrous. The blades are oblong to oblanceolate or elliptic, 1.5-4" long, 0.5-1.25" wide, apex acute and glandular apiculate, base cuneate, upper surface dark green and with a few glandular hairs, lower surface white due to the many white hairs and with scattered glandular hairs, and margins serrulate with glandular teeth. The flowers are produced before the leaves and are glabrous, mostly pink with occasional white petals. The capsule is ellipsoid-cylindric, narrowed above, pubescent, and about 0.75" long. Along the edges of small streams (FACW) in acid pH soils in the pine regions of the state. The current records include 35 parishes.

WHITE AZALEA (***Rhododendron oblongifolium***) is also known as Texas Azalea, Bush Azalea, Thicket Azalea, and White Bush Honeysuckle. The specific epithet, "*oblongifolium*", refers to the leaf shape. It is a shrub that reaches 5 ft tall. The bark is smooth and gray to reddish brown. The twigs are gray to red brown and glabrous. The blades are oblong, obovate, or elliptic, 1-2" long, 0.5-1" wide, base cuneate, apex acute, upper surface dull green, lower surface paler and pubescent along veins, and margins entire, slightly revolute and ciliate with white hairs. The flowers are formed after the leaves are mature and are white and very viscid on the outside due to the many glandular hairs. The capsule is elongate, 0.4-0.75" long, covered with glandular hairs, and persists on the stem. Along the edges of bogs and baygalls (FACW) in acid pH soils. The current reports include seven parishes all west of the Mississippi River.

LATE AZALEA (***Rhododendron viscosum***) is also known as Swamp Azalea, Clammy Azalea, and Hammock Sweet Azalea. The specific epithet, "*viscosum*", means "sticky" in reference to the glandular flowers. It is a shrub or small tree that reaches 20 feet. The bark is smooth and brown to mottled gray. The young twigs are reddish brown and pubescent with strigose hairs and the older twigs are glabrous and brown to gray. The blades are obovate to elliptic or oblanceolate, 1.5-3.5" long, 0.5-1.5" wide, apex acute to obtuse and mucronate, base cuneate to attenuate, both surfaces lustrous green and glabrous except for scattered hairs on midrib, and margins serrulate-ciliate. The flowers are produced with mature leaves and are white to pink tinted and are very viscid on the outside due to the glands. The capsule is elongate, slenderly ovoid to ellipsoid, 0.4-0.7" long, and pubescent. Along the edges of streams in acid pH soils (FACW+) in the pine regions of the state. The current reports include 20 parishes. (syn. *Rhododendron coryi*, *Rhododendron serrulatum*)

SUMAC - *Rhus* - Anacardiaceae

The genus name, "*Rhus*", is the ancient Latin or Greek name for sumac. The genus includes shrubs to small trees. The leaves are alternate, deciduous, and compound. All members of this genus do not cause the allergic reaction as do the members of the genus *Toxicodendron* but the allergic causing species are often included in the genus *Rhus*. The flowers are small and yellowish greenish and in terminal inflorescences. The fruit is a hairy drupe.

AROMATIC SUMAC (***Rhus aromatica***) is also called Fragrant Sumac, Sweet Scented Sumac, Three Leaf Sumac, and Skunk Bush. The specific epithet, "*aromatica*", refers to the

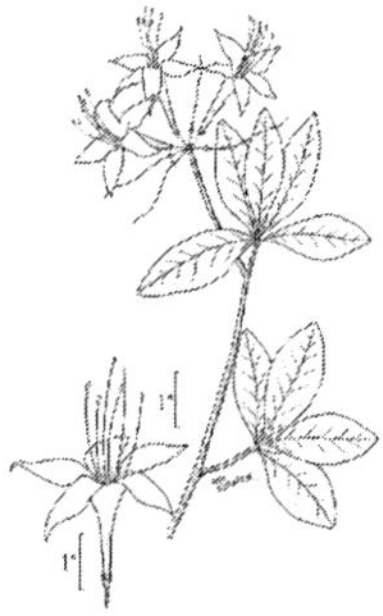

Rhododendron canescens (USDA-NRCS)

Rhododendron oblongifolium (Vines 1960)

Rhododendron viscosum (USDA-NRCS)

Rhus aromatica (Britton and Brown 1943)

aromatic leaves. It is a thicket-forming shrub that reaches up to 9 ft. The twigs are slender, brown, and pubescent when young but glabrous later. The leaves are trifoliolate and 1-4" long. The leaflets are oval, ovate, or obovate, glabrous to pubescent, and margins crenate-dentate, sometimes incised. The lateral two leaflets are sessile, base rounded or truncate and asymmetrical, 0.75-1.75" long, and 0.6-1.3" wide. The terminal leaflet has a short petiolule, base cuneate, apex acute to acuminate, 1.5-3" long, and 0.75-2.5" wide. The flowers are in a spike-like cluster. The drupe is red to orange, conspicuously hairy, and about 0.3" diameter. The fruits are eaten by several birds including wild turkey and flicker, and mammals including raccoon, opossum, and the twigs browsed by whitetail deer. It is reported to be a strong diuretic. In sandy soils west of the Mississippi River. The current records include 20 parishes.

WINGED SUMAC ***(Rhus copallinum)*** is also known as Shining Sumac, Wing Rib Sumac, Flame Sumac, Mountain Sumac, Smooth Sumac, Flame Leaf Sumac, Black Sumac, Dwarf Sumac, and Upland Sumac. The specific epithet, "*copallinum*", means "copal gum" and refers to the resemblance of the dried sap to that of the Copal Tree. It is a shrub or small tree that reaches up to 35 ft. The almost smooth, papery bark has horizontal splashes of light and dark gray, many small excrescence and cracks at irregular intervals exposing the reddish-brown beneath. The twigs are thick, reddish-brown, finely hairy, with sticky yellowish sap. The leaves are one odd pinnately compound, with 7-17 leaflets, 5-12" long, and with a broadly winged rachis. The leaflets are lanceolate, 1-3" long, base asymmetrical and rounded, apex acute or acuminate, margins entire or with a few teeth, upper surface shiny-glabrous, lower surface paler, and turning dark red to purple in late Summer and Fall. The flowers are in a thick panicle about 5" long. The drupes are about 0.1" diameter, dark red when ripe, and turning black and remaining through the winter. The panicle of fruits droops at maturity. The reddish brown, light, coarse-grained, soft wood weighs 32 lb per cu ft and is used for posts. The bark and leaves is rich in tannin and used for curing leather and manufacturing dyes. The crushed fruits are used to make lemonade and are mixed with Elderberry for jelly. The Creek Indians used the wood for dysentery. The fruit is used by 20 species of birds and whitetail deer and rabbit browse the young twigs occasionally. In mesic to well drained soils throughout the state. The current reports include 42 parishes.

SMOOTH SUMAC ***(Rhus glabra)*** is also called Scarlet Sumac, Red Sumac, White Sumac, Shoe Make, Vinegar Tree, Senhalanac, Pennsylvania Sumac, Upland Sumac, and Sleek Sumac. The specific epithet "*glabra*" refers to the shiny hairless leaves. It is a shrub to small tree that reaches 20 feet. The bark is brown and smooth or becoming scaly. The twigs have a whitish bloom and are thick and gray or brown. The leaves are 12-20" long, one odd pinnately compound with 11-31 leaflets and with a red non winged rachis. The leaflets are elliptic, lanceolate, or oblong, apex acuminate, base subcordate to cuneate and asymmetrical, upper surface bright green, lower surfaces white to gray, and margins sharply serrate. The flowers are in a thick panicle 5-9" long. The drupes are about 0.1" diameter, bright red when ripe, and turning a dull red and remaining through the winter. The panicle of fruits is erect at maturity. The wood is orange with whitish sap. The leaves are reported to have been mixed with tobacco and smoked. Records show that the date of earliest cultivation was 1620. In well drained soils throughout the state but absent from floodplains and the marsh. The current records include 38 parishes.

GOOSE BERRY - *Ribes curvatum* - Grossulariaceae

Goose Berry is also known as Granite Goose Berry, Drooping Goose Berry, Georgia Goose Berry, Southern Currant, and Slender Goose Berry. The genus name, "*Ribes*", is from an old

Rhus copallinum (USDA-FS)

Rhus glabra (USDA-FS)

Ribes curvatum (Vines 1960)

Danish name "ribs" for the red currant. The specific epithet, "*curvatum*", refers to the drooping twigs. It is an armed shrub to 9 ft. The bark is gray to red-brown or purplish. The twigs are red-brown to gray, glabrous but slightly pubescent when young, recurved or drooping, and bearing slender red stipular spines. The leaves are alternate or fascicled, simple, deciduous, and with palmate major veins. The blades are suborbicular, 0.3-1" long and wide, 3-5 lobed with lobes toothed and terminal lobe often mucronate, base subcordate to truncate to broadly cuneate, upper surface lustrous light green and lower surface pale green. The white flowers are in racemes. The fruit is a green globose berry that is 0.25-0.4" diameter. The fruit is sometimes used for preserves and is relished by many birds and small mammals. The plant was introduced into cultivation about 1898. In sandy soil in extreme northwestern Louisiana with current uncultivated records only from Caddo and DeSoto parishes. (syn. *Grossularia campestris, Grossularia curvata, Grossularia texensis*)

CASTOR BEAN - *Ricinus communis* - Euphorbiaceae

Castor Bean is also known as Castor Oil Plant, Castor Plant, Lepo, and Palma Christi. The genus name, "*Ricinus*", is the classical name of the tick, in reference to the tick shape of the seed. The specific epithet, "*communis*", means common. It is a monoecious shrub to 15 ft in temperate climates. The stems are hollow, gray to brown or red, ridged and with a distinct ring at the nodes. The leaves are alternate, simple, evergreen, and with palmate major veins. The petiole is peltately attached to the blade. The blades are nearly orbicular, 0.3-1.5 ft wide and long, palmately 6-11 lobed, glabrous and glaucous on both surfaces, lobe apex acute to acuminate, and lobe margins serrate. The male flowers are yellow and the female flowers are pink to pinkish red. The fruit is a red, green, or bluish, spiny, burr-like capsule. The fruits split at maturity to expose white to tan, attractive, but very poisonous, seeds which can be fatal if ingested. In the past, the seeds were used to control moles by placing the seeds into the mole tunnel. The oil of the seeds is sold in foreign countries to keep away moles and malaria. The seeds yield the castor oil of commerce and regarded by some as the best of all purgatives and especially favored for children and the aged. It is also used externally for itch and ringworm. The oil is known in French as *huile de ricin,* in German as *Castarol*, in Italian as *olio di ricino,* and in Spanish as *aceite de castor.* Planted as an ornamental and persisting and also at dump sites (FACU) throughout the state. The current records include 15 parishes.

LOCUST - *Robinia* - Fabaceae

The genus name, "*Robinia*", honors Jean Robin and his son, Vespasian Robin, herbalists to Henry IV of France, who first cultivated the Locust-tree in Europe. The genus includes armed shrubs or trees. The leaves are alternate, deciduous, and one odd pinnately compound. The flowers are in axillary drooping racemes. The fruit is a legume.

BRISTLY LOCUST (***Robinia hispida)*** is also known as Rose Acacia Locust. The specific epithet, "*hispida*", refers to the dense bristly hairs on the foliage and inflorescence. It is a shrub that reaches 18 feet. The young twigs are brown and densely reddish pubescent while the older twigs are gray to brown, striated, and glabrous to pubescent. The hairs are rather stiff and could be considered "armed". Stipular spines are observed in a few plants. The leaves are 5-9" long with 7-15 leaflets. The leaflets are suborbicular, ovate, or oblong, apex mucronate, base and apex rounded or broadly cuneate, upper surface dark green, and lower surface paler. The rachis is often reddish pubescent. The flowers are reddish purple to pink. The legume is densely hairy, cylindrical, and 2-3" long. Planted as an ornamental and persisting and perhaps spreading with current reports from 20 parishes.

Ricinus communis (USDA-FS)

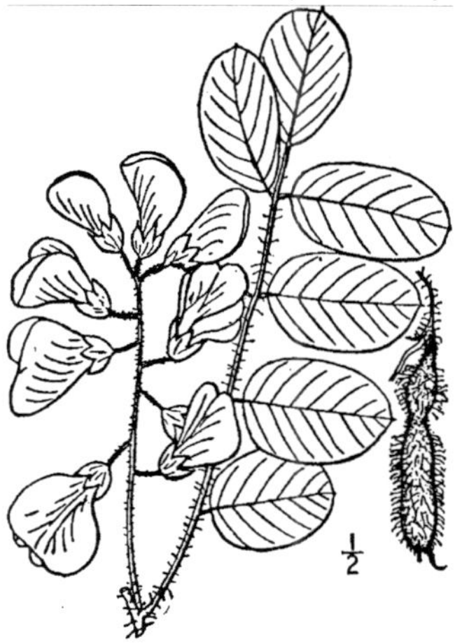

Robinia hispida (Britton and Brown 1943)

BLACK LOCUST (***Robinia pseudoacacia)*** is also known as White Locust, Yellow Locust, Red Locust, Red Flowering Locust, Green Locust, Silver Locust, Honey Locust, False Acacia, Bastard Acacia, Common Locust, Post Locust, Pea Flower Locust, Silver Chain, and False Acacia Locust. The specific epithet, "*pseudoacacia*" refers to its resemblance to an acacia. It is a shrub to small tree to 50 ft tall. The bark is gray to reddish brown, thick, rough, ridged, deeply furrowed, and sometimes twisted. The inner bark is pale yellow. The twigs are zigzag, thick, greenish brown, glabrous, and with paired stipular spines at the nodes. The leaves are 8-14" long with 7-19 leaflets. The leaflets are elliptic, 0.5-2" long, 0.5-1" wide, upper surface medium green and glabrous, lower surface paler and pubescent along veins, and apex and base rounded to obtuse. The flowers are white. The legume is flat, glabrous, narrowly oblong, straight or slightly curved, 2-5" long, and about 0.5" wide. The wood is greenish yellow or light brown, durable, hard, heavy, strong, stiff, and close-grained. The sapwood is white and narrow. The wood weighs about 45 lb per cu ft. It is shock resistant, shows little shrinkage, and is very resistant to decay and is used for fenceposts. Perhaps not native to Louisiana but now widely distributed due to plantings and spreading from seeds. The habitat is variable from mesic (UPL) to somewhat moist sites. The current reports include 48 parishes.

ROSE - *Rosa* - Rosaceae

The genus name, "*Rosa*", is the ancient Latin name. It includes armed sprawling shrubs or climbing vines. The twigs have stipular spines and prickles. The leaves are alternate and odd one pinnately compound. The stipules are usually conspicuous and adnate to the petioles. The flowers are large, distinctly radially symmetrical and with many stamens. The fruits are achenes that are enclosed in the floral cup or hypanthium that is termed the "hip". There are two native species (***Rosa carolina*** and ***Rosa setigera***) with many other cultivated species and cultivars. We include the three most widespread cultivated and escaping species but many other species and cultivars are planted and could persist or escape. The hips are edible by wildlife and humans; are reported to be an excellent source of vitamin C (60 times the concentration of lemons). The flowers are an excellent nectar source for bees.

MACARTNEY ROSE (***Rosa bracteata)*** is also known as Prairie Rose, Evergreen Rose, Cherokee Rose, Hedge Rose, Wild Rose, and Rosa Blanda. The specific epithet, "*bracteata*", refers to the bracts of the flower. It is a climbing to trailing semierect shrub to 20 ft. The stems are reddish brown to gray, densely pubescent, with many red glandular tipped hairs, and the stipular spines and prickles are stout, red, flattened, and often recurved. The stipules are pectinate fringed. The leaves are evergreen with 5-11 leaflets. The leaflets are obovate or elliptic, 0.5-1" long, 0.25-0.5" wide, apex rounded to retuse or truncate and mucronate, base cuneate or rounded, upper surface lustrous dark green, lower surface pale and glabrous to pubescent, and margins crenate-serrulate. The flowers are solitary or clumped with slightly notched white petals and large, reflexed, pubescent, toothed sepals. The hip is orange-red, subglobose to pyriform, 0.5-1" long and densely pubescent. It was first introduced into cultivation in 1893. A native of China that is widely planted and persists and escapes throughout the state. The current reports include 46 parishes.

CAROLINA ROSE ***(Rosa carolina)*** is also known as Pasture Rose. The specific epithet, "*carolina*", refers to North or South Carolina. It is a rhizomatous semi-upright shrub to 3 ft tall. The twigs are green-brown and glabrous. The stipular spines and prickles are short and terete. The adnate portion of stipules is linear and the free portion is lanceolate with glandular margins. The leaves are deciduous and the 3-9 leaflets are elliptic, oblong, or ovate-lanceolate, 0.6-1.3" long, 0.3-0.8" wide, apex acuminate, base acute, dull green above and glabrous, paler and

Robinia pseudoacacia (USDA-FS)

Rosa bracteata (Britton and Brown 1943)

Rosa carolina (USDA-NRCS)

pubescent on the veins beneath, and the margins are serrate. The flowers are few to solitary with pink slightly notched petals and glandular reflexed sepals. The hip is subglobose, red, about 0.3" diameter, and glandular hispid. Native in well drained soils (FACU) throughout the pine regions of the state. The current reports include 24 parishes.

CHEROKEE ROSE ***(Rosa laevigata)*** is a high climbing vine to 30 ft. The specific epithet, "*laevigata*", refers to the smooth leaflets. The stems are green, glabrous, slender, and the stipular spines and prickles are stout, flattened, recurved, and reddish brown. The stipules are short pectinate along fused portion but entire along the larger free portion. The leaves are evergreen with three, rarely five, leaflets. The leaflets are ovate-lanceolate or elliptic-ovate, 1-3" long, 0.3-1" wide, upper surface lustrous dark green, lower surface paler and reticulate, margins distinctly serrate. The flowers are solitary on a long pubescent pedicel, white, rarely pink, slightly notched at apex, and erect to reflexed sepals. The hip is pyriform or obovoid, dark red, very bristly with long hairs, and 1-1.5" long. This introduced species has been designated the state flower of Georgia, a surprising designation for an exotic. A native of China that is widely planted and persists and escapes throughout the state. The current reports include 23 parishes.

MULTIFLORA ROSE (***Rosa multiflora***) is also known as Japanese Rose. The specific epithet "multiflora" refers to the profuse flowering clusters. It is a climbing or sprawling shrub to 15 ft. The stems are red to green, glabrous, and the stipular spines and prickles are stout and recurved. The stipules margins are distinctly pectinate. The leaves are deciduous with 5-11 leaflets. The leaflets are ovate or elliptic or oblong, 1-1.6" long, 0.5-1" wide, base cuneate, apex acute or rounded, dark green and glabrous above, paler and often slightly pubescent beneath, and margins serrate to dentate. The flowers are numerous in a terminal panicle and white, pink, or red. The sepals are pubescent, ovate, acute, and deciduous. The hip is red, globular to ovoid, glabrous, and about 0.3-0.5" long. Native to Asia and widely cultivated (UPL) and persisting and escaping with reports from 28 parishes.

PRAIRIE ROSE (***Rosa setigera***) is also known as Cherokee Rose and Climbing Rose. The specific epithet, "*setigera*", means "bristle-bearing". It is a high climbing vine to 15 ft. The stems are green to red, glabrous, and with stout, broad based straight to slightly curved stipular spines and prickles. The stipules are adnate for more ½ of length and short pectinate along fused portion. The leaves are deciduous with three leaflets, or 5-7 on vigorous stems. The leaflets are lanceolate to oblong, 1.5-4" long, base acute, apex acute to acuminate or rounded, upper surface lustrous dark green, lower surface paler and often glandular along veins, and margins sharply serrate. The terminal leaflet has a long petiolule, 0.5" or longer, while the lateral leaflets are sessile or nearly so. The flowers are several in a corymb with pink, rarely white, petals, and lanceolate glandular pubescent sepals. The hip is globose to ellipsoid, red, often glandular pubescent, and about 0.5" diameter. In well drained soils (FACU) with non cultivated reports from Bossier, Caddo, Morehouse, and Red River parishes in north Louisiana.

BLACK BERRY, DEW BERRY - *Rubus* - Rosaceae

The genus name, "*Rubus*", is the old Roman name for Black Berry or perhaps for the red color of the fruit or stems. The genus includes armed and trailing to erect and arching shrubs. The aboveground stems live for two years; the first year stems (primocanes) are sterile and erect and the two year old stems (floricanes) produce flowers, are erect to sprawling or trailing. Both kinds of stems have recurved prickles scattered along the internodes. The stipules are green, linear, and with an acuminate apex. The leaves are alternate, deciduous, and palmately compound with five or three leaflets. The leaflets are elliptic to narrow ovate or slightly obovate,

Rosa laevigata (Vines 1960)

Rosa multiflora (Radford, Ahles, and Bell 1968)

Rosa setigera (Britton and Brown 1943)

base attenuate to acute or truncate, upper surface dark green, lower surface paler, 2-4" long, 1-2" wide, and margins serrate. Prickles are often found on the petiole, rachis, and petiolule. The flowers are white, radially symmetrical, and with many stamens. The fruits are an aggregate of black oblong drupelets. The fruits are eaten fresh or made into jams, jellies, pies, cobblers, and other desserts. Many wildlife, especially birds, eat the fruits. The young shoots can be eaten or dried to make tea. All parts of the plant have been used medicinally to treat chronic diarrhea, dysentery, cholera infantum, ulcers, sore throat, swelling of limbs, dropsy, summer complaint, and cholera morbus. The roots and rhizomes are used as an astringent, tonic and diuretic. Most species in this genus occupy open disturbed areas and cannot tolerate too much shade. Note: many species of this genus have been described and quite a number are reported for Louisiana. We are opting to describe and illustrate the two most common species.

BLACK BERRY (***Rubus argutus***) is also known as Louisiana Black Berry, Sharp Toothed Black Berry, and Sawtooth Black Berry. The specific epithet "*argutus*" refers to the sharp toothed leaves. The primocanes are light green and glabrous. The floricanes are dark green to brown, erect or arching to 7 ft and glabrous. The leaves are glabrous. Black Berries flower later than Dew Berries usually from April to May. The fruits are oblong, longer than wide. Very common in many habitats (FACU+) throughout the state. It probably occurs in all with current reports from 52 parishes. (syn. *Rubus abundiflorus*, *Rubus betulifolius*, *Rubus floridus*, *Rubus louisianus*, *Rubus rhodophyllus)*

DEW BERRY (***Rubus trivialis***) is also known as Southern Dew Berry and Mississippi Dew Berry. The specific epithet, "*trivialis*", means common or ordinary. The primocanes are red to green and pubescent. The floricanes are red, brown, or dark green, glandular pubescent and trailing and often rooting at the nodes. The petioles and petiolules are glandular pubescent. Dewberries flower earlier than Black Berries usually in February or March. The fruits are globose, about as wide as long. Very common throughout the state from mesic to dry sites (FAC). It occurs in all with current reports from 59 parishes. (syn. *Rubus mississippianus*, *Rubus rubrisetus*)

PALMETTO - *Sabal* - Arecaceae

The genus name, "*Sabal*", is of uncertain origin but possibly an American common name. The genus includes shrubs or trees. The leaves are evergreen, with a long unarmed petiole, and a fan shaped blade with the lower portion fused and upper portion dissected into narrow linear strips. The 3-parted white numerous flowers are in a panicle on a scape that is up to 8 ft tall. The fruits are globose black drupes. The leaves are used for fans, roofs, and religious ceremonies. The fruits were eaten by Indians and by wildlife including robins, fish crows, and raccoons. Cattle occasionally browse the leaves. The flowers are a source of honey.

PALMETTO (***Sabal minor***) is also known as Dwarf Palm, Bush Palmetto, Blue Palm, Blue Stem, Dwarf Palmetto, and Swamp Palm. The specific epithet, "*minor*", refers to this palm's dwarf or trunk less stature. The stem is underground with only leaves above ground. The blades are fan shaped, 2-3 ft long, 3.5-4.5 ft wide, divided into about 30 segments, with few to no fibers along the edge of the segments, and the petiole is 2-6 ft long. The drupes are 0.25-0.5" diameter. In wet (FACW) to mesic hardwood forests, especially common in the floodplains of large streams. It probably occurs in all with current reports from 55 parishes. (syn. *Sabal louisiana*)

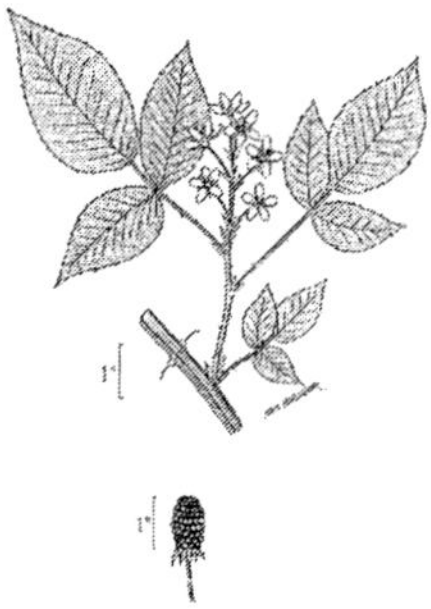

Rubus argutus (USDA-NRCS)

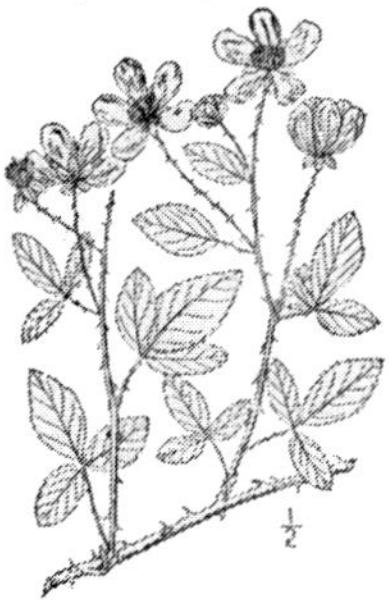

Rubus trivialis (Britton and Brown 1943)

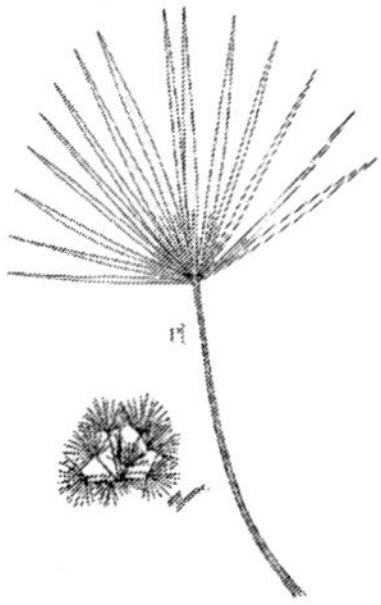

Sabal minor (USDA-NRCS)

CABBAGE PALMETTO (***Sabal palmetto***) is also known as Sabal Palm. The specific epithet "*palmetto*" is probably traced to an American common name for palms. It is a shrub or tree that reaches a height of 90 feet but nowhere near that tall in Louisiana. The blades are "V" shaped, 3-6 ft long and wide, divided into 40 to 90 segments, with numerous fibers along the edge of the segments, and the petiole is 3-7 ft long. The drupes are 0.4-0.6" diameter. It is the state tree of Florida and South Carolina. Early Florida Indians and later settlers were known to harvest the large leaf buds from the top of this plant. It is said the have the look and taste of cabbage. These growing tips were relished as a food source. However, removing the bud kills the tree. The identification of this species in the state has been a problem but there are four parishes (Calcasieu, Orleans, Richland, amd St. Charles) with herbarium records. Its wetland designation is FAC indicating mesic habitats.

WOODY GLASS WORT - *Salicornia virginica* - Chenopodiaceae

Woody Glass Wort is also known as Sapphire, Lead Grass, and Perennial Salt Wort. The genus "*Salicornia*" is from "sali" for salt and "cornu" for horn, in reference to the horn like branches. The specific epithet "*virginica*" is for the state of Virginia. It is a low mat forming shrub. The twigs are green, gray, or brown and thick and fleshy. The leaves are evergreen, small, opposite, scale like, connate and fused to stem. The flowers are in terminal spikes with leaves similar to those along the stem. The fruit is a one seeded utricle. In sandy soil (OBL) along the coast. The current records include six coastal parishes. (syn. *Salicornia europea*).

WILLOW - *Salix* - Salicaceae

The genus "*Salix*" is the old Latin name for the willows. The genus includes dioecious trees or shrubs. The leaves are alternate, simple, deciduous, and with pinnate major veins. The younger leaves often have small stipules that do not persist. The flowers are naked and in catkins. The fruit is a capsule with many cottony tufted seeds.

COASTAL PLAIN WILLOW (***Salix caroliniana***) is also known as Carolina Willow and Ward Willow. The specific epithet "*caroliniana*" is for North or South Carolina. It is a shrub or small tree to 20 ft. The bark is reddish brown to gray, with broad ridges and deep fissures, and distinctly checkered. The twigs are slender, yellow to reddish brown or gray, and pubescent when young but becoming glabrous with age. The blades are lanceolate, sometimes falcate, 2-7" long, 0.5-3.5" wide, upper surface bright green and glabrous, lower surface glaucous and white pubescent but becoming glabrous with age, apex acute to acuminate, base acute to rounded, and margins serrulate. The catkins are produced before the leaves and are 3-4" long. The capsules are ovoid-conic, about 0.25" long, and apex abruptly pointed. Along the edges of streams and other bodies of water (OBL) mostly in southern Louisiana with reports from Calcasieu, Franklin, Iberia, St. Mary, and St. Tammany parishes.

PRAIRIE WILLOW (***Salix humilis***) is also called Small Pussy Willow, Upland Willow, and Gray Willow. The specific epithet "*humilis*" refers to the low stature of the plant. It is a shrub to 9 ft. The twigs are slender, yellow to brown, and pubescent when young but becoming glabrous with age. The blades are oblanceolate to obovate, 0.4-1.2" long, 0.4-1" wide, base attenuate to cuneate, apex obtuse to acute, upper surface grayish green and glabrous, lower surface gray or white pubescent, and margins entire and revolute. The catkins are produced before the leaves and are 0.5-1" long. The capsules are pubescent, lanceolate, and about 0.25" long. In pine forests (FACU) sporadically throughout the state. The variety ***humilis*** is reported from Caddo, Calcasieu, Morehouse, Ouachita, Tangipahoa, West Carroll, and Winn parishes with the variety ***tristis*** reported only from St. Helena Parish. The var. ***tristis*** is smaller, mostly

Sabal p lmetto (Godfrey 1988)

Salicornia virginica (Vines 1960)

Salix caroliniana (USDA-NRCS)

Salix humilis (USDA-NRCS)

shorter than 3 ft and is often merged with var. ***humilis***.

NARROW LEAF WILLOW (***Salix interior***) is also called Sandbar Willow, Coyote Willow, Dusky Willow, Slender Willow, Basket Willow, and Gray Sandbar Willow. The specific epithet "*interior*" mean "inner" and could be traced to its distribution in the interior of the United States. It is a shrub or small tree to 15 ft. The bark is grayish brown, longitudinally fissured, and older trunks becoming furrowed. The twigs are slender, glabrous, and reddish brown to gray. The blades are linear to linear lanceolate, often falcate, base attenuate, apex acuminate, 1.5-4" long, mostly narrower than 0.25", upper surface grayish green and glabrous, lower surface grayish green and pubescent to glabrous, and margins with remote large subdentate glandular teeth (fewer than ten per inch) or sometimes entire. The catkins are produced after the leaves are mature and are 1-2.5" long. The capsules are ovoid-lanceolate, with a long tip, and about 0.25" long. Along the edges of streams (OBL), especially the Mississippi River, throughout the state. (syn. *Salix exigua*)

BLACK WILLOW ***(Salix nigra)*** is also known as River Willow, Catkins Willow, Scythe Leaved Willow, Swamp Willow, and Pussy Willow. The specific epithet "*nigra*" means black and refers to the bark. It is a fast growing tree often branched at base that reaches 125 feet. The bark is light brown to black, rough, and deeply fissured. On old trunks, the bark becomes quite shaggy. The twigs are brittle, reddish brown, and are easily rooted. The blades are narrowly lanceolate, falcate, 4-6" long, 0.5-0.75" wider, apex long acuminate, base acute to rounded, upper surface medium green, lower surface slighter paler than upper, and margins serrate with small glandular teeth that are closely spaced (more than ten teeth per inch). The catkins are produced before the leaves and are 1-3" long. The capsules are conic-ovoid, apex pointed, and about 0.2" long. The wood is soft, light (27 lb per cu ft), weak and not durable but does not warp, check, or splinter. It is used for artificial limbs, charcoal, toys, doors, fuel, cheap furniture, boxwood, excelsior, paper pulp and baskets. The bark is rich in tannin and used for pain relief, expectorant, hemostatic, astringent, and a tonic to break up fever. The bark contains salicylic acid, which is a chemical closely related to aspirin. In olden times, the inner bark was used in making a light purple dye and as a substitute for quinine. During Revolutionary Days, charcoal from Black Willow was used for gunpowder. The twigs are commonly browsed by deer. The tree is valuable for pollen and considerable honey has been reported from willows. A very common tree, especially in wet areas (OBL), but also invading dry open area. In all with current reports from 63 parishes.

WEEPING WILLOW (***Salix x sepulcralis***) is also known as Garb Willow, Napoleon Willow, and Babylon Weeping Willow. The specific epithet "sepulcralis" = cemetery; and was named because it was abundant in Hungarian-Austrian cemeteries. It is reported to be a hybrid between ***Salix alba*** and ***Salix babylonica***. It is a tree reaching a height of 50 ft with conspicuously drooping branches. The bark is dark gray and deeply furrowed. The twigs are slender, glabrous, elongate, and green when young but turning yellow or brown with age. The blades are narrowly lanceolate, often falcate, 3-7" long, 0.25-0.5" wide, apex long acuminate, base acute to attenuate, upper surface medium green, lower surface glaucous and midrib yellow, and margins serrate with translucent teeth. The catkins are produced before the leaves and are 0.5-1.5" long. The capsules are ovoid conic, glabrous, and about 0.2" long. This species is native to China and widely planted throughout the state and persists and perhaps escapes into moist habitats (FACW). The current records include 24 parishes. (syn. *Salix babylonica*)

Salix interior (USDA-NRCS)

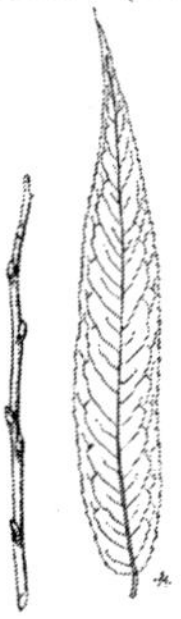

Salix nigra (USDA-NRCS)

Salix X ***sepulcralis*** (USDA-FS)

ELDER BERRY - ***Sambucus canadensis*** - Caprifoliaceae

Elder Berry is also known as American Elder, Common Elder, Sweet Elder, Pie Elder, and Elderblow. The genus name, "*Sambucus*", is the classical Latin name, and the specific epithet, "*canadensis*", refers to Canada, where the plant grows at its most northern limit. It is a large, deciduous shrub to small tree that reaches up to 30 feet. The bark is gray to red when young and as it gets older it becomes reddish yellowish or gray usually with warty projections. The large stem of this plant is dotted with small cork-like spots and the pith is large, soft and white. The leaves are opposite, deciduous, 4-11" long, and one odd pinnate compound with 5 to 11 leaflets. The leaflets are elliptic to oval or lanceolate, margins sharply serrate, glabrous to hirtellous beneath, 1.6-5" long, 0.8-2.5" wide and stipular at base. The upper surface is lustrous and bright green. The small delicate white fragrant flowers are in dense flat-topped or umbrella-shaped clusters and are produced from May to July. The fruit is a deep purple or black drupe. The leaves are used to make a hot poultice to relieve pain and to promote the healing of sprains and bruises. It is also used as an insecticide. The flowers are used to flavor drinks, fried, flavor wine, facial cream, mixed with mint, stimulant, carminative, diuretic, diaphoretic, eye lotions, and tea for dyspeptic. The fruits are used to make pies, wine, jelly, fritters, cold drink, and cough syrup. It is a good source of vitamin C. The stems are used as a tap for maple, whistle, flute, and as the barrel for popguns. The popgun was very popular in years past and was made by cutting an eight to ten inch section of a straight elderberry stem and then removing the soft pith. A plunger was whittled to fit into the elderberry barrel usually from an oak branch. The ammo to go into the popgun was the fruits of China Berry. The dried leaves of elderberry have been used as an insecticide, and the bark in preparing a black dye. The leaves, stem, roots, unripe berries, and bark are poisonous. They cause nausea, vomiting, and diarrhea. Numerous birds and animals eat the fruits and seeds. White-tail deer and cattle browse it. Butterflies are attracted to the flowers. To make Elderberry Face Cleaner, take a cup of buttermilk and add four tablespoons of Elderberry flowers to it. Bring to a light boil for about twenty minutes. Remove from heat and let soak for about two hours. Strain and use a cloth to clean the face with the liquid by light rubbing. The cleaner is gentle to the skin and very effective. It is widely distributed in Louisiana, found most frequently on moist sites (FACW) along margins of woods, roads, and fields and has been reported from all 64 parishes. *(syn. Sambucus simpsonii, Sambucus nigra* ssp. *canadensis)*

WESTERN SOAP BERRY - ***Sapindus saponaria*** var. ***drummondii*** - Sapindaceae

Western Soap Berry is also known as Soap Berry, China Berry, Wild China Berry, Wild China Tree, Indian Soap Plant, Mexican Soap Berry, and Drummond Soap Berry. The genus "*Sapindus*" is from a Latin phrase meaning "soap of the Indes". The specific epithet "*saponaria*" refers to a soaplike substance called saponin found in the fruits and the "*drummondii*" honors Thomas Drummond, who collected plants in Texas for Joseph Hooker. It is a large shrub to small tree to 40 ft. The bark is yellowish-gray and smooth on younger trunks and becomes gray and divided into scaly plates on older trunks. The twigs are thin, yellow-green, and finely hairy. The leaves are alternate, deciduous, and one odd pinnate compound. The leaves are 5-18" long with 11-19 leaflets. The leaflets are lanceolate, falcate, 1.5-4" long, 0.5-0.75" wide, apex acuminate, base asymmetrical and cuneate to obtuse, yellowish green, upper surface glabrous, lower surface with distinct veins and glabrous to pubescent, and margins entire. The yellowish white flowers are in a large terminal panicle. The fleshy drupe-like fruit is globose, translucent, white to yellow or black, and 0.5-1" diameter. The wood is light brown or yellowish, close-grained, hard, strong, and weighs 51 lb per cu ft. The wood is used to make baskets or frames or as fuel. The fruit contains large quantities of saponin, once used as a fish poison and to make soap. The fruit is reported to be poisonous but also used medicinally to treat

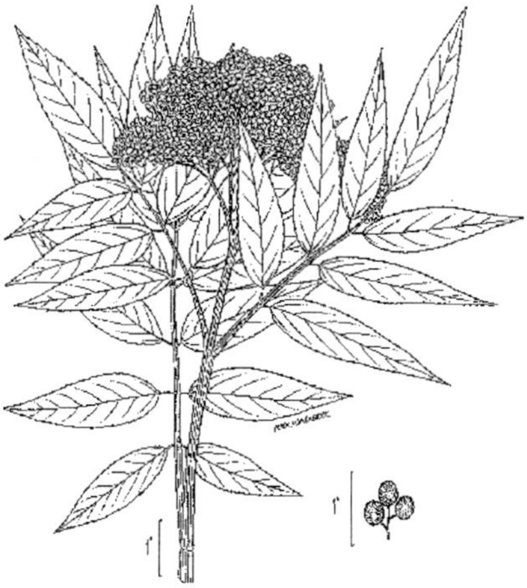

Sambucus canadensis (USDA-NRCS)

Sapindus saponaria var. ***drummondii*** (USDA-FS)

renal disorders, rheumatism, and fevers. Buttons and necklaces are made from the seeds. In well drained soils (FACU-) mostly in western Louisiana with some cultivated plants. The current records include 24 parishes.

SASSAFRAS - *Sassafras albidum* - Lauraceae

Sassafras is also known as Ague Tree, Saxifrax Tree, White Sassafras, Sassafac, Cinnamon Wood, Smelling Stick, Saloop, and Gumbo File. The genus name, "*Sassafras*", is derived from the word salsafras from the early French settlers, with reference to sassafras's medicinal properties. The specific epithet, "*albidum*", is derived from "alba" which refers to the light-colored condition of the wood or perhaps the underside of the leaf. It is a small to medium dioecious tree that reaches up to 90 feet. The bark is reddish brown to gray with fissures and broad flat ridges. The inner red bark is aromatic. The leaves and young twigs are bright green and quite mucilaginous. The leaves are alternate, simple, deciduous, and with pinnate major veins. The blades are 4-6" long, 1.5-4" wide, ovate or elliptic, margins entire or one-or-two mitten shaped lobed, apex of blade or lobe acute to acuminate, base cuneate, upper surface bright green, and lower surface glaucous and paler green. The crushed blade has an aromatic odor. The leaves become bright yellow, red and orange in the fall. The flowers are yellowish green, clustered, and appear from March to April before the leaves. The fruit is a blue-black drupe with the pedicel and thick calyx-lobes bright red. The aromatic, durable, dull orange brown with thin, light yellow sapwood, soft, weak, brittle wood weighs 32 lb per cu ft. It is used for posts, poles, rails, buckets, cabinets and interior finish, ox-yokes, cooperage, light boats, barrels, dugout canoes, and fuel. The dried leaves are used for file which is popular in Louisiana for the thickening of gumbo; the dried leaves have an odor of dried orange peels and the best time to gather the leaves for file is reported to be August. The leaves are also reported to produce a soft yellow-tan dye. The bark is used to disinfect root canals, flavoring in root beer, sarsaparilla, chewing gum, candies, toothpaste, mouthwashes, flavoring tobacco, soaps, perfumes, and the source of artificial heliotrope. It is used for home remedies as sudorific for colds, spring tonic "to thin the blood", and for high blood pressure of old people. Godfrey's cordial is a mixture of opium and sassafras. It is used as a diaphoretic, stimulant, diuretic, carminative, and treatment for bronchitis. One of the first exports from Jamestown by Captain John Smith was Sassafras. It is also considered by many authorities to have provided the first of the American medicinal plant drugs to reach Europe in 1564. It has been in cultivation since 1630. Twenty-eight species of birds, wild turkey, raccoons, and squirrels eat the leaves and the fruit and twigs are eaten by white-tail deer, marsh rabbit, and black bear. Warning: Sassafras roots have recently been reported to be carcinogenic. In well drained soils (FACU), especially in open areas and in particular along fence rows. It is planted in some areas with current records from 49 parishes.

GULL FEED - *Scaevola plumieri* - Goodeniaceae

Gull Feed is also called Beach Berry, Ink Berry, and Scaevola. The genus "*Scaevola*" means toward the left and honors a Roman hero, who was left-handed, Caius Mucius Scaevola. The specific epithet "*plumieri*" is probably in reference to the plume like flowers. It is a shrub to 5 ft. The branches are drooping and the twigs are green to brown, striated, and hollow. The leaves are opposite, evergreen, simple, and with pinnate major veins. The blades are fleshy, oblanceolate, 1-3" long, 0.5-1.5" wide, apex acute to rounded, base attenuate, upper surface dark green, lower surface pale to yellow green, and margins entire. The flowers are axillary, pink to white, fan shaped with 5-6 lobes all one side, and produced year round. The fruit is a shiny black ovoid drupe that is 0.3-1" long. Along the coast in sandy soils (FAC) of islands. The only parishes with records are Jefferson and Plaquemines.

Sassafras albidum (USDA-FS)

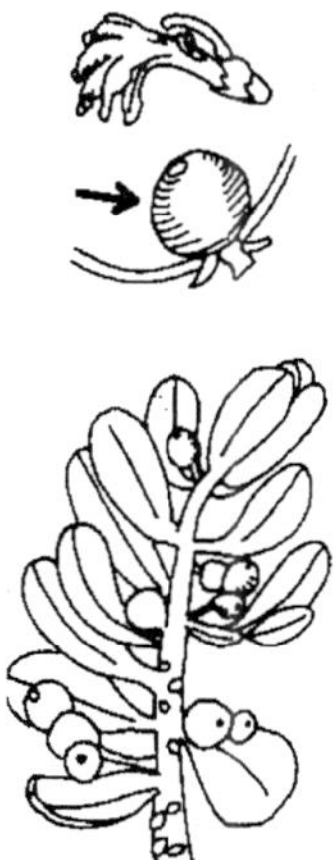

Scaveola plumieri (Hall 1993)

BAY STAR VINE - *Schisandra glabra* - Schisandraceae

Bay Star Vine is also called Carolina Magnolia Vine, Climbing Magnolia, Wild Sarsaparilla, and Schisandra. The genus "*Schisandra*" is from the Greek schizein (to cleave) and aner-ados (man stamen) referring to the cleft anther cells. The specific epithet "*glabra*" is probably in reference to the glabrous leaves. It is a monoecious sprawling shrub or vine that climbs by twining. The twigs are gray to brown, glabrous, and striated. The leaves are alternate but often clustered, simple, deciduous, and with pinnate major veins. The blades are ovate to oval or elliptic, 2-6" long, 1-3" wide, apex acute to abruptly acuminate, base cuneate to truncate, upper surface bright green, lower surface paler, texture thin, and margins entire or sparsely toothed. The crushed blade has a sweet odor. The crimson flowers are in axillary drooping clusters. The fruit is a red globose berry that is about 0.25" diameter. In well drained hardwood slope forests sporadically throughout the state but never common. The current reports include 10 parishes. (syn. *Schisandra coccinea*)

CANDLE BERRY - *Sebastiania fruitcosa* - Euphorbiaceae

Candle Berry is also known as Sebastian Bush, Gulf Sebastian Bush, and False Gum. The genus name, "*Sebastiania*", honors Antonio Sebastiani, a writer on Roman plants. The specific epithet "*fruticosa*" means shrubby. It is a monoecious shrub that reaches 12 feet. The young twigs are slender, green to brown, and glabrous and the older twigs are gray to brown and glabrous. The leaves are alternate, simple, tardily deciduous, and with pinnate major veins. The blades are elliptic to oblong or oval, 0.5-3" long, 0.5-1.5" wide, apex acute or short acuminate, base cuneate, upper surface dull green, lower surface paler, and margins entire. The flowers are in yellow green terminal spikes with the male terminal and the female at the base. The fruits are light brown, three lobed globose capsules with three grayish brown seeds. Along the edges of streams, especially on the natural levee, throughout the pine regions of the state. The current reports include 19 parishes. (syn. *Sebastiania ligustrina)*

SAW PALMETTO - *Serenoa repens* - Arecaceae

The genus name, "*Serenoa*", honors Sereno Watson (1826-1892), distinguished American botanist. The specific epithet, "*repens*", refers to the creeping habit of this species. It has horizontal creeping stems with leaves to 7 ft. The leaves are evergreen, with a long armed petiole, and a fan shaped blade with the lower portion fused and upper portion dissected into narrow linear strips. The terminal strips are further divided into two strips. The petiole is usually longer than the blade and with two rows of recurved prickles. The 3-parted white numerous flowers are in a panicle on a scape that is up to 5 ft tall. The fruits are oblong blue-black drupes that are 0.5-1" long. The leaves are used for Christmas decorations. Current medical research, especially in Europe, suggests that the plant's drupes might be beneficial in the treatment of prostate cancer. Another interesting attribute is the belief, at least by some in the Far East, that the plant's fruit is a powerful aphrodisiac. These latter two uses have led to a thriving international trade for saw palmetto fruit. During 1995, the price per pound of Saw Palmetto drupes soared from 10 cents to $3.50 per pound, allowing some south Florida gatherers to gross as much as $300.00 per day for their summer harvest. It is much more common to the east of Louisiana but is rare in pine flatwoods (FACU) in southeastern Louisiana. The only records are from Orleans and St. Tammany parishes.

Schisandra glabra (Vines 1960 as ***Schisandra coccinea***)

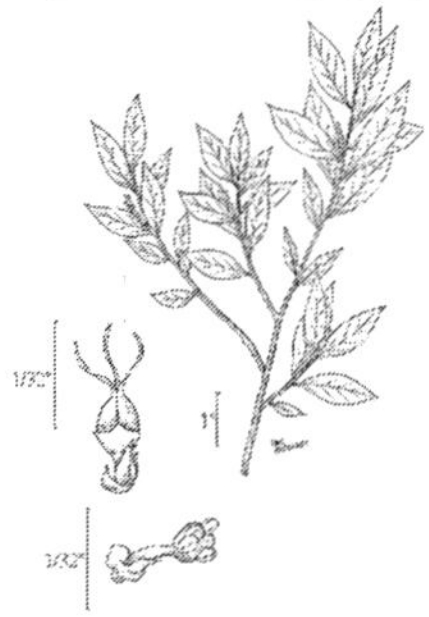

Sebastiana fruticosa (USDA-NRCS)

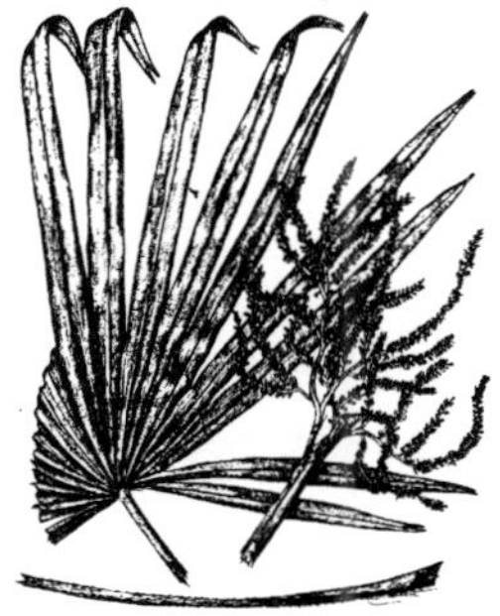

Serenoa repens (Vines 1960)

POISON BEAN, RATTLE BOX - *Sesbania* - Fabaceae

The genus "*Sesbania*" is the Latinized version of the old Adansonian name, Sesban, which is presumably of Arabic origin. The genus includes many herbaceous species but others that are woody shrubs to small trees with a large white pith. The leaves are alternate, deciduous, and one even pinnate compound. The flowers are in racemes. The fruit is an oblong drooping legume that is four sided and four winged and with a long acuminate apex and 2-3" long.

POISON BEAN (***Sesbania drummondii***) is also known as Rattle Bush, Rattle Box, Drummond Rattle Box, Siene Bean, Coffee Bean, and Senna. The specific epithet "*drummondii*," honors Thomas Drummond, an English botanist, who collected in Texas from 1833 to 1834. It is a shrub or small tree to 20 ft. The young stems are green and smooth but turning light brown with age and becoming somewhat scaly. The 12-60 leaflets are narrowly oblong to elliptic, 0.3-1" long, base cuneate, apex rounded and mucronate, upper surface dull green and glabrous, lower surface paler and glaucous, and margins entire. The flowers are yellow to sometimes orange yellow. The legume is light brown. The wood is green to yellowish or white but of no commercial value. The seeds are poisonous to livestock, especially in sheep and goats, and death may follow if a quantity is eaten. The symptoms are diarrhea, extreme weakness, and lethargy from one to two days after eating. However, it is reported that the seeds were used as a substitute for coffee during the Civil War period. In wet roadside ditches and along the edges of streams and other water bodies (FACW) throughout the state. The current reports include 39 parishes. (syn. *Daubentonia drummondii*)

BRAZILIAN RATTLE BOX (***Sesbania punicea***) is also known as Sesban, Daubentonia, Locust, Red Siene Bean, Spanish Gold, and Purple Sesban. The specific epithet "*punicea*," refers to the puniceous, or bright red flowers. It is a shrub to 10 ft. The young stems are green to reddish brown with white scattered hairs and the older stems are reddish brown to gray and glabrous. The 12-40 leaflets are linear to narrowly oblong, 0.3-1" long, base cuneate, apex rounded to truncate and mucronate, upper surface pale green and glabrous, lower surface paler, glaucous, and finely reticulate veined, and margins entire. The flowers are bright red to orange red. The fruit is dark brown. There is not much information available for this species but it is closely related to ***Sesbania drummondii*** and probably has the same poisonous qualities etc. Along roadsides and edges of forests (FAC+) throughout the state. The current reports include 35 parishes. (syn. *Daubentonia punicea*)

CHITTIM WOOD - *Sideroxylon* - Sapotaceae

The genus *Sideroxylon* is from Greek "sidero = iron" and "xylon = wood" in reference to the hard wood. The members of this genus are small trees or shrubs with alternate, simple, deciduous leaves with pinnate major veins. The blade margins are entire. The leaves are often clustered on spur shoots that often end in a thorn. The flowers are in fascicles with five sepals and five white petals. The fruit is a black to dark purple drupe like berry with one large seed. Various authors state that the fruit is edible, it is bitter so one would have to be hungry to eat this fruit.

CHITTIM WOOD (***Sideroxylon lanuginosum***) is also known as Gum Bully, Gum Bumelia, Wooly Buckthorn, Gum Elastic, Chittam Wood, and False Buckthorn. The specific epithet "*lanuginosum*" refers to the wooly hairs on the under surface of the leaves. It is a small tree that reaches up to 60 feet. The bark is gray or dark brown gray. It is divided into narrow, irregularly ridges and covered with scales. The twigs are slender and gray or rusty-hairy. The blades are wedge shaped but rounded at tip, gradually narrowed toward the base, 1-3" long, 0.5-

Sesbania drummondii (Vines 1960)

Sesbania punicea (Vines 1960)

Sideroxylon lanuginosum (USDA-FS)

1" wide, oblanceolate, apex rounded or obtuse, upper surface shiny and dark green, and lower surface rusty to white pubescent. The berry is ellipsoid and about 0.4" long. The wood is heavy (40 lb per cu ft), rather soft, not strong, close-grained, light brown or yellow with thick, lighter colored sapwood. This species is found in a number of habitats from well drained areas to the clay soils of inland prairies and to the sandy soils of the coast (FAC). It is most abundant in the prairie soils like those at Copenhagen, Keiffer or Fort Polk. The current records include 39 parishes. (syn. *Bumelia lanuginosa)*

SWAMP CHITTIM WOOD ***(Sideroxylon lycioides)*** is also known as Buckthorn, Buckthorn Bully, Mock Orange, Iron Wood, Chittam Wood, Gum Elastic, Southern Buckthorn and Buckthorn Bumelia. The specific epithet, "*lycioides,*" applies to the *Lycium*-like fruit, which in turn is named for the country of Lycia. It is a much-branched shrub or tree up to 40 feet tall. The bark is reddish brown to gray but on older trees may develop loose plates or scales. The twigs are shiny reddish brown to gray. The blades are elliptic to oblanceolate, 2-6" long, 0.5-2" wide, base cuneate to attenuate, apex acute to acuminate, upper surface bright green and glabrous, and lower surface paler and glabrous and reticulate veined. The berry is oval-oblong and 0.3-0.75" long. The wood is brown to yellow, close-grained, heavy, hard, not strong, and weighs 46 lbs per cu ft. The leaves are browsed occasionally by livestock, and the fruit eaten by a number of species of birds. This species is found along the edges of streams and lakes (FACW) throughout the state. The current records include 41 parishes. (syn. *Bumelia lycioides, Bumelia smallii)*

COASTAL CHITTIM WOOD (***Sideroxylon reclinatum***) is also known as Florida Bully and Smooth Bumelia. The specific epithet "*reclinatum*", is from "recline", to lean or bend back probably in reference to the habit of the plant. It is a shrub to 10 ft. The blades are oblanceolate, spatulate, or obovate, 0.5-2" long, 0.5-1" wide, apex rounded sometimes emarginate, base cuneate, upper surface dark green and glabrous, and lower surface paler and glabrous. The berry is ovoid oblong and 0.2-0.4" long. Rare with only one report from an elevated area (FAC) in Cameron Parish. (syn. *Bumelia reclinatum*)

GREEN BRIAR, SAW BRIAR - *Smilax* - Liliaceae

The genus name, "*Smilax*", is from the Greek word for an evergreen oak. There is a town in Kentucky named Smilax. This genus includes woody vines that climb with tendrils and are usually armed with prickles on the stem and in some species also on the leaves. The young twigs are green but can change with age in some species. The leaves are alternate, simple, and with palmate major veins. The tender tips of the stems of most species are edible raw or parboiled. The flowers are yellowish green, and radially symmetrical with three sepals and three petals. The fruit is a berry with 1-3 seeds. The plants are rhizomatous and are usually tuberous. The tubers are filled with starch and were harvested by Native Americans and early settlers for food. The tubers were beaten to a pulp, strained to remove the woody fibers, and then ground to make a meal, which was made into a dough or mush, or used, in a cooling drink. Many animals eat the fruits and deer browse the tender tips.

BEACH GREEN BRIAR ***(Smilax auriculata)*** is also known as Bamboo Vine, Horse Brier, and Arrow Leaf Smilax. It reaches heights of 30 ft. The specific epithet, "*auriculata*", is for the ear-like leaf bases. The stems are slender, zigzag, rounded or angled, finely lined, and often pinkish in the lower portion. The leaves are evergreen and the blades are oval, ovate, elliptic, and lanceolate or hastate, with the more typical form broadly extended into two lateral

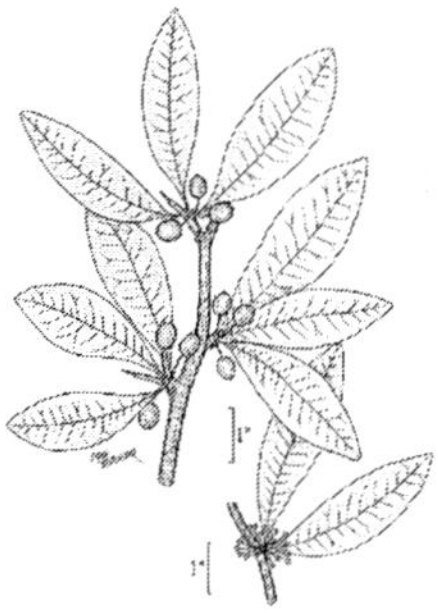

Sideroxylon lycioides (USDA-NRCS)

Sideroxylon reclinatum (Godfrey 1988 as ***Bumelia reclinata***)

Smilax auriculata (Vines 1960)

auriculate lobes. The apex is abruptly pointed and mucronate or rounded to notched. The margin is entire and if examined from a lens from the underside the marginal thickening is double with a narrow groove between (a good way to distinguish it from species such as ***Smilax bona-nox***). The leaves are thin, leathery, and prominently reticulate veined. The primary veins usually five in larger leaves or fewer in smaller leaves. The surfaces are glabrous and lustrous green above and beneath, or slightly glabrous and lustrous green above and beneath, or slightly glaucous beneath. The blades are 1-5" long and 0.5-3" wide. The petiole is about 2" long and often with a bifid coiled tendril as an appendage. The umbel stalk is shorter than the petiole. The berries are 0.25-0.5" diameter, sub-globose, flattened, or sometimes slightly 3-lobed, and reddish to purplish or finally black. In sandy soil (FACU) along the coast with reports only from Plaquemines, St. Bernard, and St. Tammany parishes.

FIDDLE LEAF GREEN BRIAR (***Smilax bona-nox***) is also known as Green Briar, Saw Green Brier, China Brier, Tramp's Trouble, Stretch Berry, Fringed Green Brier, Cat Brier and Bull Brier. The specific epithet, "*bona-nox*", means "good night" and originates from the Spanish buenos noches. It is tough and wiry and reaches a height of 25 ft with the lower portion usually square in cross section. The prickles are stiff, flattened at base, black-tipped and clustered at the nodes and internodes. The leaves are tardily deciduous and the blades are hastate, deltoid, or ovate, 1-5" long.4" wide, with 3-5 veins from base, apex acute or abruptly cuspidate, base rounded, truncate, cordate, or with lobes, lower surface with prickles especially along midvein, and a distinct vein along the margins. The umbel stalk is distinctly longer than the leaf petiole. The globular or ovoid berries are black and about 0.25" diameter. In many habitats (FAC) but especially in bottomland hardwoods throughout the state. It probably occurs in all with current reports from 59 parishes.

CAT GREEN BRIER (***Smilax glauca***) is also known as Wild Sarsaparilla, Cat Brier, Glaucous Leafed Green Brier, Sow Brier, and Sarsaparilla Vine. The specific epithet, "*glauca*", means bluish-gray and refers to the color of the undersides of the leaves. It is slender, terete, and climbs by coiled tendrils at the nodes. The young stems are glabrous or somewhat glaucous, prickle less or with a few prickles. The prickles are stiff and broad based. The older stems are green to dark brown, and glabrous. The leaves are tardily deciduous and the blades are mostly broad ovate but can vary all the way to narrowly lanceolate, 1.5-4" long, 1-3" wide, apex acute to acuminate or cuspidate, base rounded to truncate or almost cordate, with 3 main veins, upper surface medium green, and lower surface distinctly glaucous. The umbel stalk is distinctly longer than the leaf petiole. The subglobose berries are black or dark blue with a glaucous bloom and about 0.3" diameter. In mesic to dry (FAC), especially pine forests, throughout the state. The current reports include 45 parishes.

LOWLAND BAMBOO VINE (***Smilax laurifolia***) is also known as China Brier, Blaspheme Vine, Bay Leaf Smilax, Laurel Green Briar, Laurel Brier and Bamboo Vine. The specific epithet, "*laurifolia*", is for the laurel-like leaves. The lower portion of the stem is black and with many prickles. It is high-climbing vine to 25 ft with numerous upright, often twisted, evergreen leaves. The blades are leathery, often conduplicate, oblong, lanceolate, elliptic or linear, 3-5" long, 0.5-1.5" wide, apex cuspidate and acute to acuminate, base cuneate to rounded, with three main veins and the midvein distinctly larger, and both surfaces pale lustrous green, glabrous, and often glaucous. The umbel stalk varies from shorter to longer than the petiole. The berries are persistent, glaucous, subglobose, black, and about 0.3" diameter. In acid baygalls and other moist areas (FACW+) in the pine regions of the state. The current reports include 33 parishes.

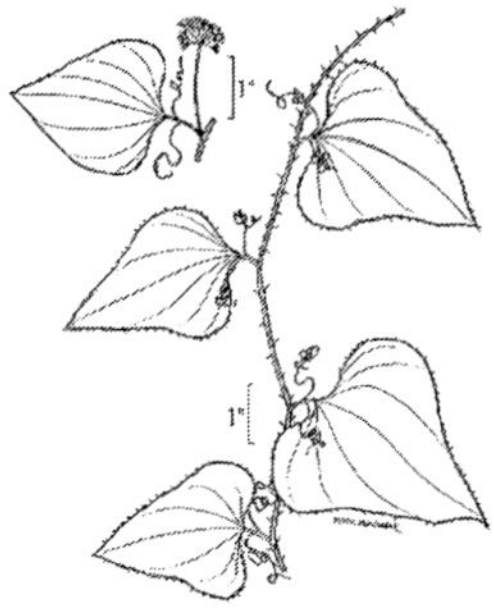

Smilax bona-nox (USDA-NRCS)

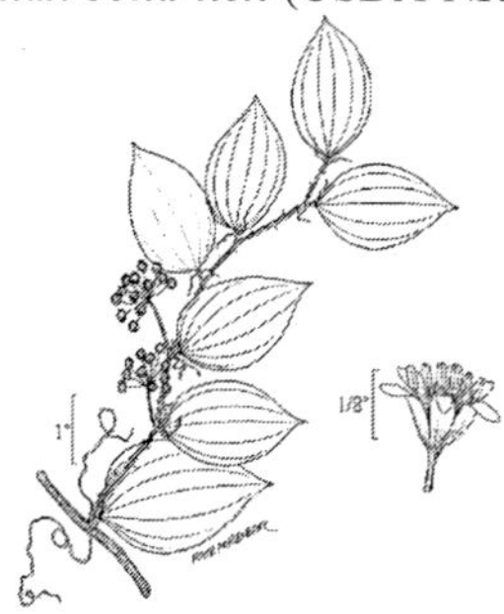

Smilax glauca (USDA-NRCS)

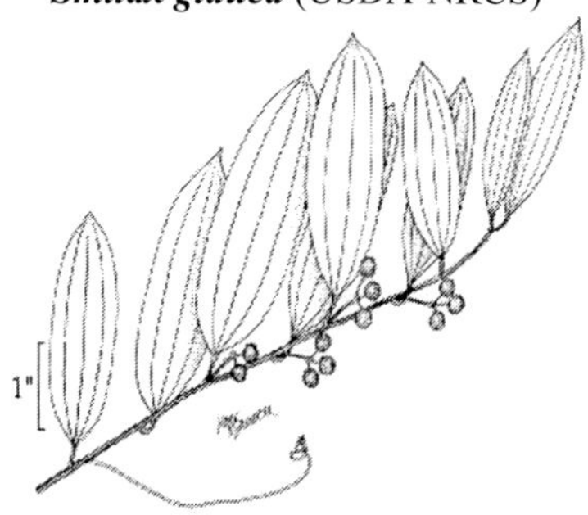

Smilax laurifolia (USDA-NRCS)

SARSAPARILLA VINE (***Smilax pumila***) is also known as Dwarf Green Briar, Hairy Green Brier, Prostrate Green Brier, Brier Vine, Ground Brier, and Wild Sarsaparilla. The specific epithet, "*pumila*", is for its dwarf habit. The stem is slender, unarmed, brown, wiry, often sprawling, and reaches a height of 3 ft. The leaves are deciduous to tardily deciduous. The blades are ovate to oval or heart-shaped, 2-4" long, 1.5-3" wide, apex obtuse to acute and often mucronate, base cordate, with 3-5 main red veins, upper surface lustrous dark green, and lower surface reticulate veined and densely pubescent with gray to brown hairs. The flowers are produced in October-November with the umbel stalk shorter than the petiole. The berries are orange, ovoid, with an obtuse point, and about 0.3" long. In well drained often shady habitats in the pine regions of the state. The current records include 33 parishes.

COMMON GREEN BRIAR (***Smilax rotundifolia***) is also known as Cat Brier, Bull Brier, Horse Brier, Sow Brier, Hungry Vine, Bread and Butter, Devil's Hop Vine, Bamboo Brier, Wait-a-Bit, and Biscuit Leaves. The specific epithet, *"rotundifolia,"* is for the rounded shape of leaves. It is tough and wiry vine that reaches to a height of 20 ft and is round in cross section. The prickles are stiff and flattened. The leaves are deciduous to tardily deciduous and the blades are ovate to orbicular or heart shaped, rarely lanceolate, 2-6" long, 1-6" wide, apex often apiculate, obtuse to rounded or acute, base cordate, with 5 main veins, upper surface dark green, and lower surface slightly paler. The umbel stalk is about the same length to shorter than the petiole. The berry is ovoid to globose, black to bluish black, and about 0.25" diameter. In a wide range of habitats (FAC) from dry to moist and open to shade. It probably occurs in all with current records from 60 parishes.

UPLAND BAMBOO VINE (***Smilax smallii***) is also known as Lance Leaf Green Brier, Coral Green Brier, Jackson Vine, Jackson Brier, Bamboo Vine, and Thornless Smilax. The specific epithet "*smallii*" honors John Kunkel Small (1869-1938), noted botanist and student of the flora of the southeastern United States. It is a high climbing vine that has few or no prickles. The leaves are evergreen and the blades are lanceolate to elliptic, often conduplicate, 2-5" long, 0.75-2" wide, apex acute or acuminate, base broadly cuneate to truncate, with 5-7 main veins, upper surface shiny deep green, and lower surface paler with indistinct veins. The umbel stalk is about the same length as the petiole. The berries are subglobose, dull red or reddish brown, and about 0.3 diameter. In well drained soils (FACU), especially pine forests throughout the state. The current records include 47 parishes.

HELLFETTER (***Smilax tamnoides***) is also known as Bristly Green Brier, Hag Brier, Devil Green Brier, and Wild Sarsaparilla. The specific epithet, "*tamnoides*", translates to "resembling *Tamnus*", a genus in the Dioscoreaceae. It is stout and reaches a height of 40 ft. The upper stem is often unarmed but the lower portion has many straight, slender brown to black prickles. The underground tubers are unusual in having prickles. The leaves are deciduous and the blades are ovate to heart-shaped, or elliptic-ovate, 2-6" long, 2-5" wide, apex acute to obtuse or abruptly cuspidate, base rounded or cordate, with 5-7 main veins, upper surface is lustrous dark green, lower surface paler green, and the blade dries ashy green. The umbel stalk is longer than the petiole. The berries are bluish-black, subglobose, and about 0.25" diameter. In mesic (FAC+) habitats mostly in shade situations. It may occur in all with current reports from 59 parishes. (syn. *Smilax hispida*)

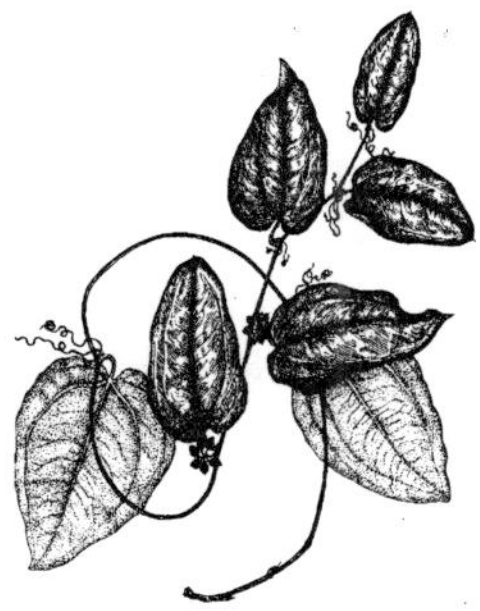

Smilax pumila (Vines 1960)

Smilax rotundifolia (Britton and Brown 1943)

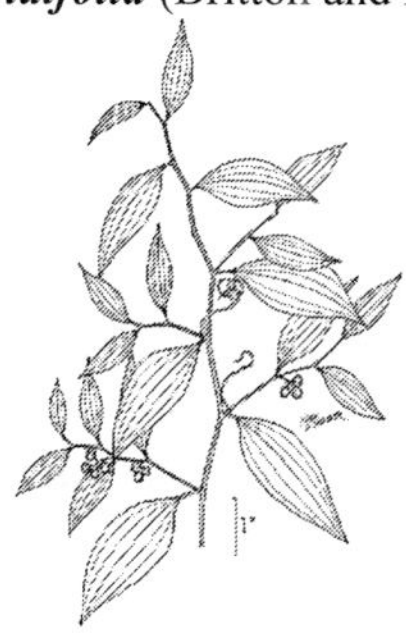

Smilax smallii (USDA-NRCS)

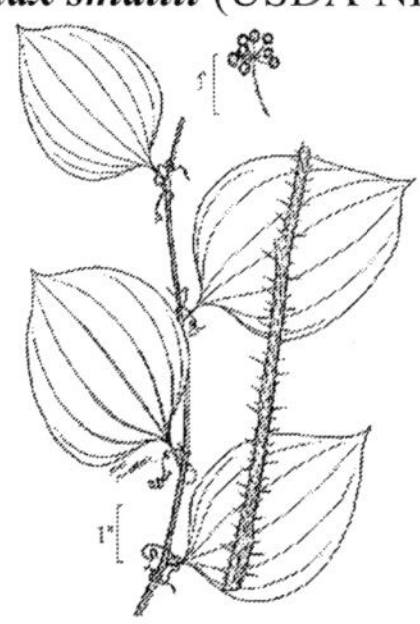

Smilax tamnoides (USDA-NRCS)

RED BERRY GREEN BRIAR (***Smilax walteri***) is also known as Red Berry Bamboo, Sarsaparilla, Coral Green Briar and Redbead Green Briar. The specific epithet "*walteri*", honors Thomas Walter, 1740-1789, early student of the flora of the Carolinas. It does not form tubers. The above ground stem is slender, terete, mostly without prickles, and up to 10 ft. The leaves are deciduous and the blades are ovate to lanceolate, 1.75-5" long, 0.75-3" wide, apex mucronate and obtuse to acute, base rounded to truncate, with 5 to 7 major veins, and both surfaces lustrous dark green but lower surface distinctly reticulate veined. A distinct characteristic is the red petiole. The umbel stalk is usually shorter than the petiole. The berries are globose to obovoid, red to orange, and about 0.3" diameter. Along the edges of lakes, streams, and other wet areas (OBL), especially in the pine flatwoods of southeastern and southwestern Louisiana. The current records include 30 parishes.

JERUSALEM CHERRY - *Solanum*- Solanaceae

The genus "*Solanum*" is from Latin "solamen" meaning "solace" or "quieting", probably referring to the medicinal qualities of some of the species in the genus. It includes many herbaceous species and in Louisiana, two woody shrub species. The stems are green to brown. The leaves are simple, alternate, deciduous, and with pinnate major veins. The blades are narrowly elliptic to spatulate, 1-2.5" long, 0.3-1" wide, apex acute to rounded, base cuneate, and margins entire to sinuate. The radial symmetrical white flowers are axillary in groups of 1 to 3. The fruit is an orange to red, globose berry that is 0.5-0.75" diameter. The leaves and unripe fruits are reported to be poisonous. Both species are native to Brazil.

FALSE JERUSALEM CHERRY (***Solanum capsicastrum***) is also called Christmas Cherry and Tomatito of Christmas. The specific epithet "*capsicastrum*" means "pepper aster" (capsica = pepper like and astrum = aster like). The major characteristic is the pubescence on the young stems and the underside of the leaves. Widely cultivated and escaping and spreading mostly in southern Louisiana. The current reports include 16 parishes. (Not Illustrated)

JERUSALEM CHERRY (***Solanum pseudocapsicum***) is also called Winter Berry. The specific epithet "*pseudocapsicum*" means false caspsicum thus false pepper. The major characteristic is the glabrous nature of all plant parts. Widely cultivated and escaping and spreading throughout the state. The current reports include 24 parishes. (Not Illustrated)

EVE'S NECKLACE - *Sophora affinis* - Fabaceae

Eve's Necklace is also known as Texas Sophora, Pink Sophora, Coral Bean, and Necklace Tree. The genus name, "*Sophora*", is from the Arabic word "Sophero", which was applied to some tree of the same family. The specific epithet, "*affinis*", means "related to", probably in reference to the genus name. It is a shrub or small tree that reaches 35 ft. The bark is gray to reddish brown and broken into small, thin, oblong scales. The twigs are slender, green to brown and often streaked with lighter brown, and often swollen at the nodes. The leaves are alternate, deciduous, one odd pinnately compound, and 3-9" long. The 7 to 13 leaflets are 0.75-1.5" long, about 0.5" wide, base rounded or cuneate, apex acute to obtuse and often retuse and mucronate, upper surface lustrous dark to yellow green and glabrous, lower surface slightly paler and pubescent, and margins entire. The fragrant flowers are in drooping axillary racemes and are pink to white. The fruits are shiny black legumes with narrowed or constricted pods that create a beaded or necklace like appearance. The seeds are red. The wood is light red, heavy, strong, hard, and the sapwood is yellow. In clay soils in northwest Louisiana with records from Caddo Parish and with other, probably cultivated, records from Jefferson, Lafayette, and Orleans parishes.

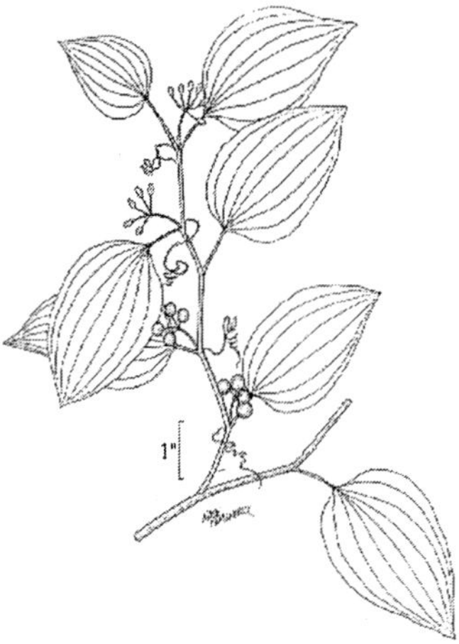

Smilax walteri (USDA-NRCS)

Sophora affinis (USDA-FS)

MEADOW SWEET - *Spiraea* - Rosaceae

The genus "*Spiraea*" is from the Greek word "speira = wreath or band or a plant used in garlands". The plants are shrubs with arching branches. The leaves are alternate, simple, deciduous, and with pinnate, palmate in one species, major veins. The small flowers are white or pink but attract attention with the concentration of flowers in umbels, racemes or panicles. The fruit is a follicle. There are several species native to North America as well as a number that are introduced. This genus is in need of closer examination in the state.

REEVES MEADOW SWEET (***Spiraea cantoniensis***) is a native of China and Japan. The specific epithet "*cantoniensis*" is probably for the Canton Province of China. The stems are glabrous. The blades are lanceolate to elliptic, 1-3" long, glabrous, apex acute, base cuneate, upper surface dark green, lower surface bluish green, and margins doubly serrate. The white flowers are in simple corymbs and produced before and with the leaves. Cultivated and persisting with reports from Caddo, DeSoto, East Baton Rouge, and Franklin parishes. (Not Illustrated)

JAPANESE SPIREA (***Spiraea japonica***) is a native of China and Japan. The specific epithet "*japonica*" is for Japan. It reaches a height of 6 ft. The stems are reddish brown, ridged, and pubescent. The blades are narrowly ovate to oblong-lanceolate, 1-3" long, 0.3-1" wide, apex acuminate, base attenuate, upper surface medium green, lower gray green, and margins serrate. The white or pink flowers are in a terminal compound corymb and produced with the leaves. Cultivated and persisting in Natchitoches Parish.

BRIDAL WREATH SPIREA (***Spiraea prunifolia***) is a native of Korea and China. The specific epithet, " *prunifolia*", refers to the leaves being plum like. It reaches a height of 9 ft. The stems are slightly zig-zag and pubescent. The blades are oblong to ovate, 1-3.5" long, 0.5-1" wide, apex acute to obtuse, base obtuse, lustrous dark green above, slightly paler below and pubescent, and margins serrate. The white flowers are double in simple corymbs and produced before the leaves. Cultivated and persisting in 10 parishes.

THUNBERG'S MEADOW SWEET (***Spiraea thunbergii***) is also called Baby's Breath Spirea and Thunberg Spirea. The specific epithet "*thunbergii*" honors Carl Peter Thunberg (1743-1828), Dutch physician and botanist who introduced many Japanese plants. It is a native of China and Japan that reaches a height of 5 ft. The stems are brown and thin. The blades are linear, 1-2" long, narrower than 0.25", apex and base acuminate, pale green, glabrous, and margins sharply serrate. The white flowers are small in simple corymbs and produced before the leaves. Cultivated and persisting in 13 parishes. (Not Illustrated)

VANHOUTTE SPIREA (***Spiraea x vanhouttei***) is a hybrid of ***Spiraea triloba*** and ***Spiraea cantoniensis***. The hybrid was produced in 1862 in France and the specific epithet "*vanhouttei*" honors Louis van Houtte, a noted Belgian Horticulturist. It reaches a height of 8 ft. The twigs are reddish brown. The blades are ovate, with 3 palmate major veins, 0.75-2" long, apex acuminate, base attenuate, green, and margins with large coarse teeth, almost lobes, in the upper half but entire at base. The white flowers are in simple corymbs and produced before the leaves. Cultivated and persisting in 19 parishes.

Spiraea japonica (Britton and Brown 1943)

Spiraea prunifolia (Steyermark 1968)

Spiraea vanhouttei (Bailey 1949)

AMERICAN BLADDER NUT - ***Staphylea trifolia*** - Staphyleaceae

American Bladder Nut is also known as Bladder Nut. The genus "*Staphylea*" is a Greek word meaning cluster of grapes in reference to the clusters of flowers. The specific epithet "*trifolia*" refers to the three leaflets. It is a shrub to small tree to 25 ft. The bark gray, smooth but becoming fissured on older stems.The twigs are green to brown or gray, smooth, curved, and often striped. The leaves are opposite, deciduous, and odd pinnate compound with three leaflets. The leaflets are elliptic, ovate, or obovate, 1.5-4" long, base obtuse to cuneate, apex acuminate to obtuse, upper surface bright green and pubescent along the veins, lower surface paler and pubescent, and margins serrulate. The white flowers are in drooping racemes. The fruits are three lobed, bladder like, green to brown capsules with distinct veins. The only records (Natchitoches and Rapides parishes) are old and perhaps from cultivated specimens.

VIRGINIA STEWARTIA - ***Stewartia malacodendron*** - Theaceae

Virginia Stewartia is also known as Silky Camellia and Round Fruited Stewartia. The genus name, "*Stewartia,*" honors of John Stuart (1713-1792), Earl of Bute and patron of botany. The specific epithet, "*malacodendron,*" is from the Greek *malakos* ("soft") and *dendron* ("tree"), referring to the silky pubescence. It is a shrub to small tree to 20 ft. The twigs are slender, light to dark brown or gray, and pubescent when young but glabrous with age. The leaves are alternate, simple, deciduous, and with pinnate major veins. The blades are oval to elliptic or ovate, 2-4.5" long, 1-2" wide, base acute, apex acute to short acuminate, upper surface light to medium green and glabrous, lower surface light green and pubescent, and margins sharply serrate with mucronate teeth. The flowers are large, axillary, radially symmetrical, and with five white spreading silky pubescent petals and many stamens. The fruit is a globose capsule that is 0.5-0.75" diameter. In mesic soils, especially along small streams, in southeastern and central Louisiana. The current reports include 11 parishes.

SNOWBELL - ***Styrax*** - Styracaceae

The genus name, "*Styrax,*" is the Greek name of the Old World tree producing storax. It includes shrubs to small trees. The bark is smooth and reddish brown to gray. The buds are double and the crushed twigs do not have a strong odor. The leaves are alternate, simple, deciduous, and with pinnate major veins. The petioles are short. The flowers are white with five recurving petals and yellow anthers. The flowers are in racemes and produced late February to early April with the leaves. They are fragrant and resemble orange blossoms. The fruits are green to dark brown one-seeded drupes that are subglobose to obovoid and about 0.25" long. The fruits are eaten by wood ducks. A gummy substance, storax, may be extracted from the plant and the scientific name ***Styrax*** is derived from this fact.

SMALL SNOWBELL (***Styrax americanus***) is also known as Spring Orange, Snowbell, Little Snowbell, American Snowbell, and Storax. The specific epithet, "*americanus*", refers to North America. It can reach a height of 10 ft. The blades are oval, elliptic, or oblong, 0.75-4" long, apex acute or short acuminate, base cuneate, upper surface dark green and glabrous, lower surface paler and pubescent to glabrous, margins entire to serrate or remotely large toothed. In bottomland hardwood forest and along the edges of streams (FACW) throughout the state. The current records include 47 parishes. (syn. *Styrax pulverulentus*)

Staphylea trifoliata (Britton and Brown 1943)

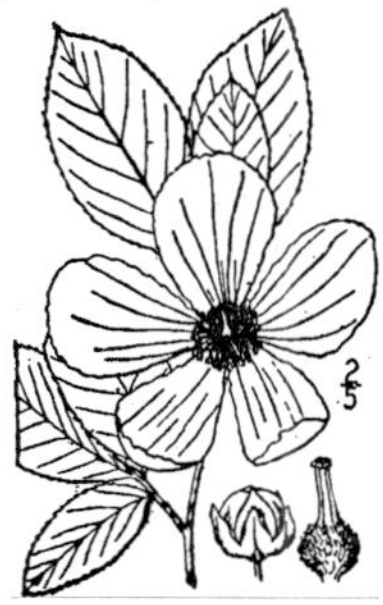

Stewartia malacodendron (Britton and Brown 1943)

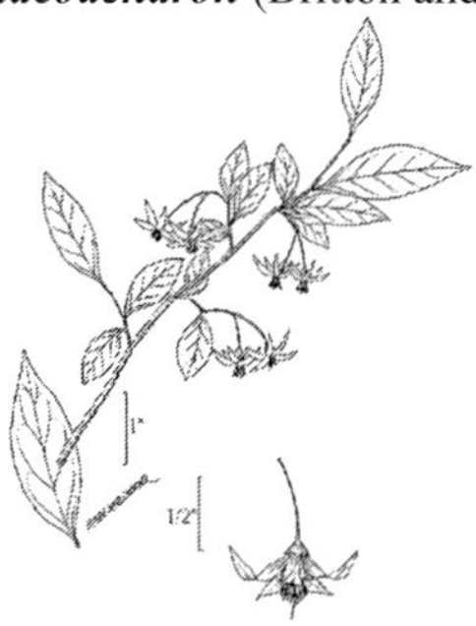

Styrax americanus (USDA-NRCS)

LARGE SNOWBELL *(Styrax grandifolius)* is also known as Big Leaf Snowbell, Large Leaved Storax, Coast Snowbell, Storax, and Mock Orange. The specific epithet, *"grandifolius,"* refers to the large leaves. It can reach a height of 25 feet The blades are oval, obovate, or elliptic, 2-6" long, 1-3" wide, upper surface bright green, lower surface paler and pubescent, and margins entire or remotely serrate. In well drained forests (FACU-), especially hardwood slopes, in the pine regions of the state. The current records include 37 parishes.

SNOW BERRY - ***Symphoricarpos orbiculatus*** - Caprifoliaceae
Snow Berry is also known as Buck Rush, Coral Berry, Indian Currant Snow Berry, Snap Berry, Buck Berry, Wolf Berry, Wax Berry, Turkey Bush, and Indian Current. The genus name, "*Symphoricarpos*", comes from the Greek "*sympherein*" ("to bear together") and "*karpos*" ("fruit"). The specific epithet, "*orbiculatus*", refers to the rounded leaf. It is an erect, much branched shrub up to 6 feet. The bark is shreddy and brown. The young twigs are slender, brown and pubescent while the older twigs are gray and with shreddy bark. The leaves are opposite, simple, tardily deciduous, and with pinnate major veins. The blades are oval, ovate, elliptic, or orbicular, 1-2.5" long, 0.5-2" wide, apex acute, obtuse, or rounded, base rounded or cuneate, upper surface dull green and glabrous, lower surface paler and glabrous to pubescent and often slightly glaucous, and margins entire. The flowers are small, bell-shaped, greenish to pinkish, and borne in dense clusters in the axils of upper leaves. The fruit is a round to ellipsoid, red to pink drupe. Several birds eat the fruits and deer browse the foliage. The slender stems are sometimes woven into baskets. It was first introduced into horticulture in 1727. Along roadsides and edges of forests (FAC-) in northern Louisiana. The current records include 14 parishes. (syn. *Symphoricarpos symphoriocarpos)*

SWEET LEAF, HORSE SUGAR - ***Symplocos tinctoria*** - Symplocaceae
Sweetleaf is also known as Horse Sugar, Yellow Wood, Wild Laurel, Florida Laurel, and Dye Leaves. The genus name, *"Symplocos,"* is from the Greek "symploke" (a connection) and refers to the united stamens. The specific epithet, "*tinctoria,*" comes from the tincture, or dye, that can be made from the leaves and roots. It is a shrub or small tree up to 35 feet. The bark is gray to reddish, smooth, but with age becoming a little rough and with warty bumps. The twigs are slender, terete, gray to reddish brown, and pubescent when young and becoming glabrous with age. The twigs have chambered pith. The leaves are alternate, simple, evergreen to tardily deciduous, and with pinnate major veins. The blades are lanceolate to elliptical, 5-7" long, 1-2" wide, apex acute to acuminate, base cuneate, upper surface dark green, lower surface paler, and margins entire or remotely serrate. The yellow fragrant flowers are in dense clusters. The fruit is a drupe that is 0.4" long and is green most of the year and at maturity is orange-brown to purple. The reddish-brown wood is close-grained, brittle, weak, soft, light, and weighs about 33 lb per cubic ft. The leaves have a sweet apple peel taste and can be used to relieve thirst. Livestock can also feed on the leaves. The leaves are reported to have medicinal properties and early settlers were reported to use it as a tonic. In late Winter to early Spring, yellow gall like structures are formed on the twigs. These are edible and are reported to be caused by bacteria. A very common plant in mesic to dry habitats (FAC) throughout the pine regions. It is currently reported from 39 parishes.

Styrax grandifolius (USDA-FS)

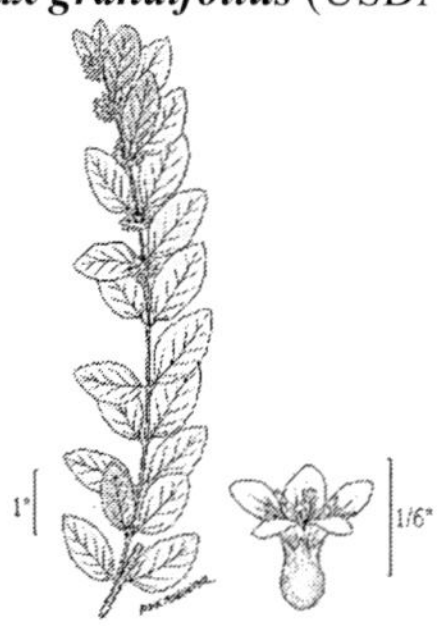

Symphoricarpos orbiculatus (USDA-NRCS)

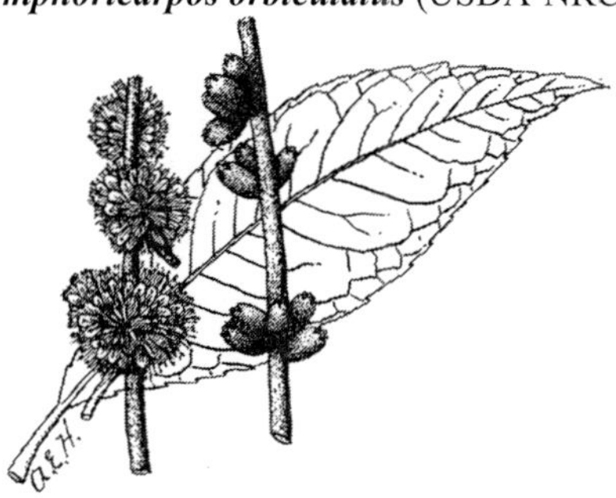

Symplocos tinctoria (USDA-FS)

SALT CEDAR, TAMARISK - *Tamarix* - Tamaricaceae

The genus name, *"Tamarix,"* is the ancient name, probably with reference to the Tamaracine people of southern Europe, where the plant grew or for the Tamaris River in Spain. It includes shrubs to small trees. The branches are often drooping, glabrous, often glaucous, and red to gray. The leaves are simple, evergreen, alternate but with very short internodes, and with obscure but pinnate major veins. The leaves are tiny, scalelike, appressed, overlapping, and sessile. The blade apex is acute to obtuse, base wide and clasping the stem, brownish green, and with scales. The branches with the appressed, small, evergreen, leaves are very much "cedar like" hence the common name. The flowers are white or pink and in terminal clusters of racemes. The fruit is a capsule with numerous small seeds that have a tuft of hairs at the apex. The wood is light-colored, close-grained, takes a high polish, often twisted of knotty. Most species can tolerate saline conditions hence the word "salt" in their common names. Some species produce manna, a sweet hardened exudate, resulting from the puncture of the stem by scale insects. This is considered by some to be the manna of the Bible. Some species produce galls from which tannins are obtained for tanning leather. There are four species reported for the state with the separation of the species very difficult and based primarily on flower characteristics. All four species are introduced in Louisiana and are the most common along the coast but can be found in sandy soils along streams or disturbed sandy areas inland. The most common species is ***Tamarix gallica*** with the common names of French Tamarisk, Salt Cedar, Manna Bush, Athel, Eshel, Asul, Athul, and Atle. The specific epithet "*gallica*" refers to France. This species is rated FACW with regard to wetland status. The current reports include 16 parishes. The second most common species is ***Tamarix ramosissima*** with the common names of Salt Cedar and Tamarisk. The specific epithet "*ramosissima*" means most densely branched. It is native to Asia and southeastern Europe. The other two species are ***Tamarix canariensis*** (Canary Island Tamarisk) reported from Cameron, St. Bernard, and Vermilion parishes and ***Tamarix africana*** (African Tamarisk) (Not Illustrated) reported from Cameron and Vermilion parishes. The specific epithet "*canariensis*" refers to the Canary Islands and "*africana*" to Africa where each species was native.

BALD CYPRESS, POND CYPRESS - *Taxodium* - Taxodiaceae

The genus "*Taxodium*", is from the Greek and means "yewlike", in reference to the leaves. It includes small to large monoecious trees that are often buttressed at base. The knees are brownish structures with shreddy bark that project from the roots and used to stabilize the plant in the very wet soil. The bark is silvery to cinnamon-red and finely divided by numerous longitudinal fissures. On old trees, the bark often forms loose flakes. The twigs are very thin, flat, green, and deciduous. The leaves are deciduous, needlelike, sessile, entire, apex acute and apiculate, crowded on thin twigs (giving a feather like appearance), 0.4-0.8" long, dull pale green, and remaining attached to the twig which drops off in the late Fall. The pollen or male cones are drooping reddish brown panicles and produced in late winter to early Spring but do not persist. The seed or female cones are solitary or paired at the ends of the branches, globular, about 1" diameter, and composed of several 4-angled flattened scales which fall separately at maturity. The seeds are two-winged and about 0.25" long. The heartwood is variable, yellowish to light or dark brown and sometimes reddish and is extremely decay resistance and is called "wood eternal" or "wood everlasting". The wood is light or dark brown, soft, easy to work, straight-grained, moderately hard, not strong, very durable, and weighs 28-32 lb per cu ft. The wood has a multitude of uses, from hand-made "shakes", fenceposts, construction timbers, siding, greenhouses, hotbed lumber, interior trim, fancy paneling, shingles, posts, poles, coffins, cross-ties, boats, shipbuilding, window frames, doors, to furniture and fine cabinet construction. The knees are sometimes made into souvenirs and the cone resin used as an analgesic for

Tamarix gallica (Vines 1960)

Tamarix ramosissima (USDA-NRCS)

wounds. The seeds are eaten by a number of species of birds, including wood ducks and squirrels.

BALD CYPRESS (***Taxodium distichum***) is also known as White Cypress, Gulf Cypress, Southern Cypress, Tidewater Red Cypress, Yellow Cypress, Red Cypress, Black Cypress, Swamp Cypress, and Sabino Tree. The specific epithet, "*distichum*", means "two-ranked", and refers to the leaves. It is a large tree to 125 ft. The leaves are spreading from the branches and in two rows or distichous. Bald Cypress was designated by the 1963 State Legislature as our state tree. It was chosen because it characterizes the swamps and lowlands for which Louisiana is well known. In wet sites (OBL) such as swamps, edges of bayous and lakes and is also planted in many drier sites. It occurs in all parishes with current reports from 62 parishes.

POND CYPRESS (***Taxodium ascendens***) is also called Black Cypress. The specific epithet "ascendens" means rising "upright" in reference to the leaves. The leaves are not distinctly distichous, as they are in Bald Cypress, but more or less spirally attached and appressed to the branchlets. However, this distinction is not always clear-cut. It is also usually smaller in stature only reaching a height of 80 ft. In wet pine flatwoods or savannahs (OBL) in extreme southeastern Louisiana with reports only from St. Tammany Parish.

BASS WOOD, LINDEN - *Tilia americana* - Tiliaceae

Basswood is also known as Linden, Lime Tree, White Wood, Bast Tree, Bee Tree, Blacklime Tree, White Lind, Southern Lind, Whistle Wood, Spoon Wood, Daddy Nuts, Monkey Nuts, and Wickup. The genus name, "*Tilia*", is the classical Latin name for lime or linden. The specific epithet "*americana*" is for North America. It is a large tree to 130 ft. The bark is gray with shallow fissures and flat ridges. The twigs are gray to brown and glabrous. The leaves are alternate, simple, deciduous, and with five palmate major veins. The blade is broad ovate, 5-6" long, 3-4" wide, apex acute or acuminate, base asymmetrical, cordate to truncate, upper surface dark green and glabrous, lower surface paler green and often with axillary tufts of hairs, and margins coarsely serrate with gland tipped teeth. The yellowish white flowers are in drooping cymes that are attached to a leaf-like bract. The fruits are globose to ovoid, dry, densely brown pubescent drupes that are about 0.25" diameter. The wood is soft, light, easy to handle, and the heartwood is pale brown to reddish while the sapwood is white to brownish. The wood is straight-grained, rather soft, weak in endwise compression, low in shock resistance, works well, and has a specific gravity of 0.40. It takes paint well, finished smoothly, glues well, has poor nail-holding ability, seasons well, and holds shape fairly well. It is used for piano keys, Venetian blinds, handles, rollers, cabinets, laundry appliances, caskets, crates, furniture, and all types of veneer. The flowers make good honey. The dried flowers produce a great tea. It has been cultivated since 1752. The var. ***caroliniana*** (Carolina Bass Wood) and var. ***heterophylla*** (White Bass Wood) are reported for Louisiana. The variety name "*caroliniana*" is for North or South Carolina and "*heterophylla*" means "variable leaf". The main differences include: pubescent twigs and leaves with larger tufts of hairs for Carolina Bass Wood compared to glabrous twigs and leaves with smaller tufts of hairs for White Bass Wood. Both are in well drained, often hardwood slope, forests throughout the state. Carolina Bass Wood is reported from 39 parishes and White Basswood from Iberia, Natchitoches, Rapides, and St. Mary parishes.

Taxodium distichum (USDA-FS)

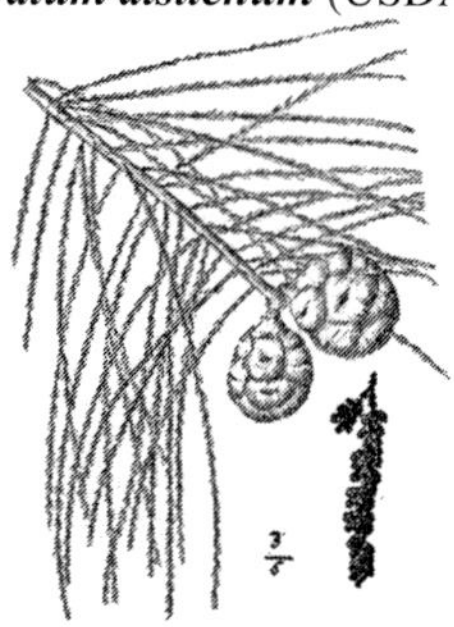

Taxodium ascendens (Britton and Brown 1943)

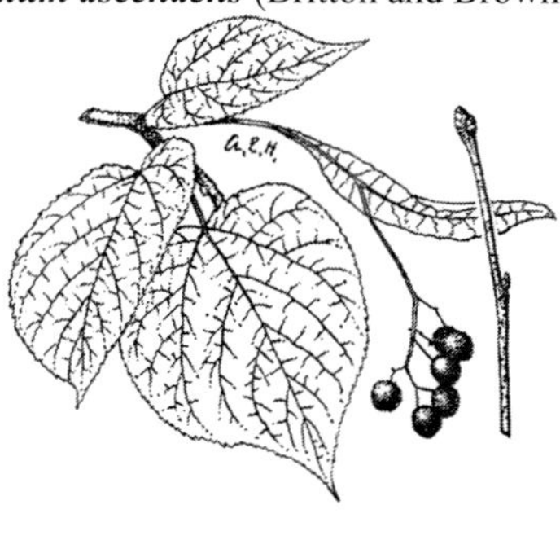

Tilia americana (USDA-FS)

POISON IVY, OAK, SUMAC - *Toxicodendron* - Anacardiaceae

The genus name, "*Toxicodendron*", is from the ancient Greek and means "poison tree". It includes vines, shrubs, and small trees. The leaves are alternate, deciduous, and compound. The flowers are greenish-white and in axillary clusters. The fruits are globose to subglobose greenish white drupes. All species can cause allergic reactions in many people including blisters, eruptions, itching, and burning as a result of a nonvolatile, poisonous oil known as toxicolendrol or urushiol. Probably, the best method of removing the oil before inflammation begins is to wash with laundry soap. Also, do not scratch the irritated area since this helps to spread the toxic oil. Even in winter months or in water, all plant parts can cause an allergic reaction. Breathing the smoke can cause irritation of the respiratory tract. The sap has been used as a tincture to treat rheumatism, ischia, neuralgia, fever, and eye hypertrophy of the heart. It is sometimes used for making varnish. The Indians used the powdered root as a poultice to cause a swelling to open. Goats that eat poison ivy get dry milk. Ants and 75 species of birds eat the fruit. Deer browse the foliage in spring and summer.

POISON OAK (***Toxicodendron pubescens***) is also known as Oak Leaf Ivy, Oak Leaf Poison Ivy, Poison Wood, Poison Weed, Scratch Ivy, Poison Ivy, and Poison Sumac. The specific epithet "*pubescens*" refers to the pubescent leaves. It is a low, erect shrub to 3 ft. The leaves are trifoliolate. The leaflets are ovate to obovate, 2-3.5" long, 1.5-2.75" wide, apex acute to obtuse, base cuneate to obtuse, upper surface dark green, lower surface paler and densely pubescent, and margins dentate to 3-7 lobed. The lateral leaflets are slightly smaller, asymmetrical and usually entire apically and lobed basally. The drupes are pubescent when young but becoming glabrous with age and about 0.2" diameter. In dry sandy soil (FACU) in the pine regions of the state. The current reports include 25 parishes. (syn. *Toxicodendron quercifolium, Toxicodendron toxicarum*, *Toxicodendron toxicodendron, Rhus toxicodendron, Rhus toxicarum*)

POISON IVY (***Toxicodendron radicans***) is also known as Three Leaf Ivy, Climbing Ivy, Poison Oak, Mercury, Black Mercury Vine, Markry, Mark Weed, and Picry. The specific epithet, "*radicans*", refers to its climbing habitat. It varies from an erect shrub to a high climbing vine that climbs with aerial roots. These high-climbing vines may attain a diameter of 6" and sometimes will display more foliage than the supporting tree. The twigs are glabrous to pubescent but become very pubescent on large stems. The leaves are alternate, pinnate compound (3 leaflets), and deciduous. The blades are oval to lanceolate, 1-8" long, 0.5-5" wide, apex acuminate, base cuneate to rounded, upper surface dull green and glabrous, lower surface paler and glabrous to slightly pubescent, and margins entire to dentate or lobed. The lateral leaflets are slightly smaller and asymmetrical. The drupes are waxy, 0.2-0.4" diameter, and glabrous to pubescent. In moist to mesic or even dry habitats (FACU) throughout the state. It occurs in all with current records from 61 parishes. (syn. *Rhus radicans)*

POISON SUMAC (***Toxicodendron vernix)*** is also known as Poison Dogwood, Poison Elder, Poison Ash, Poison Tree, Poison Wood, and Thunder Wood. The specific epithet, "*vernix*", means "varnish", erroneously referring to the Japanese Lacquer Tree. It is a shrub or small tree up to 25 feet tall. The bark is thin and light gray. The twigs are stout, brown to orange, and glabrous. The leaves are alternate, deciduous, and one odd pinnately compound. The 7-13 leaflets are oblong, oblong-ovate, or oval, 2.5-4" long, 1-2" wide, apex short acuminate, base cuneate or rounded, upper surface dark green, lower surface paler and pubescent, and margins entire. The petiole, rachis, and petiolules are often red. The flowers are in panicles up to 8" long. The drupes are glabrous, and 0.2-0.3" diameter. In acid baygalls

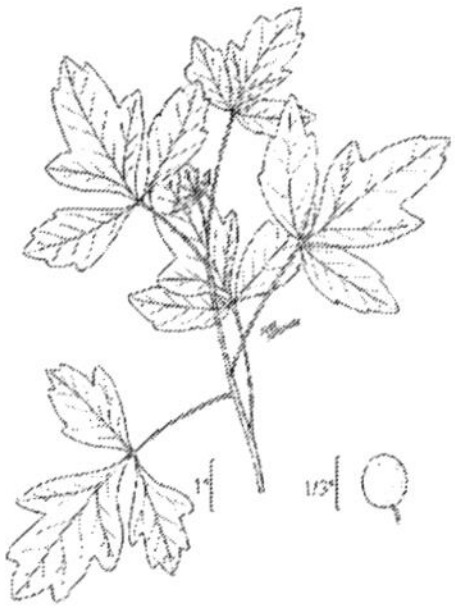

Toxicodendron pubescens (USDA-NRCS)

Toxicodendron radicans (USDA-NRCS)

Toxicodendron vernix (USDA-FS)

(OBL) in southeastern, southwestern and central Louisiana with reports from 12 parishes. (syn. *Rhus vernix*)

CLIMBING DOGBANE - ***Trachelospermum difforme*** - Apocynaceae

Climbing Dogbane is also known as Climbing Star Jasmine, Double Pod, and Southern Jasmine. The genus name, "*Trachelospermum*", means "neck-seeded" perhaps in reference to the tufts of hairs on the seed. The specific epithet, "*difforme*", means "differing shape" compared to other species in the genus. It is a vine that is herbaceous above and woody basally with milky sap that climbs by twining. The stem is slender, twining, green to reddish brown and pubescent in the upper portion but glabrous basally. The leaves are opposite, deciduous, and with pinnate major veins. The blades are lanceolate to broadly elliptic or oval, 1-3" long, 0.5-2" wide, apex acute or acuminate, base cuneate to rounded, upper surface dark green and glabrous to pubescent, lower surface paler and pubescent, and margins entire. The groups of small, cream colored to yellowish, fragrant flowers have fine-lined brown markings inside and are located near the main stems. The fruits are elongate "pod like" paired slender terete follicles, 5-9" long, and smaller in diameter than 0.2". The numerous seeds are linear-oblong and with tufts of fluffy hairs at one end. In moist habitats (FACW), especially bottomland hardwood forests, throughout the state. The current reports include 54 parishes.

CHINESE TALLOW TREE - ***Triadica sebifera*** - Euphorbiaceae

Chinese Tallow Tree is also known as Chicken Tree, Pop Corn Tree, Money Tree, and Vegetable Tallow Tree. The genus name, "*Triadica*", means "three divisions or parts", probably in reference to the capsule. The specific epithet, "*sebifera*", refers to the vegetable tallow, or wax, of the fruit. It is a small monoecious tree with milky sap that reaches up to 30 feet. The bark is brownish gray with shallow fissures. The young twigs are slender and green and the older twigs are grayish brown with many small lenticels. The leaves are alternate, simple, deciduous, and with pinnate major veins. The blades are rhombic-deltoid, 1-3.5" long, 1-3" wide, apex long acuminate, base widely cuneate to obtuse, upper surface dark green, lower surface paler, and margins entire. The petiole is about as long as blade and has two green glands on the upper side at the base of the blade. In the Fall, the leaves turn red. The flowers are yellowish green with the male in terminal erect to drooping spikes and the female at the base. The fruit is a capsule with three persistent white seeds. The seeds contain wax that is used for soap, clothes, and cloth dressing. The milky sap is reported to be poisonous. Everywhere!!!! This species can tolerate wet to dry (FAC), open to shade, and all soils. It is introduced from China but has and is spreading rapidly. It occurs in all with current records from 58 parishes. (syn. *Sapium sebiberum*)

ELM - ***Ulmus*** - Ulmaceae

The genus, "*Ulmus*", is the ancient Latin name. It includes trees. The leaves are alternate, simple, deciduous, and with pinnate major veins. The lateral veins are distinctly parallel and obvious. The leaf margins of the native species are doubly serrate while the leaf margins of the two introduced species are singly serrate. The base of the leaf is distinctly asymmetrical (unequal) but is nearly symmetrical in the introduced species ***Ulmus pumila***. The leaves turn a golden yellow in the fall. The flowers are perfect, small, and inconspicuous. The petals are absent and the sepals are green, brown, or red. The fruit is a samara and are produced in fascicles or racemes. Dutch elm disease can affect any elm species and is caused by the fungus *Ophiostoma ulmi*, which invades the water-conducting vessels of elms. The leaves of infected trees wilt, turn yellow or brown, and then fall. Another diagnostic feature is the formation of brown or green streaks in the infected sapwood. This discoloration is visible when the bark is peeled back on symptomatic branches. The fungus is spread by elm bark beetles which feed on

Trachelospermum difforme (Britton and Brown 1943)

Triadica sebifera (Main and Mills 1981)

and breed in elm trees. Dutch Elm Disease was discovered in 1930 in Ohio and has killed many elms in the north and east.

WINGED ELM (***Ulmus alata***) is also known as Mountain Elm, Small Leaved Elm, Cork Elm, Water Elm, Wahoo Elm, Red Elm, and Witch Elm. The specific epithet "*alata*" refers to the corky wings on the twigs. It is a small to medium-sized tree that reaches up to 80 feet and is the most common of the elms found in the state. The bark is reddish brown to gray, fissured into irregular furrows and rows and breaking up into scales. The twigs are gray, glabrous and usually with two corky wings. The blades are oblong lanceolate to elliptic, sometimes falcate, 1-3" long, 0.5-1.5" wide, apex acute to acuminate, base cuneate to subcordate, upper surface dark green and smooth, lower surface paler and pubescent, and margins doubly serrate. The flowers and fruits are produced before the leaves in January to March. The samara is elliptic, 0.25-0.3" long, with ciliate margins and a pubescent surface. The heavy, difficult to split, light brown with thick, lighter colored sapwood, close-grained, hard but not as strong as other elms wood weighs 47 lb per cu ft. The wood is used for tool handles, vehicle parts, agricultural implements, woodenware, baskets, flooring, veneer, furniture, cooperage, cabinets, sporting goods, stock stoves, boxes, crates, framework, trunks, toys, construction ships, fuel, and hubs of wheels. The Indians used for the wood for canoes and the bark for rope. Birds use the seeds and buds. It occupies the highest and driest sites of all the native elms in Louisiana. It can be found throughout the state on dry to mesic sites but not usually in bottomlands (FACU+). The current reports include 53 parishes.

AMERICAN ELM (***Ulmus americana***) is also known as River Elm, White Elm, Water Elm, Soft Elm, Rock Elm, Common Elm, and Swamp Elm. The specific epithet "*americana*" refers to North America. It is a tree that reaches up to 120 ft and is a much loved and famous American tree known sometimes to reach the age of 300 years. It is the largest and most important of the six elms native to eastern North America. The bark is light to dark gray, fissured, and breaking up into scales. The twigs are brown and pubescent to glabrous and the buds are dark brown, glabrous, and pointed. The blades are oval, obovate or oblong, 4-6" long, 2-3" wide, apex acute or abruptly acuminate, base cordate on one side and round to cuneate on the other side, upper surface dark green and smooth to slightly scabrous, lower surface paler and pubescent, and margins doubly serrate. The flowers and fruits are produced from January to March before the leaves. The samaras are oval-obovate, deeply notched at the apex, 0.4-0.6" long, with ciliate margins but glabrous on the surface. The wood is light brown, with thick, somewhat lighter colored sapwood. The tough, heavy (35 lb per cu ft), hard, strong, durable, difficult to split wood is used for tool handles, vehicle parts, agricultural implements, woodenware, baskets, flooring, veneer, furniture, cooperage, cabinets, sporting goods, stock stoves, boxes, crates, framework, trunks, toys, construction ships, fuel, and hubs of wheels. The Indians used the wood for canoes and the bark for rope. The bark is also used for ox whips. The Iroquois Indians of western New York used the dark, ashy, gray bark with crisscrossing flaky ridges to line their canoes. The fruit is used by gallinaceous birds and the twigs and leaves by deer, opossum, and cottontail. American Elm is the state tree of North Dakota, Nebraska, and Massachusetts. The separation of American and Slippery Elm is very difficult when the fruits are not present. American Elm usually has leaves that are not as scabrous, glabrous twigs, and will be on slightly wetter sites (FACW) than Slippery Elm (FAC). The current reports include 59 parishes.

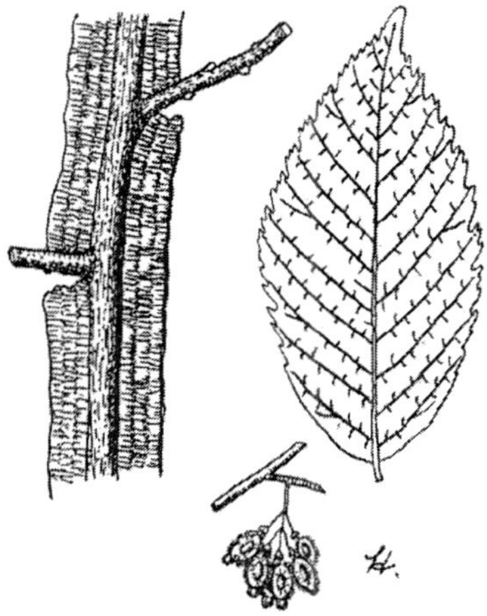

Ulmus alata (USDA-FS)

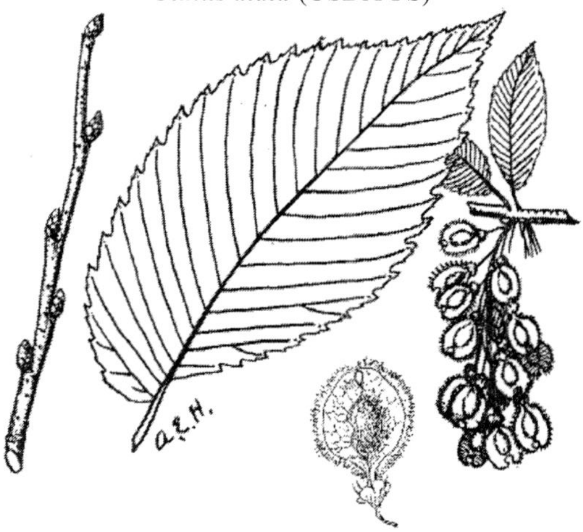

Ulmus americana (USDA-FS)

CEDAR ELM (***Ulmus crassifolia***) is also known as Cork Elm and Small Leaved Elm. It is a tree that reaches a height of 80 feet. The specific epithet "*crassifolia*" refers to the rough thick leaves. The bark of medium-sized trees is broken into broad, flat, scaly ridges. The twigs are red brown, pubescent, and occasionally have corky wings. The blades are broadly elliptic to ovate, 1-2" long, 0.75-1" wide, apex acute to obtuse, base rounded or cuneate, upper surface dark green and scabrous, lower surface paler and pubescent, and margins doubly serrate. The flowers appear from late September into October with the leaves. The samaras are oval-elliptic or oblong, 0.25-0.5" long, and pubescent. The wood is reddish brown, brittle, with a thick layer of lighter colored sapwood. It is primarily on clay soils (FAC) in Louisiana only west of the Mississippi River. The current reports include 37 parishes.

CHINESE ELM (***Ulmus parvifolia***) and **SIBERIAN ELM** (***Ulmus pumila***) are both native to Asia and have been widely planted throughout Louisiana. The specific epithet "*parvifolia*" refers to the small leaves and the specific epithet "*pumila*" mean small, perhaps in reference to the plant size. Chinese Elm is also called Lace Bark Elm and is native to China, Japan, and Korea. Siberian Elm is also called Dwarf Elm, Asiatic Elm, and Chinese Elm and is native to China and Korea. These two species are reported to be resistant to Dutch elm disease. Both are small to medium size trees reaching a height of 50 to 70 feet. The bark on Siberian Elm is brownish and furrowed while the bark of Chinese Elm is thin and flaky and often with orange to salmon colored patches. The leaves are smaller than American Elm and Slippery Elm; mostly shorter than 2" long and have singly serrate margins. The base of Siberian Elm is often nearly symmetrical. The tip of the leaves in Siberian Elm is acute to acuminate while the leaf tip in Chinese Elm tends to be acute only. Siberian Elm has slightly wider leaves, mostly wider than 1", compared to 1" or less in Chinese Elm and the lateral veins fork only three times between the midrib and the margins in Siberian Elm and five or more branches can be found in Chinese Elm. The teeth in Siberian Elm also tend to be sharper than those of Chinese Elm. Siberian Elm flowers in the spring (mostly March) with the fruits remaining until leaf development. Chinese Elm flowers in the fall (mostly October) and the fruits remain on the tree into early winter. The samaras of Chinese Elm are nearly orbicular while Siberian Elm's fruits are broadly elliptic. Widely planted and persisting with incomplete records on the distribution of the two species in Louisiana.

SLIPPERY ELM (***Ulmus rubra***) is also known as Red Elm, American Elm, Indian Elm, Moose Elm, Rock Elm, and Sweet Elm. It is a short-lived medium-sized tree that reaches a height of 80 feet. The specific epithet "*rubra*" refers to the reddish wood. The bark is dark gray to reddish brown, divided by shallow fissures, and covered with thick, loose scales. The twigs are gray and densely pubescent when young but more toward glabrous with age. The buds are brown, red-hairy, large, and blunt. The blades are obovate, ovate, or oblong, 4-8" long, 2-3" wide, apex acute or abruptly acuminate, base cordate on one side and round to cuneate on the other side, upper surface dark green and very scabrous, often also pubescent, lower surface paler and pubescent, and margins doubly serrate. The dried leaves have a medicinal odor. The flowers and fruits are produced from February to March before the leaves. The samaras are nearly orbicular, not deeply notched at the apex, about 0.4" long, with glabrous margins but pubescent on the surface. The wood is dark reddish-brown, with thin, lighter-colored sapwood, heavy (37 lb per cu ft), tough, hard, strong, close-grained, durable and easy to split. It is used for crossties, fence posts, hubs of wheels, agricultural implements and ribs for small boats. Pioneers and Indians used the inner bark to make tea, to make a nutritious flour, to counter thirst, and to treat sore throats. The inner bark is mucilaginous and was used to treat coughs, ulcers, dysentery, tumors, gunshot wounds, frost bite, burns, skin diseases, erysipelas, felons, and

Ulmus crassifolia (USDA-FS)

Ulmus parvifolia (Li 1963)

Ulmus pumila (USDA-FS)

Ulmus rubra (USDA-FS)

diarrhea. Slippery Elm is found along slopes and higher elevations in bottomlands throughout the state. The separation of Slippery Elm and American Elm is very difficult when the fruits are not present. Slippery Elm usually has leaves that are very scabrous and one test is to slap the leaf to your clothing and if it sticks and holds, it is Slippery Elm but if it falls off, it is American Elm. The twigs of Slippery Elm are usually pubescent and it will be on slightly drier sites (FAC) than American Elm (FACW). The current reports include 52 parishes. (syn. *Ulmus fulva)*

HUCKLE BERRY, BLUE BERRY - *Vaccinium* - Ericaceae

The genus name, "*Vaccinium*", is thought to be from the Latin vaccinus ("of cows") or bacca for berry. This genus includes shrubs to small trees with alternate, simple, deciduous to tardily deciduous or evergreen leaves with pinnate major veins. The flowers are white to pinkish and usually in racemes. The fruit is a globose to subglobose false berry. The false berries provide food for many wildlife, especially birds. The false berries of some species are relished by humans.

WINTER HUCKLE BERRY (***Vaccinium arboreum)*** is also known as Tree Huckle Berry, Farkle Berry, Tree Sparkle Berry, Sparkle Berry, Whortle Berry, and Goose Berry. The specific epithet, "*arboreum*", is a Latin word which calls attention to the fact that this shrub often grows to tree size. It is a large shrub to small tree that reaches up to 30 feet and usually gnarled. The bark is thin, gray to grayish brown, with narrow ridges, and shredding into large plates The twigs are light brown to dark brown and glabrous to pubescent. The leaves are tardily deciduous to evergreen and the blades are oval to obovate, or elliptic, 1-3" long, 0.5-1" wide, apex acute to rounded, base cuneate, upper surface dark glossy green, lower surface dull green and often sparsely pubescent, and margins entire to slightly dentate. The winter weather turns the leaves from shiny green to deep maroon. The flowers are pink to white, bell-shaped, about 3/8" long, resembling Lily of the Valley and produced from March to May after the new leaves have been produced. The false berries are long-stemmed, shiny, globose, and 0.3-0.5" diameter. It has lots of seeds and is not very sweet but ripen in late summer to early fall and persist through the winter. In the past these false berries were highly sought after during the late fall and winter when no other fruits were available. And, the fruits probably tasted very good to humans or animals who had never tasted a snickers bar. The hard, very close-grained, brown wood weighs 48 lb per cu ft and is used for tool handles, tobacco pipes, woodenware, and novelties. The bark is used for tanning leather. The root bark is used to treat diarrhea. The fruit and foliage is used by a number of species of birds and mammals. It is commonly found in the understory of pine-oak forests (FACU) near streams and lakes but is absent from the floodplain of the Mississippi River. The current reports include 40 parishes.

HIGH BUSH BLUE BERRY (***Vaccinium corymbosum)*** is also known as Whortle Berry, Common High Bush Blue Berry, Swamp Blue Berry, and Northern High Bush Blue Berry. The specific epithet, "*corymbosum*", refers to the corymbose, or clustered flowers. It is a shrub to 12 ft. The twigs are warty, yellowish green when young but turning brown to gray with age. The leaves are deciduous and the blades are elliptic to ovate, 1.5-3" long, 0.5-0.75" wide, apex acuminate to acute, base cuneate, upper surface shiny glaucous green, lower surface paler and pubescent especially on the veins, and margins entire to serrate. The white rarely pink urceolate flowers appear after the leaves from May to June and are in a large dense corymb. The false berries are blue to blue-black with a glaucous bloom, 0.2-0.5" diameter, are in a large dense corymb, and mature from June to August. In mesic to dry areas (FAC) in pine forests sporadically throughout the state. The current reports include nine parishes.

Vaccinium arboreum (USDA-FS and USDA-NRCS)

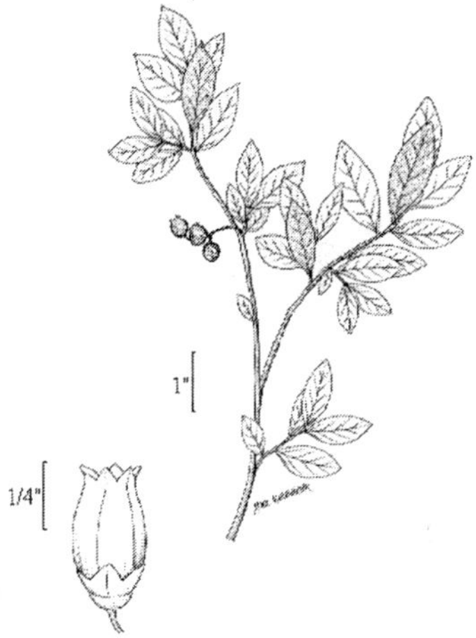

Vaccinium corymbosum (USDA-NRCS)

DWARF BLUE BERRY ***(Vaccinium darrowii)*** is also known as Darrow Evergreen Blue berry and Glaucous Blue berry. The specific epithet, "*darrowii*", honors George McMillan Darrow (b. 1889), who first brought the plant into cultivation in 1940. It is a low shrub to 1.5 ft. The leaves are evergreen and the blades are elliptic to spatulate, 0.3-0.5" long, apex acute to obtuse, base cuneate, upper surface dark green, sometimes glaucous, lower surface grayish to bluish green and pubescent, and margins entire. The flowers are white to pink, urceolate, and in small axillary clusters. The false berry is glaucous blue and about 0.25" diameter. In dry sandy pine forests (FACU) in southeastern Louisiana. The current reports include St. Helena, St. Tammany, Tangipahoa, and Washington parishes.

SUMMER HUCKLE BERRY (***Vaccinium elliottii)*** is also known as Elliott's Blue berry, Spring Huckle Berry, and May Berry. The specific epithet, "*elliottii*", honors Stephen Elliot (1771-1830), an early student of the flora of South Carolina. It is a tall shrub that grows up to 12 feet. The twigs are green and glabrous. The leaves are deciduous and the blades elliptic to ovate, 0.5-2" long, 0.25-0.6" wide, apex acute or obtuse, base cuneate or rounded, upper surface dark green, lower surface paler, and margins entire. The flowers are pinkish or white, cylindric to narrowly urceolate, and are produced before the leaves in January to March. The false berries are black, lustrous, and 0.2-0.3" diameter. The false berries ripen from May to June. The fruits are edible and were highly sought after for pies, jams, and cobblers. As a youngster in St. Helena Parish in the late 1950's, Charles Allen picked many of these fruits for the making of the "Huckleberry Pie". Some were also frozen so as to have the delicacy during the off season. In mesic to well drained soils (FAC+) throughout the state except for the large floodplains and marsh. The current records include 37 parishes.

BAYGALL BLUE BERRY (***Vaccinium fuscatum)*** is also known as Thick Leaf Blue Berry and Black High Bush Blue Berry. The specific epithet, "*fuscatum*", refers to the brownish hairs. It is a shrub to 9 ft. The twigs are slender, light green to brown and pubescent when young but brown and glabrous with age. The leaves are evergreen to tardily deciduous and the blades are coriaceous, lanceolate to elliptic or ovate, 1.3-2" long, 0.5-1" wide, apex acute, base rounded, upper surface dark green, lower surface paler with distinct veins and glabrous to pubescent, and margins entire. The flowers are urceolate, pink to reddish, fascicled, and produced January to March before the leaves. The false berry is dull dark blue to black, about 0.3" diameter, anad matures May to June. In baygalls (FACW) in acid pH areas of the state. The current reports include 22 parishes. (syn. *Vaccinium arkansanum)*

DEER BERRY (***Vaccinium stamineum***) is also known as Buck Berry, Squaw Huckle Berry, High Bush Huckle Berry, and Southern Goose Berry. The specific epithet, "*stamineum*" refers to the prominent stamen structure. It is a shrub to 6 ft. The twigs are green to red and pubescent when young but turning green to brown and glabrous with age. The leaves are deciduous and the blades oblong to elliptic or oval, 0.75-2.75" long, 0.3-1" wide, apex acute to short acuminate, base cuneate or rounded, upper surface shiny light green, lower surface dull green and pubescent with distinct veins, and margins entire and ciliate. Small leaves often occur at the base of the pedicels. The white, bell-shaped flowers have long, straight, orange stamens that project past the corolla. The flowers are produced in distinct racemes from April to June with the leaves. The false berries are green to yellowish, about 0.4" diameter, and mature from May to September. The false berries are found from May to September. In mesic to dry soils (FACU), especially pine forests throughout the pine regions of the state. The current records include 37 parishes. (syn. *Vaccinium caesium)*

Vaccinium darrowii (Vines 1960)

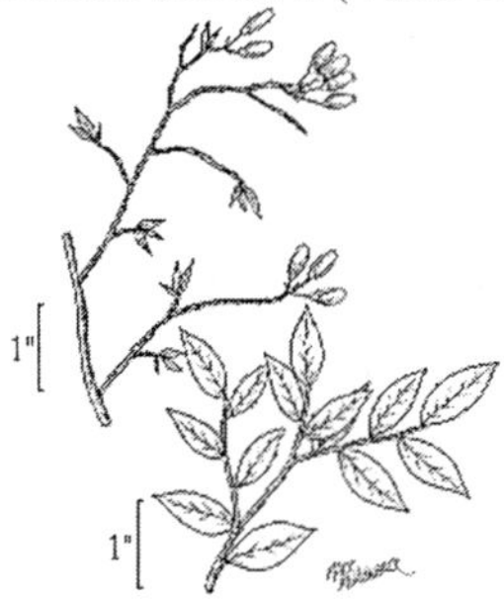

Vaccinium elliottii (USDA-NRCS)

Vaccinium fuscatum (Vines 1960)

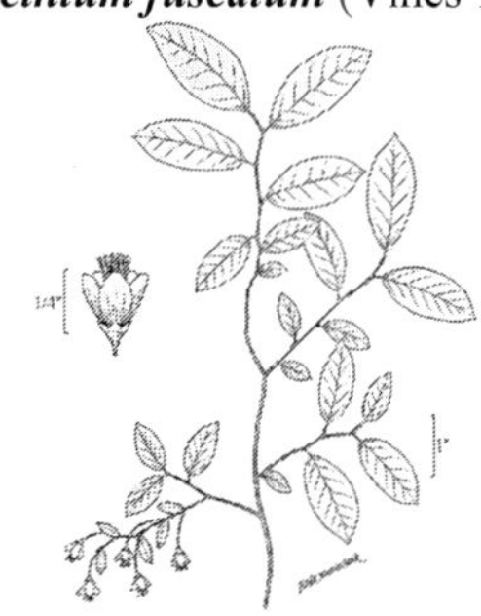

Vaccinium stamineum (USDA-NRCS)

LARGE CLUSTER BLUE BERRY (***Vaccinium virgatum***) is also known as Rabbit-eye Blue Berry and Medium Cluster Blue Berry. The specific epithet, "*virgatum*", refers to the slender twigs. It is a shrub that reaches 3 feet tall. The twigs are slender, green to reddish-brown and pubescent but becoming brown or gray and glabrous with age. The leaves are deciduous and the blades are spatulate to narrowly elliptic or oblanceolate, 1.2-2" long, 0.4-0.8" wide, apex acute to acuminate, base cuneate, upper surface dark green, lower surface paler and pubescent, and margins serrulate. The flowers are clustered, cylindric urceolate, deep pink, and produced January to March before the leaves. The false berries are black, lustrous, 0.3-0.4" diameter, and mature May to June. In mesic to dry pine forests (FACU) in the pine regions of the state. The current records include 30 parishes.

TUNG OIL TREE - *Vernicia fordii* - Euphorbiaceae

Tung Oil Tree is also known as China Wood Oil Tree and Candle Nut Tree. The genus name, "*Vernicia*", comes from vernic referring to varnish. The specific epithet, "*fordii*", honors C. Ford, former superintendent of the Hong Kong Botanic Garden. It is introduced from Central Asia and becomes a small tree to 30 ft. The bark is smooth and light to dark gray. The leaves are alternate, deciduous, large, and with palmate major veins. The margins are entire or 3-5 lobed and the base is rounded or semi-cordate. The upper surface is dull green with fine brown hairs and the lower surface is paler. The veins are sparingly hairy or glabrous. The petiole is as long as the blade with two glands on the petiole at the base of the blade. The flowers are borne in large conspicuous cymes and appear when the leaves start to unfold. The petals are white and streaked with red at the base. The fruit is a 2-3" diameter drupe that becomes ripe in September. The seeds are thick shelled and poisonous. Oil from the seed is used in making paint, varnish, soap, linoleum and other items. In Asia, the oil is used for fuel and for treating woodenware and cloth. This species was once cultivated in the south in orchards for its oil. Some remnants of the orchards still persist, in particular, in southeastern Louisiana. In other places throughout the state, the plant is now cultivated as an ornamental and sometimes escapes or persists. The current records include 27 parishes. (syn. *Aleurites fordii)*

HAW, VIBURNUM - *Viburnum* - Caprifoliaceae

The genus "*Viburnum*" is an old Latin name for *Viburnum lantana.* This genus includes shrubs to small trees. The leaves are opposite, simple, deciduous, and with pinnate major veins. The white flowers are in flattened to rounded terminal cymes. The fruits are drupes.

MAPLE LEAF VIBURNUM (***Viburnum acerifolium)*** is also known as Maple Leaf Arrow Wood, Dockmackie, Arrow Wood, Possum Haw, Squash Berry, and Guelder Rose. The specific epithet, "*acerifolium*", refers to the maple-like foliage. It is a shrub to 7 ft. The twigs are tan to reddish brown and stellate pubescent when young but becoming gray and glabrous with age. The blades are suborbicular to ovate, 1-5" wide and long, apex acuminate, base rounded or cordate, upper surface dull green and pubescent, lower surface dull green, pubescent and yellow and black dotted, and margins coarsely dentate and often 3-lobed apically. The drupes are purplish-black, ovoid to oblong, and about 0.3" long. Deer and rabbits eat the twigs and leaves. Several birds eat the fruit. The plant has been in cultivation since 1736. In well drained hardwood slope forests (FACU) west of the Missippi River. The current reports include 12 parishes.

Vaccinium virgatum (Britton and Brown 1943)

Vernicia fordii (Vines 1960 as ***Aleurites fordii***)

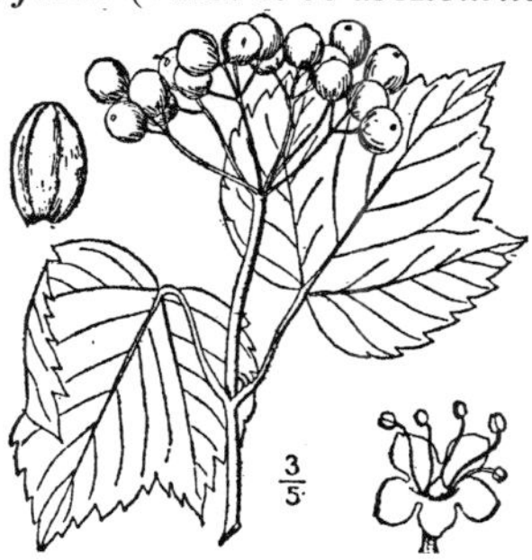

Viburnum acerifolium (Britton and Brown 1943)

ARROW WOOD ***(Viburnum dentatum)*** is also known as Southern Arrow Wood, Mealy Tree, Withe Rod, and Withe Wood. The specific epithet, "*dentatum*", refers to the coarsely toothed margins of the leaf. It is a large shrub to small tree to 20 ft. The bark is gray to grayish brown or reddish brown. The twigs are elongate, straight, gray to grayish brown or reddish brown, and pubescent when young but becoming glabrous with age. The blades are ovate-lanceolate to rotund, 1-4.5" long, apex acute to short acuminate, base rounded to subcordate, upper surface medium green and pubescent to glabrous, lower surface with 5-11 pair of straight or nearly so veins, and margins distinctly dentate. The drupes are bluish-black fruits, subglobose, and 0.2-0.4" diameter. The straight and not tapering branches are used for arrow shafts by the Indians. Wildlife utilizes the foliage, stems and fruit. In mesic to dry (FAC) areas throughout the state, especially in the pine regions. The variety ***dentatum*** is separated from variety ***scabrellum*** based on broader leaves and pubescent flowers in ***scabrellum***. The varietal name "*scabrellum*" means somewhat rough. The current reports for variety ***dentatum*** include 43 parishes and 28 parishes for var. ***scabrellum***.

POSSUM HAW ***(Viburnum nudum)*** is also known as Withe Rod, Swamp Haw, Shonny Haw, Southern Wild Raisin, Smooth Withe Rod, Naked Withe Rod, Bil Berry, Nanny Berry, and Swamp Haw Viburnum. The specific epithet, "*nudum*", refers to the rather naked stems with remote leaves. It is a shrub or small tree to 20 ft. The bark is smooth and gray to brown. The young twigs are green, gray, or reddish brown, lustrous, finely grooved, and rusty pubescent and the older twigs are glabrous. The blades are oblong, elliptic or narrowly ovate, 2-5" long, 1-2" wide, apex acute, acuminate, obtuse, or rounded, base cuneate or rounded, upper surface dark green to olive green and lustrous, lower surface paler and with rusty brown scales, and margins entire. The drupes are pink when young but maturing to a glaucous blue, ellipsoid, and 0.25-0.5" diameter. The variety ***nudum*** is separated from variety ***cassinoides*** by narrow leaves in var. ***cassinoides***. The varietal name "*cassinoides*" refers to its resemblance to ***Ilex casssine***. In moist (FACW) acid pH areas, especially baygalls, in the pine regions of the state. The current records for variety ***nudum*** is 27 parishes and for variety ***cassinoides*** is eight parishes mostly in southeastern and southwestern Louisiana.

SMOOTH BLACK HAW ***(Viburnum prunifolium)*** is also known as Southern Black Haw, Sheep Berry, Nanny Berry, Sweet Haw, Sweet Sloe, Stag Bush, Arrow Wood, and American Sloe. The specific epithet, "*prunifolium*", means "plum-leaved". It is a shrub or small tree to 40 ft. The bark is gray and broken by fine ridges into short sections like Dogwood. The twigs are slender, rigid, green to reddish or brown, and sometimes with spur branches. The blades are oval to ovate or oblong, 1-3" long, 0.5-2" wide, apex acute to obtuse or rounded, base obtuse to rounded, upper surface dull green lower surface paler, and margins serrulate. The drupes are bluish black, ellipsoid to ovoid, and 0.3-0.5" diameter. The wood is reddish brown, hard, weighs 52 lb per cu ft, and is of no commercial importance. The plant has been cultivated since 1727. The fruit is eaten by several species of birds and mammals. Deer browse the twigs and foliage during the summer months. The foliage is considered high in humus building capacity. A drug prepared from the powdered bark of the root or stem is officially known as Fluidex-tractum Viburni Prunifolii; at one time it was used in the treatment of uterine colic and as a general antispasmodic. In meisc to dry areas (FACU) soils in the western part of the state. The current records include 17 parishes, all west of the Mississippi River.

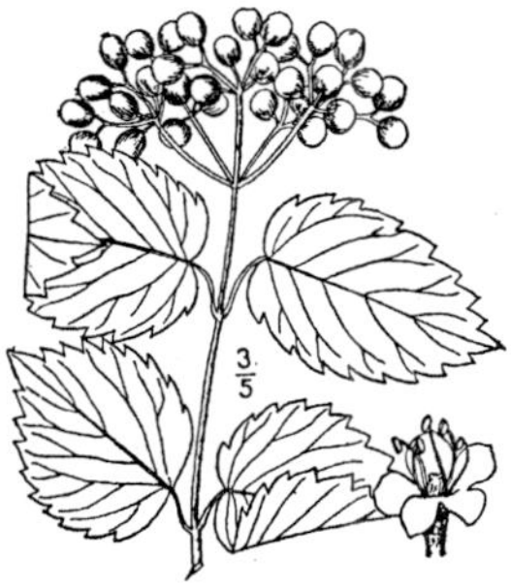

Viburnum dentatum (Britton and Brown 1943)

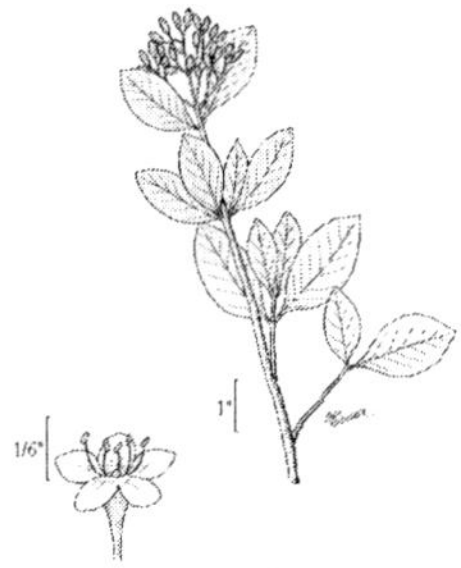

Viburnum nudum (USDA-NRCS)

Viburnum prunifolium (USDA-NRCS)

RUSTY BLACK HAW (***Viburnum rufidulum***) is also known as Rusty Black Haw Viburnum, Rusty Nanny Berry, Black Haw, Sweet Haw, Southern Nanny Berry, and Southern Black Haw. The specific epithet, "*rufidulum*", refers to the rufous-red hairs on young parts. It is a shrub or small tree that reaches up to 40 feet. The bark is rough with ridges that are narrow and rounded breaking into dark reddish brown plates (similar to Persimmon and Flowering Dogwood). The young twigs are gray with reddish hairs and the older twigs are glabrous and reddish brown. The blades are elliptic to obovate or oval, 1.5-4" long, 1-2.5" wide, apex acute to obtuse, base cuneate or rounded, upper surface glossy dark green, lower surface paler with red hairs on veins, and margins serrate. The drupes are glaucous bluish black, oblong to obovoid, and 0.3-0.5" diameter. The wood is hard, heavy, strong, fine-grained, and with a disagreeable odor. The powdered bark of roots or stems is used for medicine for uterine, colic, general antispasmodic, astringent, tonic, sedative, abortive, diuretic, and nervine. The fruits are edible and are used by wildlife. In mesic hardwood slope and Inland Prairie Forests (FACU) throughout the state. The current records include 30 parishes.

PERIWINKLE - *Vinca* - Apocynaceae

The genus "*Vinca*" is from Latin "vincio = to bind", in reference to the use of the vines in wreaths. It includes sprawling vines or shrubs with milky sap. The twigs are green to light brown, glabrous, squarish in cross section, and ridged. The leaves are opposite, simple, evergreen, and with pinnate major veins. The blade upper surface is dark green and pubescent along veins, the lower surface paler and pubescent along veins, and the margins ciliate. The blue flowers are solitary and large. The fruits are paired follicles.

BIG LEAF PERIWINKLE (***Vinca major***) is also called Large Periwinkle. The specific epithet "major" means "larger". The blades are round ovate, 0.5-3" long, 0.5-2.5" wide, apex obtuse to acute, and base truncate. The flowers are 1-2" long. The follicles are 1-2" long. It is a native of Eurasia that is cultivated throughout the state. It persists in many locations but is spreading in some. The current records include 32 parishes.

COMMON PERIWINKLE (***Vinca minor***) is also known as Running Myrtle. The specific epithet "minor" means "smaller". The blades are oblong to ovate, 0.5-1.5" long, 0.3-1" wide, apex acute, and base rounded to truncate. The flowers are 0.3-1" long. It is a native of Europe that is cultivated throughout the state. It persists and is currently reported from Caddo, Iberia, Natchitoches, and Ouachita parishes.

CHASTE TREE - *Vitex* - Verbenaceae

The genus "*Vitex*" is the classical Latin name for *Vitex agnus castus*. It includes small trees. The bark is smooth, light to dark gray on young branches and is gray with broad ridges and shallow fissures on old trunks. The twigs are squarish in cross section, green to reddish brown or gray, pubescent, and ridged. The leaves are opposite, deciduous, and palmate compound with 3 to 9 (mostly 5) leaflets. The leaflets are 1-5" long, dark dull green on the upper surface and grayish to white pubescent on the lower surface. The leaflet apex is acuminate, base cuneate, and margins entire to revolute. The crushed stem and leaves are aromatic. The pubescent flowers are in terminal panicles and blue to purplish or white. The fruit is a brown to black globular drupe that is smaller than 0.2" diameter.

Viburnum rufidulum (USDA-FS)

Vinca major (Bentham 1865)

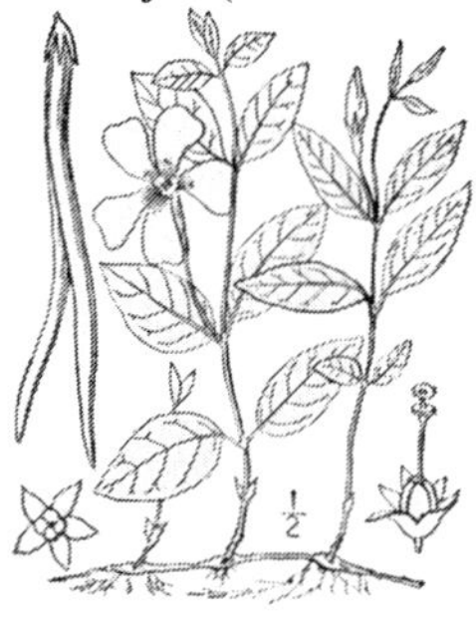

Vinca minor (Oregon State Univ. 1914)

LILAC CHASTE TREE (***Vitex agnus-castus***) is also known as Monk's Pepper Tree, Wild Pepper, Indian Spice, Abraham's Balm, Hemp Tree, Sage Tree, Wild Lavender, Common Chaste Tree, True Chaste Tree, Tree of Chastity, and Chaste Lamb Tree. The specific epithet, "*agnus-castus*", is from "agnus" ("lamb") and "castus" ("pure, holy, or chaste"). The leaflets are elliptic and narrower than Negundo Chaste Tree and the inflorescence is not as dense. The seeds are reported to be a sedative. In Brazil, a perfume is made from the flowers, and the aromatic leaves are used to spice food. The fruits have been used as a pepper substitute. It is a native of Europe that is cultivated and persists throughout the state. The current reports include 21 parishes.

NEGUNDO CHASTE TREE (***Vitex negundo***) is also known as Chaste Tree and Vitex. The specific epithet "*negundo*" is a Malay word for Chaste Tree. The leaflets are ovate and wider than Lilac Chaste Tree and the inflorescences are denser. The leaves are smoked to treat headaches, and infusion of the leaves is used to treat ulcers and abscesses. The flower buds are used to treat pneumonia, the roots as an expectorant and tonic, and the oil is rubbed on rheumatic joints. The roots, leaves, and fruits are used to treat colds, coughs, and bacterial dysentery, and to prevent malaria. It is a native of China that is cultivated and persists. The current reports include Evangeline, Richland, and St. Mary parishes. (Not Illustrated)

GRAPE, MUSCADINE - *Vitis* - Vitaceae

The genus name, "*Vitis*", is the classical Latin name for grape. It includes vines that climb with tendrils. The leaves are simple, alternate, deciduous, and with palmate major veins. The base of the blade is cordate. The flowers are small, green, and fragrant. The fruit is a globose berry. The fruits are eaten by many animals including birds. Humans eat the fruits raw and also make jelly and wine with them. The young leaves were reported to be cooked as a green and the mature leaves are used to wrap foods for cooking. Some species have been brought into cultivation.

SUMMER GRAPE (***Vitis aestivalis***) is also known as Pigeon Grape and Bunch Grape. The specific epithet, "*aestivalis*", refers to the summer blooming. The stems have rough, shreddy, and dark gray to brown bark. The twigs are terete, reddish brown, pubescent when young but becoming glabrous with age. The pith is brown and interrupted at the nodes with the gap greater than 1 mm. The tendrils are bifurcate near the end. The blades are cordiform to orbicular, 3-8" long and wide, upper surface dull green and glabrous, lower surface glaucous and with cobwebby hairs, and margins entire to crenate or dentate or 3-shouldered or 3-5 lobed. The lobe apices are acute and the sinuses rounded to acute. The flowers and berries are in drooping panicles to 6" long. The berries are black or purple, 0.2-0.6" diameter, and ripen in the fall. The variety ***aestivalis*** is found throughout the state in mesic to well drained sites (FAC-). The current reports include 40 parishes. The second variety is ***lincecumii*** and the varietal name "*lincecumii*" honors Gideon Lincecum (1793-1874), early collector of Texas plants. Common names for this variety include Post Oak Grape, Turkey Grape, and Big Summer Grape. It differs from variety ***aestivalis*** in having more densely pubescent twigs, a larger fruit (larger than 0.5"), and is restricted to very sandy habitats. The current records for this variety include 12 parishes in the pine region of western Louisiana.

Vitex agnus-castus (Vines 1960)

Vitis aestivalis (USDA-NRCS)

GRAY GRAPE (***Vitis cinerea***) is also known as Grayback Grape, Winter Grape, Downy Grape, Ashy Grape, and Pigeon Grape. The specific epithet, "*cinerea*", refers to the gray tomentum of the leaves. The stems have rough, shreddy, and dark gray to brown bark. The twigs are slightly angled, pubescent, and green when young but becoming pale brown to gray with age. The pith is brown and interrupted at the nodes with the gap greater than 1 mm. The tendrils are bifurcate near the end. The blades are cordiform, 4-8" long and wide, upper surface dark green and pubescent but becoming glabrous with age, lower surface densely gray pubescent and margins entire to crenate or dentate, or 3-shouldered or sometimes 3 lobed. The apex is acute to acuminate. The flowers and berries are in drooping panicles to 10" long. The berries are black or purple, 0.2-0.4" diameter, and ripen in the Fall. In mesic to moist habitats (FAC+) throughout the state. It probably occurs in all with current reports from 59 parishes.

FOX GRAPE (***Vitis labrusca***) is also known as Northern Fox Grape, Plum Grape, Northern Muscadine, Swamp Grape, and Wild Vine. The specific epithet, "*labrusca*", was the classical name of a wild vine. The common name "Fox" as applied to this grape originated from the old verb "to fox," meaning to intoxicate. The colonists from Europe were interested in the wine-making quality of New World grapes, more than in the fruit product. Some popular accounts variously ascribe the name "foxy" to the reddish brown hairs of the lower leaf surface resembling a fox's pelt, or to the lobed leaves looking like a fox's track, or to the fruit having a foxy pungent odor or taste. The stems have rough, shreddy, and light to dark brown bark. The twigs are slightly angled, yellow to red pubescent, and green when young but becoming terete and light to dark brown with age. The pith is brown and interrupted at the nodes with the gap greater than 1 mm. The tendrils are numerous, and bifurcate or trifurcate near the end. The young twigs and leaves often have spinose pubescence with glandular tips. The blades are cordiform, 3-8" long and wide, upper surface dull green and glabrous, lower surface white to yellow due to dense pubescence, and margins entire to crenate or dentate and often 3-shouldered. The apex is acute. The flowers and berries are in drooping panicles to 12" long. The berries are black, larger than 0.5" diameter, and ripen in the fall. Cultivated and perhaps native in hardwood slope forests (FAC+) in the northern part of the state. The current records include 12 parishes.

MUSTANG GRAPE (***Vitis mustangensis***) has an interesting common name that may have been the source of the scientific name. The specific epithet, "*mustangensis*", is derived from "mesteño", Spanish for strayed or feral, although the term is usually applied to horses. The stems have fibrous dull grayish brown bark. The twigs are slightly angled and densely pubescent with white hairs. The pith is brown and interrupted at the nodes with the gap greater than 1 mm. The tendrils are bifurcate or trifurcate near the end. The blades are cordiform to reniform, often concavely folded, 2.5-6" long and wide, upper surface white pubescent when young but becoming dark green and glabrous with age, lower surface densely covered with white pubescence, and margins entire or with shallow, broad, and obtuse teeth or 3-shouldered or sometimes deeply 3-5 lobed. The apex is acute to obtuse. The flowers and berries are in drooping panicles to 5" long. The berries are black or rarely dark red, larger than 0.5" diameter, and ripen in the summer. On dry sandy soils in western Louisiana. The current reports include eight parishes. (syn. *Vitis candicans*)

CAT GRAPE (***Vitis palmata***) is also known as Cat Bird Grape. The specific epithet, "*palmata*", refers to the palmately lobed leaves. The stems are chestnut to red, striated, and separating into thin plates. The twigs are slender and herbaceous like, tip with few leaves, angled, red, and with scattered white pubescence. The pith is brown and interrupted at the nodes with the gap greater than 1 mm. The tendrils are bifurcate and red when young. The blades are

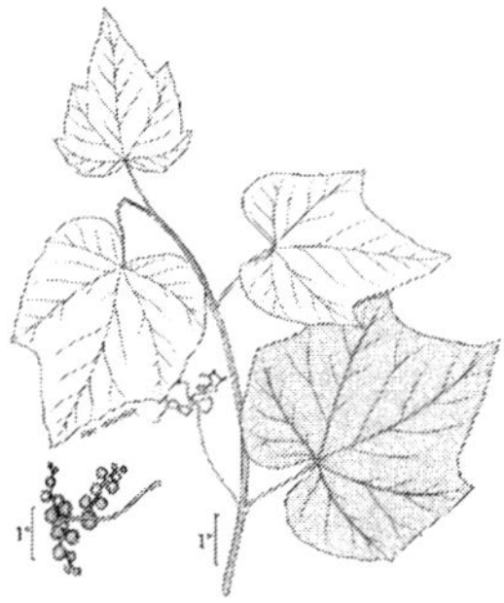

Vitis cinerea (USDA-NRCS)

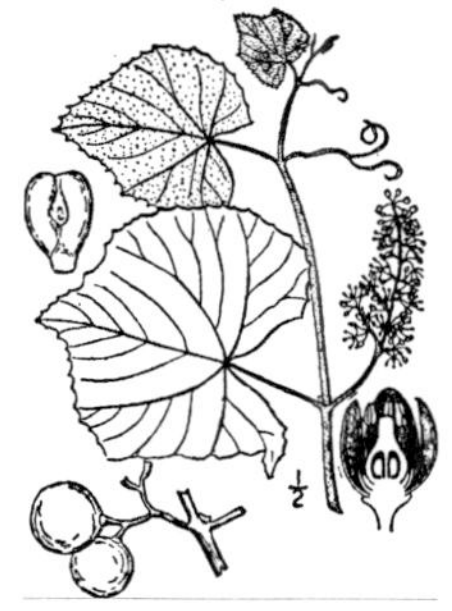

Vitis labrusca (Britton and Brown 1943)

Vitis mustangensis (Vines 1960 as ***Vitis candicans***)

Vitis palmata (USDA-NRCS)

cordiform, 2.5-4.5" long and wide, upper surface dark glossy green and glabrous, lower surface paler green and with a few hairs along the veins, and margins dentate-serrate and deeply 3, rarely 5, lobed. The lobe apex is acuminate and the sinuses are rounded to acute. The flowers and berries are in drooping panicles to 6" long. The berries are bluish black, smaller than 0.25" diameter, and ripen in late fall. It flowers later, June to July, than other species and thus the fruit matures later. It occupies the wettest habitats of all grape species. In bottomland hardwood forests (FACW-) throughout the state and very common in the Mississippi River floodplain. The current records include 33 parishes mostly west of the Mississippi River. (syn. *Vitis rubra*)

RIVERBANK GRAPE (***Vitis riparia***) is also known as Fox Grape, Winter Grape, Chicken Grape, Raccoon Grape, Frost Grape, and August Grape. The specific epithet "*riparia*" refers to the streambank habitat where it is often found. The stems are reddish and with weakly fibrous plates. The twigs are terete, green to red, and glabrous to pubescent. The pith is brown and interrupted at the nodes with the gap shorter than 0.5 mm. The tendrils are bifurcate. The blades are cordiform, 3-5" long and wide, upper surface light to yellow green and glabrous, lower surface paler green and with a few hairs along the veins or glabrous, and margins dentate-serrate and 3-shouldered or shallowly 3-lobed. The apex and/or lobe apex are acuminate. The flowers and berries are in drooping panicles to 5" long. The berries are black, glaucous, and smaller than 0.5" diameter. In moist habitats along streams (FACW) throughout the state. The current records include 19 parishes.

MUSCADINE (***Vitis rotundifolia***) is also known as Scuppernong, Southern Fox Grape, Bullace Grape, Bullet Grape, and Bull Grape. The specific epithet, "*rotundifolia*", refers to the rounded leaves. The stems are reddish brown to gray and smooth. Some plants produce aerial rootlets. The twigs are green to reddish brown, angled, and glabrous to slightly pubescent. The pith is brown and continuous through the nodes. The tendrils are unbranched. The blades are cordiform to reniform, 1.5-5" long, 2-4" wide, upper surface lustrous dark green and glabrous, lower surface paler green and with a few hairs along the veins or glabrous, and margins crenate to dentate. The apex is short acuminate. The flowers and berries are in globose clusters to 3" diameter. The berries are black or purple, rarely bronze, glaucous, with lenticels on the covering, and 0.5-1" diameter. The fruits ripen from July to September. In mesic to dry habitats (FAC) throughout the state but absent from large floodplains. The current records include 54 parishes.

FROST GRAPE (***Vitis vulpina***) is also known as Fox Grape, Winter Grape, Chicken Grape, Raccoon Grape, Riverbank Grape, and August Grape. The specific epithet, "*vulpina*", pertains to the fox, which eagerly devours the fruit. The stems are reddish and with weakly fibrous plates. The twigs are terete, green to red, and glabrous to pubescent. The pith is brown and interrupted at the nodes with the gap longer than 1 mm. The tendrils are bifurcate. The blades are cordiform, 3-5" long and wide, upper surface light to yellow green and glabrous, lower surface paler green and with a few hairs along the veins or glabrous, and margins dentate-serrate and 3-shouldered or shallowly 3-lobed. The apex and/or lobe apex are acute to short acuminate. The flowers and berries are in drooping panicles to 8" long. The berries are black, sometimes glaucous, and smaller than 0.5" diameter. In mesic to dry soils (FAC+) throughout the state. The current records include 32 parishes.

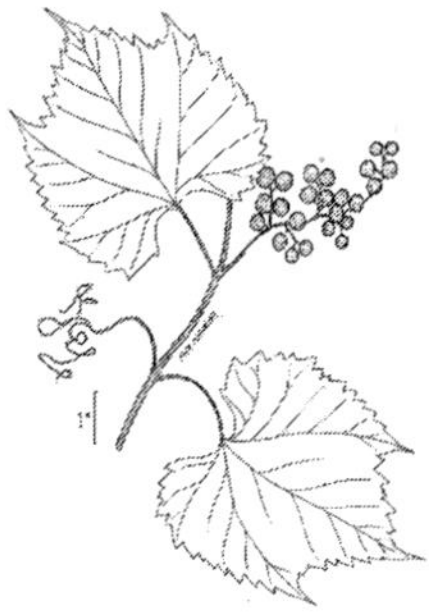

Vitis riparia (USDA-NRCS)

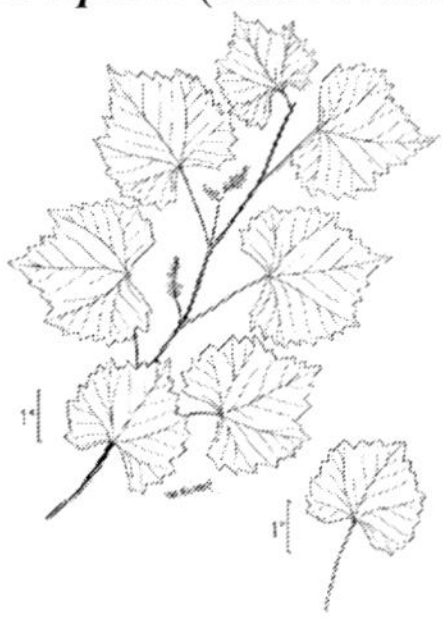

Vitis rotundifolia (USDA-NRCS)

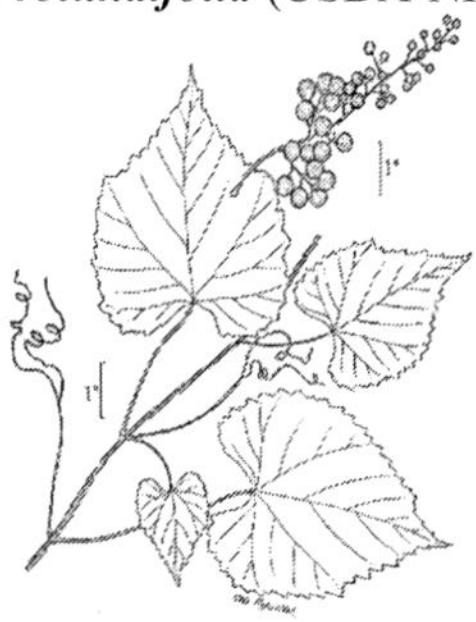

Vitis vulpina (USDA-NRCS)

WISTERIA - *Wisteria* - Fabaceae

The genus name, "*Wisteria*", honors Caspar Wistar (1761-1818), an American professor of anatomy at the University of Pennsylvania. The genus includes vines that climb by twining. The leaves are alternate, deciduous, and one odd pinnately compound. The blue, rarely white, flowers are in racemes. The fruit is a flattened linear legume.

JAPANESE WISTERIA (***Wisteria floribunda***) is also called Wisteria. The specific epithet "*floribunda*" means "profusely flowering". It is very similar to the more common Chinese Wisteria but with larger leaves and 13-19 leaflets. The racemes are slender and 10-20" long. The flowers and pedicels are each shorter than 1" long. The flowers in each raceme open gradually from the base up. A native of Japan that is widely cultivated and spreads aggressively. The current records include 13 parishes. (Not Illustrated)

AMERICAN WISTERIA (***Wisteria frutescens***) is also known as Kentucky Wisteria, Woody Wisteria, Kidney Bean Tree, and Virgin's Bower. The specific epithet "*frutescens*" refers to the woody or shrubby habit. The young stems are green and white pubescent while the older stems are reddish brown to gray and glabrous. The leaves are 5-18" long and the 5-15 leaflets are ovate to elliptic or lanceolate, 0.75-2.5" long, 0.5-1" wide, apex acute or acuminate, base rounded, upper surface dull green and glabrous, lower surface paler and pubescent, and margins entire. The racemes are 8-12" long and produced April to June with the leaves. The legumes are reddish brown, constricted between the seeds, 1.5-5" long, and glabrous. The flowers can be dipped into batter and fried in oil to make excellent fritters. However, the seeds are quite poisonous and two are enough to harm a child. It was first cultivated in 1724. In moist habitats (FACW) along streams, especially on the natural levees, throughout the state. The current records include 38 parishes. (syn. *Wisteria macrostachya)*

CHINESE WISTERIA (***Wisteria sinensis***) is also called Wisteria. The specific epithet, "*sinensis*", denotes its Chinese origin. The young stems are green and white pubescent while the older stems are reddish brown to gray and glabrous and often contorted. The leaves are 6-12" long and the 7 to13 leaflets are oblong to elliptic, 1.5-4" long, 0.5-1.5" wide, apex acute to short acuminate, base broadly cuneate to rounded, upper surface dull green and glabrous, lower surface paler and pubescent to glabrous, and margins entire. The racemes are thick cylindric, 4-15" long and produced February to April before the leaves. The flowers and pedicels are each about 1" long. The flowers in each raceme open almost simultaneously. The legumes are velvety reddish brown, 4-6" long, and not constricted between the seeds. A native of China that is widely cultivated and spreads aggressively. The current records include 35 parishes.

YELLOW ROOT - *Xanthorhiza simplicissima* - Ranunculaceae

Yellow Root is also known as Brook Feather and Parsley Leaf Yellow Root. The genus name, "*Xanthorhiza*", is from the Greek word xanthos, meaning "yellow," and rhiza, meaning "root," which refers to the yellow roots. The specific epithet, "*simplicissima*", refers to its unbranched habit. It is a rhizomatous small shrub to 2 ft. The stems are brittle, glabrous, brown to gray, and with yellow inner bark as are the rhizomes and roots. The leaves are alternate, deciduous, and one odd pinnate compound with 3, rarely 5, leaflets. The leaflets are ovate to elliptic or oblong, 1-3" long, apex acute to acuminate, base cuneate, sessile, upper surface shining medium green, lower surface gray green, and margins toothed to incised or deeply divided. The flowers are brownish purple and in axillary drooping panicles. The fruits are curved follicles that are oblong

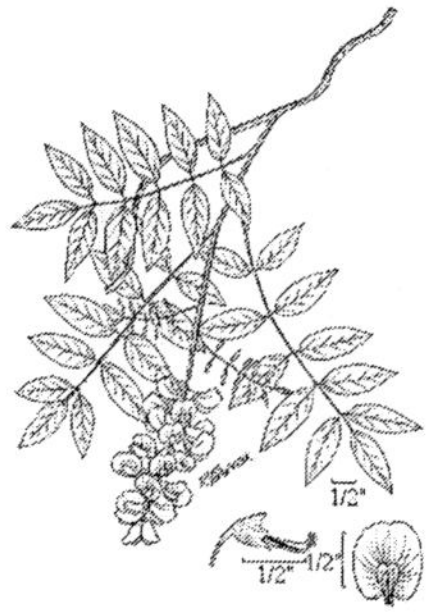

Wisteria frutescens (USDA-NRCS)

Wisteria sinensis (Vines 1960)

Xanthorhiza simplicissima (Britton and Brown 1943)

or elliptic, shorter than 0.2", and yellow to brown. The one population in Louisiana is heavily browsed by deer. Rare with only one population known along the edge of a baygall stream ((FACW-) on Fort Polk in Vernon Parish.

YUCCA, ADAMS NEEDLE - ***Yucca*** - Agavaceae

The genus name, "*Yucca*", is the native Haitian name for the Manihot genus, erroneously applied to this group of plants. The genus includes armed shrubs. The leaves are alternate or basal, evergreen, simple, and with parallel veins. The leaves are sessile and the blade is dagger like, apex long acuminate and ends in a spine tip, and the base is wrapped around stem. The white flowers are large and showy and produced in terminal panicles. The fruit is a green to black capsule. The leaves were used to hang meat for curing in the smokehouse. An easy genus to recognize but it is very difficult to separate the species.

ALOE YUCCA (***Yucca aloifolia***) is also called Spanish Bayonet and Spanish Dagger. The specific epithet, "*aloifolia*", refers to the similarity to the leaves of the aloes of South Africa. It is a shrub or small tree, with above ground stems to 25 feet. The upper leaves are green and alive while the lower leaves are dead and brown. The blades are 12-20" long, 1-1.5" wide, stiff, light green, widest in the middle, gradually narrowed toward the base and apex, flattened to concave, margins with erose teeth, and apex with dark brown spine. The capsules are 3-4" long. Native in sandy soils (FACU-) along the coast and widely planted, persisting, and escaping inland. The current reports include 27 parishes.

ADAMS NEEDLE (***Yucca filamentosa***) is also known as Yucca, Silk Grass, Bear Grass, Bears Thread, Thread and Needle, and Eves Darning Needle. The specific epithet, "*filamentosa*", means with filaments on the leaf margins. It is a stemless shrub with leaves to 4 ft and panicle to 8 ft. The blades are 1-3.5 ft long, to 2" wide, flexible and often recurved near the middle, margins with white, often twisted, threadlike fibers or entire, and apex ends in a spine. The panicle is 4-8 ft tall. The capsules are oblong and 1.5-2" long. Perhaps native to sandy soils throughout the pine regions of the state, especially in the western portion. The current reports include 26 parishes. (syn. *Yucca smalliana, Yucca concava, Yucca flaccida*)

GULF COAST YUCCA (***Yucca louisianensis***) is also known as Bear Grass, Silk Grass, Bears Thread, Thread and Needle, Eves Darning Needle, and Adams Needle. The specific epithet "*louisianensis*" refers to the state of Louisiana. It is very similar to Adams Needle but differs in having narrower leaves mostly 1" or less wide. The leaves are stiffer and tend not to be recurved in the middle. Perhaps native to sandy soils throughout the pine regions of the state, especially in the western portion. The current reports include 13 parishes. (syn. *Yucca freemanii, Yucca arkansana*)

TOOTHACHE TREE, PRICKLY ASH - ***Zanthoxylum*** - Rutaceae

The genus name, "*Zanthoxylum*", is from the Greek xanthos (yellow) and xylon (wood) referring to the yellow wood of some species. The plants are dioecious shrubs or small trees with broad based thorns, rarely unarmed. The leaves are deciduous, one odd pinnately compound, alternate, and often with a thorny rachis. The leaflets are finely crenate or entire and sessile or short-petiolate. The flowers are small and yellowish green. The fruit is a brown to reddish brown rough follicle with a shiny globose black seed, rarely two seeded. The seed persists on the outside of the follicle after dehiscence. The leaves provide food for the caterpillar of the giant swallowtail. Chewing the leaves or twigs produce first a tingling and then a

Yucca aliofolia (USDA-FS)

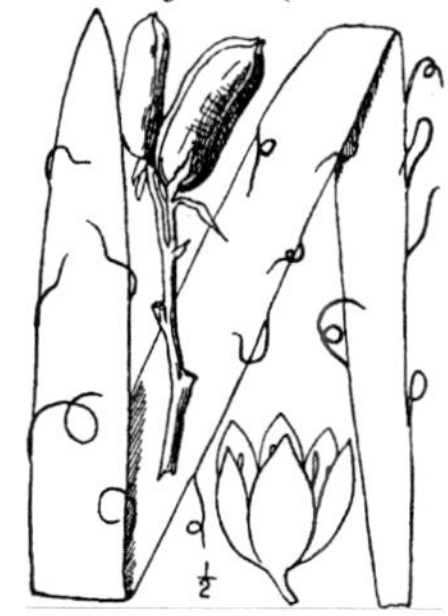

Yucca filamentosa (Britton and Brown 1943)

Yucca louisianensis (Vines 1960 as ***Yucca arkansana***)

numbness and insensibility to pain; can be used to relieve the discomfort of an aching tooth. Other medicinally uses include rheumatism, gastrointestinal stimulant in flatulence, and diarrhea. It is used as a diaphoretic, mild stimulant, and alterative agent. The fruits of a close relative are harvested in China and are used as a seasoning called Szechuan Peppercorns and is one of the spices in Five Spices. The fruits are eaten by a number of birds.

NORTHERN PRICKLY ASH (***Zanthoxylum americanum***) is also known as Common Prickly Ash, Toothache Tree, Angelica Tree, Suter Berry, and Pepper Wood. The specific epithet, "*americanum*", is for North America. It is a shrub mostly shorter than 15 feet. The bark is smooth and gray to brown while the twigs have paired flattened at the base thorns. The leaflets are elliptic to ovate or oblong, apex obtuse and slightly notched, base cuneate and symmetrical with the midvein near the center of the leaflet, 0.75-3" long, 0.5-1.5" wide, upper surface dull green, lower surface paler and pubescent especially along the veins, and margins crenate. The flowers are produced in small axillary clusters in February to April before the leaves expand. The wood is light brown, soft, not strong, and weighs 35 lb per cu ft. This species has been in cultivation since 1740. It is restricted to the clay soils of Copenhagen in Caldwell Parish.

TOOTHACHE TREE (***Zanthoxylum clava-herculis***) is also known as Prickly Ash, Southern Prickly Ash, Hercules Club, Sea Ash, Pepper Wood, Prickly Yellow Wood, Yellow Prickly Ash, Tongue Bush, Rabbit Gum, Wild Orange, Pillenterry, Wait-a-Bit, Tear Blanket and Sting Tongue. The specific epithet, "*clava-herculis*", means "club of Hercules" and refers to the trunk's thorny character. It is a small tree that reaches up to 30 ft. The bark is thin, light gray, and conspicuously roughened by corky, pyramidal or round-pointed, cone-shaped protuberances an inch or more high which often end in a stout spine. The leaflets are ovate or ovate lanceolate, often falcate, 0.5-4.5" long, apex acute or acuminate, base cuneate and asymmetrical with the midvein off center, upper surface dull green, lower surface paler and glabrous, and margins crenate-serrate. The flowers are produced in large terminal clusters from May to July with the leaves. The wood is light (31 lb per cu ft), soft, close-grained, and light brown with yellow sapwood. There is no commercial value for the wood. It is widely distributed and in a number of habitats from the sandy soils of the coast (FAC) to well drained upland soils and to the clay soil of inland prairies. The current records include 54 parishes.

COMMON JUJUBE - *Ziziphus jujuba* - Rhamnaceae

Common Jujube is also called Chinese Date. The genus "*Ziziphus*" is from an ancient Greek name which is derived from the Persian "zisafun". The specific epithet "*jujuba*" is the French common name. It is an armed tree to 50 ft. The bark is mottled gray or black, smooth on younger trunks, but roughly furrowed and peeling in loose shaggy strips on older trunks. The twigs are stout, green to gray or black, with some swollen nodes, and stipular spines. The leaves are alternate but clustered on spur shoots, simple, deciduous, and with three palmate major veins. The blades are ovate to oblong or lanceolate, 0.5-2" long, 0.3-1" wide, apex obtuse, base often asymmetrical and rounded to truncate, upper surface dark waxy green and glabrous, lower surface paler and glabrous to pubescent, and margins crenate to serrate and often with gland at base of tooth. The flowers are yellowish green and solitary or in small clusters. The fruits are subglobose to oblong drupes, 0.5-1" long, and maturing to a yellow to reddish brown or black color with a yellow pulp. The drupes are edible and are used medicinally in the Orient. It is native to Syria and was first introduced into the U.S. by Robert Chisholm in 1837 in Beaufort, North Carolina. Cultivated throughout the state and persisting. The current reports include 12 parishes.

Xanthoxylum americanum (Britton and Brown 1943)

Xanthoxylum clava-herculis (USDA-FS)

Ziziphus jujube (Vines 1960)

GLOSSARY

achene - a simple dry, indehiscent fruit with one seed that is attached to the pericarp in only one location as in the Asteraceae (Sunflower Family).
acuminate - long tapering to a sharp point, less than a 45^{0} angle.
acute - sharp pointed but less sharp than acuminate and more than obtuse, 45-90^{0} angle.
adnate - fusion or sticking together of unlike parts as stamens to petals.
aerial root - a root produced by vines that are used for climbing as in ***Toxicodendron radicans.***
aggregate fruit - a compound fruit, more than one pistil, that develops from a single flower as in the genus ***Magnolia*** which is an aggregate of follicles.
allelopathy - the production of a chemical compound that retards the growth of other plant(s).
alternate - the leaf arrangement with one leaf per node as in ***Ulmus*** and ***Prunus*** and is by far, the most common type of leaf arrangement.
androecium - the male part(s) of a flower consisting of one or more stamens.
anther - the enlarged terminal portion of the stamen that produces the pollen.
apiculate - a small distinct point from the apex.
arcuate - curved or bent.
aril - a fleshy covering over the seed as in ***Celastrus***.
aristate - a long slender projection from the apex.
armed - referring to a stem or leaf with sticker(s), spine(s), thorn(s), or prickle(s).
articulation - a joint.
bark - the outer tissue in older stems and roots of woody plants.
berry - a simple fleshy fruit with the pericarp fleshy throughout as in ***Vitis***.
bifurcate - divided into two parts.
blade - the green flattened portion of the leaf.
bract - a leaf like structure located below the flower and later the fruit.
calcareous - containing much calcium carbonate and thus an alkaline pH.
calyx - the outermost group of parts in a flower that are usually green.
capsule - a dry dehiscent fruit that splits along two or more lines to release the seeds as in ***Hibiscus***.
caryopsis - a dry indehiscent fruit with one seed that is completely fused to the pericarp as in the Poaceae.
catkin - an spike that droops and has only male flower or only female flowers as in ***Salix***.
ciliate - a margin with hairs along the edge.
clavate - club shaped or with an enlarged terminal portion.
columnar - cylindric shaped with little or no taper in diameter.
complete flower - a flower with all four major parts: sepal(s), petal(s), stamen(s), and pistil(s).
compound fruit - a fruit developed from more than one pistil.
compound leaf - a leaf that has the blade cut up into smaller segments called leaflets.
conduplicate - folded together length wise.
cone - the reproductive structure of Gymnosperms such as ***Pinus*** that is made of bracts that are not fused as in the flower of Angiosperms; it is also called a strobilus.
cordate - a base that is heart shaped.
cordiform - a heart shaped structure usually in reference to blade of a leaf.
corolla - the second group of flower parts that are usually brightly colored.

corymb - a flat topped inflorescence with stalked flowers and the stalks do not all start at the same point as in an umbel.

crenate - a margin with rounded teeth.

culm - the stem

cuneate - wedge shaped usually in reference to leaf or leaflet bases.

cyme - an inflorescence with the oldest flower produced at the top so all succeeding flowers are below the first one.

deciduous - the falling of structures from the plant, in particular, the falling of leaves in the autumn.

decumbent - reclining or lying on the ground with the tip erect or ascending, usually in reference to stems.

dehiscent - the natural opening of a fruit to release the seeds.

deliquescent - a broad widely spreading tree or shrub shape as in ***Quercus virginiana***.

deltoid - triangular or delta shaped

dentate - margins with pointed teeth that are set at right angles to the margin.

dicot - one of the two groups of Angiosperms that have two cotyledons in the seed, flower parts in 4's or 5's, netted veined leaves, and can have true woody stems.

dioecious - a plant that produces imperfect flowers and then stamen flowers on one plant and pistil flowers on a different plant. Would have "mommy" plants and "daddy" plants.

divergent - spreading from the main axis.

drupe - a fleshy fruit with the pericarp consisting of an outer somewhat tough skin or exocarp, a fleshy central portion the mesocarp, and the very hard inner layer the endocarp or pit as in plums and peaches.

elliptic - a narrow shape that is widest in the middle and tapes to both ends.

endocarp - the hard inner layer of the pericarp in a drupe, often called the pit.

entire - a margin that is smooth or not toothed no lobed.

epithet, specific - the second part to a scientific name often mistaking termed species name.

even - a pinnate compound leaves with an even number of leaflets due to the absence of a terminal leaflet

evergreen - a leaf that remains on the plant during the winter.

excurrent - a narrow pointed tree or shrub shape as in ***Pinus***.

exocarp - the outer covering or layer in a drupe.

FAC - the wetland ranking where the species is found about 50% of the time in wetlands and about 50% of the time in uplands.

FACU - the wetland ranking where the species is found most of the time (67%-99%) of the time in uplands.

FACW - the wetland ranking where the species is found most of the time in(67%-99%) of the time in wetlands.

falcate - sickle or curved shaped.

fascicles - groups or clumps.

filament - the stalk of a stamen.

fissure - a line or crease.

flower - the reproductive part that is a stem tip with four major parts possible including sepals, petals, stamens, and pistils.

follicle - a simple dry dehiscent fruit with one cavity that splits along one line at maturity as in ***Magnolia***.

fruit - the structure that matures from a pistil.

glabrous - smooth, without hairs.

glaucous - shiny, whitened with a bloom.
globose - round or spherical shaped.
glycoside - a compound with sugar derivities.
gynoecium - the female portion of the flower consisting of one or more pistils.
habit - the form or shape of a plant such as tree, shrub, or vine.
habitat - the place where a species is often found such as swamp or dry sandy area.
hastate - arrow head shaped with lobes nearly at right angles.
head - the flat or round inflorescence with the oldest flowers on the outside and each flower without a stalk.
hilum - the scar left on a seed from the attachment of the stalk (belly button)
imbricate - overlapping.
imperfect flower - a flower that has only stamen(s) or only pistil(s) but not both.
incomplete flower - a flower that is lacking one or more of the four major parts.
indehiscent - a fruit that does not open up under normal circumstances to release the seeds.
inflorescence - the arrangement of the flowers on the plant.
internode - the distance along the stem between two successive nodes.
involucre - the bracts around the flower and later the fruit.
lanceolate - lance shaped with the widest point at the base and tapering toward the apex.
lateral buds - the potential lateral stem buds that are located in the axils of leaves.
leaf - one of the three organs of the plant that consists of blade, petiole, and stipules.
leaflet - one of the parts of a compound leaf where the blade is cut up into segments.
legume - a simple dry dehiscent fruit with one cavity that splits along two lines at maturity as in the Fabaceae.
lenticels - the markings or openings on the stem produced by the stomata.
ligulate - strap shaped.
lobe - the pointed solid portion of the margin located between the sinuses.
lobed - a margin that is deeply cut into lobes by the sinuses.
major veins - the large veins at the base of the blade or leaflet.
mesocarp - the mid or central fleshy portion of the pericarp in a drupe.
midrib - the one large vein in the center of the blade with pinnate major veins and will become the rachis in a pinnate compound leaf.
monocot - one of the two groups of Angiosperms that have one cotyledon in the seed, flower parts in 3's or 6's, parallel veined leaves, and not true woody stems
monoecious - a plant that produces imperfect flowers and then its stamen flowers and its pistil flowers on the same plant.
mucilaginous - soft and sticky or gelatinouus to the touch.
mucro - the short spur or spine located on the apex.
mucronate - an apex with a short spur or spine.
multiple fruit - a compound fruit derived from more than one flower.
naked flower - a flower without petals and sepals.
needle - a long linear leaf as in ***Pinus***.
net veins - the arrangement of veins that branch and rebranch to create a net.
node - the point on the stem where a leaf is produced.
nut - a dry indehiscent fruit where the pericarp or part of the pericarp is very hard.
OBL - the wetland ranking where the species is found in wetlands almost (>99%) all the time.
oblanceolate - the opposite of lanceolate where the widest end is at the apex in a narrow shape.

oblique - unequal or oblique where the blade or leaflet base is wider on one side of the petiole or petiolule than the other side.

obovate - opposite of ovate where the widest end is at the apex in a wide shape.

obtuse - an apex that is blunt or not pointed.

ocrea - the modified stipules in the Polygonaceae that produce circles around the stem as in ***Brunnichia*** and ***Polygonella***.

odd - a pinnate compound leaves with an odd number of leaflets due to the presence of a terminal leaflet.

opposite - the leaf arrangement with two leaves per node as in ***Cornus***.

ovary - the enlarged lower end of the pistil that contains the ovule(s).

ovule - the structure(s) within the ovary that contain the egg and will become the seed.

palmate major veins - the arrangement of large veins at the base of the blade or leaflet with more than one vein, usually three or five.

palmate compound - a leaf where the palmate major veined blade is cut up into smaller segments called leaflets and all the leaflets start at the same point.

panicle - an inflorescence that has the oldest flowers at the bottom and is much branched creating a large mass of flowers.

pappus - the modified calyx in the Asteraceae and is often fluffy and used for fruit/seed dispersal.

parallel veins - as in monocots with the veins all running parallel to each other.

pectinate - a margin that is comb like where it is cut into long, linear segments.

peduncle - the stalk of a flower or inflorescence.

perfect flower - a flower that has both stamen(s) and pistil(s).

perianth - the sepals and petals collectively.

pericarp - the outer covering of the fruit that developed from the ovary or pistil wall.

petal - one segment of the corolla.

petiole - the stalk of the leaf or the structure that connects the blade to the stem.

petiolule - the stalk of a leaflet or the structure that connects the leaflet to the rachis or rachilla.

pilose - densely long pubescent.

pinnate - the arrangement of large veins at the base of the blade or leaflet with one vein only, the midrib.

pistil - the female portion of the flower that contains one or more ovules that can mature into a seed while the entire pistil or a part of it will develop into the fruit.

pith - the central core of the stem that is a softer texture than the surrounding wood and is often a different color.

pollen - the male reproductive cells that will produce the sperm.

pome - a simple fleshy fruit that has the true fruit in the center or core and is surrounded by the fleshy hypanthium which is the base of the sepals, petals, and stamens as in apples and pears.

prickle - a sticky outgrowth from the epidermis of the stem or leaf.

prostrate - lying down, usually in reference to a stem.

puberulent - pubescent with short hairs.

pubescent - with hairs.

pyrene - a drupe like fruit that has several seeds inside the pit as in the genus ***Ilex***.

pyriform - pear shaped.

raceme - an inflorescence that has the oldest flower at the bottom and each flower has a stalk as in ***Wisteria***.

rachis - the central axis of a pinnate compound leaf that develops from the midrib.

receptacle - the tip, often enlarged, of the flower stalk and is the point of attachment for the flower parts.
reticulate - distinctly net veined.
revolute - a margin that is folded under lengthwise.
rhizome - a prostrate underground stem as in ***Arundinaria***.
salverform - a corolla that has a slender tube that abruptly expands into the lobes.
samara - an achene with a wing as in ***Ulmus*** or ***Acer***.
sapling - a young immature tree.
scabrous - rough to the touch because of small projections on the surface.
scape - the stem of an inflorescence without leaves as in ***Yucca*** or ***Sabal***.
schizocarp - a fruit that splits into one seeded portions as in ***Illicium***.
seed - a packaged new plant that matures from an ovule.
sepal - one segment of the calyx
serrate - a toothed margin with sharp pointed forward pointing teeth.
serrulate - a toothed margin with small sharp pointed forward pointing teeth.
sessile - without a stalk or petiole or petiolule.
sheath - a leaf like structure that wraps around the stem.
shrub - a woody plant usually with several main stems and at maturity is shorter than 20 ft and/or smaller than 5" diameter.
sinus - the cut area of the margin located between the lobes.
spathulate - spoon shaped with the distal end very wide and the proximal portion long and narrow.
spine - a sticky structure that is located at a node where a leaf should be, thus a modified leaf.
spinescent - a margin with spines.
spur shoot - a short stumpy stem with nodes and thus leaves very close together.
stamen - the male portion of the flower that produces the pollen.
stellate - branched hairs with more than two branches thus star shaped.
stigma - the tip of the pistil that catches the pollen.
stipules - the paired structures located on the stem at the base of the leaf and can be large and leaf like, or small dots, or modified into sticky structures, stipular spines.
style - the stalk of the pistil that connects the stigma to the ovary.
stramineous - straw or light brown colored.
subsesssile - almost sessile, with s short stalk or petiole or petiolule.
suffrutescent - woody at base but herbaceous above.
synconium - a multiple fruit that is inverted with the achenes inside as in ***Ficus***.
tendril - a twining structure used by some vines for climbing.
tepals - a flower part that cannot be identified as a sepal nor as a petal as in ***Magnolia***.
terete - round or circular in cross section.
terminal buds - the potential embryonic stem located at the tip of the branch.
tessellate - with square or nearly so depression.)
thorn - a sticky structure that is located at the nodes where a branch should be, thus a modified stem.
tomentose - dense interwoven hairs.
toothed - a margin that has shallow cuts into it and is between lobed = deeply cut and entire = not cut at all.
tree - a woody plant usually with one main stem and at maturity is taller than 20 ft and/or larger than 5" diameter.
trifoliate - a compound leaf with three leaflets.
trifoliolate - a compound leaf with three leaflets as in ***Toxicodendron radicans***.

trifurcate - divided into three parts.
truncate - an apex that is squared off or almost straight across with no taper.
UPL - the wetland ranking where the species is almost always (>99%) found in non wetland or upland situations.
umbel - a flat topped inflorescence with the oldest flower on the outside and all flower stalks begin at the same point.
urceolate - urn shaped.
valvate - meeting but not overlapping.
vascular tissue - the transporting or conducting tissue in many plants.
veins - the conducting or vascular tissue in the leaf blade.
villous - long soft hairs.
vine - a plant that climbs by tendrils, twining, aerial roots, or sprawling.
whorled - the leaf arrangement with three or more leaves per node as in ***Catalpa***.
woody - term used to describe a plant with hard and woody character and/or maintaining an above ground stem during the winter or cold season.
zygomorphic - a flower that can only be divided into two equal parts as compared to one that could be divided into many equal parts in actinomorphic.

REFERENCES AND LITERATURE CITED

ALLEN, C.M., H.D. GUILLORY, C.H. STAGG, AND S.D. PARRIS. 1990. The effects of sedimentation on the flora and vegetation of baygalls in west central Louisiana. Proc. La. Acad. Sci. 53:33-41.

BAILEY, L.H. 1949. Manual of cultivated plants. The MacMillan Company, Inc., New York.

BATES, D.M., J.W. INGRAM, JR. & H.E. MOORE, JR., compilers. 1976. Hortus third, rev. ed. Macmillan Publishing Co., Inc., New York.

BENTHAM, G. 1865. Handbook of the British flora. Lovell Reeve and Company, London.

BRITTON, L.B. and A. BROWN. 1943. An illustrated flora of the Northern United States, Canada, and the British Possessions, three vol. The New York Bot. Gard. New York.

BROWN, C. A. 1945. Louisiana trees and shrubs. La. Forestry Comm. Bull. 1.

BROWN, C. A. 1956. Commerical trees of Louisiana. Louisiana Forestry Comm. Baton Rouge.

BROWN, C. A. 1957. Checklist of woody plants of Louisiana, native naturalized, and cultivated Baton Rouge, La. Forestry Comm. Bu;. No. 8. 16p.

CLEWELL, A. F. 1985. A guide to the vascular plants of the Florida Panhandle. Florida State Univ. Press, Tallahassee.

CORRELL, D. S. and M. C. JOHNSTON. 1970. Manual of the vascular plants of Texas. Texas Research Foundation, Renner.

CRONQUIST, A. 1980. Vascular flora of the southeastern United States. I: Asteraceae. Univ. North Carolina Press, Chapel Hill.

DIGGS, G.M., B.L. LIPSCOMB, AND R.J. O'KENNON. 1999. Shinners & Mahler's Illustrated flora of north central Texas. BRIT and Austin College, Fort Worth.

DUNCAN, W. H. 1967. Woody vines of the southeastern United States. Sida 3:1-76.

DYKE, R. A. 1941. Climate of Louisiana. pp. 894-903. In Climate and Man. Yearbook of Agriculture. U.S. Department of Agriculture, Washington, D.C.

FERNALD, M. L. 1950. Gray's manual of botany, 8th ed. American Book Co., New York.

FLORA OF NORTH AMERICA EDITORIAL COMMITTEE, eds. 1993. Flora of North America north of Mexico, Vol. 2. Pteridophytes and Gymnosperms. Oxford Univ. Press, New York.

FLORA OF NORTH AMERICA EDITORIAL COMMITTEE, eds. 1997. Flora of North America north of Mexico, Vol. 3. Magnoliophyta: Magnoliidae and Hamamelidae. Oxford Univ. Press, New York.

FOOTE, L.E. AND S.B. JONES. 1989. Native Shrubs and Woody Vines of the Southeast. Timber Press, Portland, OR.

FRIENDS OF HILLTOP ARBORETUM, INC. 1997. A Pocket Guide to Louisiana Native Trees. Baton Rouge.

GHANDI, K. N. & R. D. THOMAS & S. L. HATCH. 1989. Asteraceae of Louisiana. Sida, Bot. Misc. No. 4, Dallas, Tx.

GLEASON, H. A. 1968. The new Britton & Brown illustrated Flora of the northeastern United States and adjacent Canada. Hafner, New York.

GODFREY, R. K. 1988. Trees, shrubs, and woody vines of northern Florida and adjacent Georgia and Alabama. Univ. of Georgia Press, Athens.

GODFREY, R. K. & J. W. WOOTON. 1979. Aquatic and wetland plants of southeastern United States monocotyledons. Univ. Georgia Press, Athens.

GODFREY, R. K. & J. W. WOOTON. 1981. Aquatic and wetland plants of southeastern United States dicotyledons. Univ. Georgia Press, Athens.

HALL, D. W. 1993. Illustrated Plants of Florida and the Coastal Plain. Maupin House Publ., Gainesville, FL.

HARRAR, E. S. & J. G. HARRAR. 1962. Guide to southern trees. Dover Publications, New York. 709 + ix pp.

HITCHCOCK, A.S. 1950. Manual of the grasses of the United States, 2nd ed. (revised by Agnes Chase.) U.S. Dept. Agr. Misc. Publ. 200.

HOLLAND, W. C. 1944. Physiographic divisions of the Quaternary lowlands of Louisiana. Proc. Louisiana Acad. Sci. 8: 11-24.

HUNT, C. B. 1974. Natural regions of the United States and Canada. W. H. Freeman and Co., San Francisco.

HUNTER, C. G. 1989. Trees, Shrubs, & Vines of Arkansas. The Ozark Society Foundation, Little Rock, AR.

ISLEY, D. 1990. Vascular Flora of the Southeastern United States. Vol. 3. Leguminosae.

KARTESZ, J. T. & R. KARTESZ. 1994. A synonomized checklist of the vascular flora of the United States, Canada, and Greenland. Univ. North Carolina Press, Chapel Hill.

LASSEIGNE, A. 1973. Louisiana Legumes. Univ. Southwestern Louisiana Southwest Studies series 1, Lafayette.

LI, HUI-LIN. 1963. Woody Flora of Taiwan. The Morris Arboretum. Philadelphia, Pa.

LOUDON, J.C. 1838, Arbortum et fruticetum Britannicum. Longman, Orme, Brown, Green, and Longmans, London.

MACROBERTS, D.T. 1984. The vascular plants of Louisiana. Bull. Mus. Life Sci. 6, Louisiana State Univ., Shreveport.

MACROBERTS, D.T. 1988. A documented checklist and atlas of the vascular flora of Louisiana; three parts. Published by the author.

MAIN, A.C. and R.H. MILLS. 1981. Leaf Key to common trees in Louisiana. Louisiana State Univ. Extension Service Publ. No. 1669. Baton Rouge.

MILLER, H.A. 1978. How to Kow the Trees, 3rd edition. Wm. C. Brown, Dubuque, Iowa.

MOORE, D. M. 1960. Trees of Arkansas. Arkansas Forestry Comm., Little Rock.

NELSON, G. 1996. The Shrubs and Woody Vines of Florida. Pineapple Press, Sarasota, FL.

NEWTON, M. B., JR. 1972. Atlas of Louisiana, a guide for students. La. State Univ. Misc. Publ. 72-1. 196p.

OREGON AGRIC. EXP. STA. BUL. 1914. Oregon State Univ. Agricultural Experimental Station, Corvalis.

PEATTIE, D. C. 1966. A Natural History of Treees of Eastern and Central North America. Houghton Mifflin Co., Boston.

POHL, R.W. 1980. Family #15, Gramineae *in* Flora Costaricensis, William Burger Editor. Fieldiana: Botany, New Series No 4: 1-608.

RADFORD, A. E., H. E. AHLES, and R. C. BELL. 1968. Manual of the vascular Flora of the Carolinas. Univ. North Carolina Press, Chapel Hill.

REESE, W.D. 1992. Acadian flora; Native and Naturalized Woody Plants of South-Central Louisiana. Center for Louisiana Studies, Lafayette, LA.

SHINNERS, L.H. 1964. Cayratia japonica (Vitaceae) in southeastern Louisiana: new to the United States. Sida 1: 384.

SMALL, J. K. 1933. Manual of the southeastern flora. Hafner, New York

SMITH, E.B. 1994. Keys to the Flora of Arkansas. Univ. Of Arkansas Press, Fayetteville.

SOUTHWESTERN FORESTRY COLLEGE. 1991. Forestry Dept. of Yunnan Provinc. Iconographia Arborum Yunnanicorum. Yunnan, China.

STEYERMARK, J. A. 1968. Flora of Missouri. Iowa State Univ., Ames.

THOMAS, R. D. 1988. 100 woody plants of north Louisiana. Contr. Herb. Northeast Louisiana. Univ. 7, Monroe.

THOMAS, R. D. and C. M. ALLEN. 1981. A checklist of the woody plants of Louisiana.

Contr. Herb. Northeast La. Univ., Monroe (2): 1-59.

THOMAS, R. D. and C. M. ALLEN. 1982. A preliminary checklist of the Dicotyledons of Louisiana. Contr. Herb. Northeast La. Univ., Monroe (3): 1-130.

THOMAS, R. D. and C. M. ALLEN. 1984. A preliminary checklist of the Pteridosperms, Gymnosperms, and Monocotyledons of Louisiana. Contr. Herb. Northeast La. Univ. (4): 1-55.

THOMAS, R.D. and C.M. ALLEN. 1993. Atlas of the vascular flora of Louisiana, Vol. 1: ferns and ferns allies, conifers, and monocotyledons. Louisiana Department of Wildlife and Fisheries, Baton Rouge, LA.

THOMAS, R. D. and C. M. ALLEN. 1996. Atlas of the vascular flora of Louisiana, Vol. 2: Dicotyledons Acanthaceae-Euphorbiaceae. Louisiana Department of Wildlife and Fisheries, Baton Rouge, LA.

THOMAS, R. D. and C. M. ALLEN. 1998. Atlas of the vascular flora of Louisiana, Vol. 3: Dicotyledons Fabaceae-Zygophyllaceae. Louisiana Department of Wildlife and Fisheries, Baton Rouge, LA.

TOBE, J.D., K.C. BURKS, R.W. CANTRELL, M.A. GARLAND, M.E. SWEELEY, D.W. HALL, P. WALLACE, G. ANGLIN, G. NELSON, J.R. COOPER, D. BICKNER, K. GILBERT, N. AYMOND, K. GREENWOOD, AND N. RAYMOND. 1998. Florida Wetland Plants; an Identification Manual. Florida Department of Environmental Protection, Tallahassee, FL.

USDA. Forest Service Collection. Hunt Institute for Botanical Documentation, Carnegie Mellon University, Pittsburgh, PA.

USDA, NRCS. 1999. The PLANTS database (http://plants.usda.gov/plants). National Plant Data Center, Baton Rouge, LA 70874-4490 USA.

USDA. Midwestern wetland flora: Field office guide to plant species. USDA Soil Conservation Service, Midwest National Technical Center, Lincoln, Nebraska. Jamestown, ND: Northern Prairie Wildlife Research Center Home Page. http://www.npwrc.usgs.gov/resource/othrdata/plntguid/plntguid.htm (Version 16JUL97).

USDA. Northeast wetland flora: Field office guide to plant species. USDA Natural Resources Conservation Service, North National Technical Center, Chester, Pennsylvania. Jamestown, ND: Northern Prairie Wildlife Research Center Home Page. http://www.npwrc.usgs.gov/resource/1999/neflor/neflor.htm (Version 12APR99).

USDA. Southern wetland flora: Field office guide to plant species. USDA Soil Conservation Service, South National Technical Center, Fort Worth, Texas. Jamestown, ND: Northern Prairie Wildlife Research Center Home Page. http://www.npwrc.usgs.gov/resource/1999/soutflor/soutflor.htm (Version 12APR99).

USDA. Western wetland flora: Field office guide to plant species. USDA Soil Conservation

Service, West National Technical Center, Portland, Oregon. Jamestown, ND: Northern Prairie Wildlife Research Center Home Page. http://www.npwrc.usgs.gov/resource/othrdata/westflor/westflor.htm (Version 16JUL97).

VINES, R. A. 1960. Trees, shrubs, and woody vines of the southwest. Univ. of Texas Press, Austin.

VINES, R. A. 1977. Trees of east Texas. Univ., of Texas Press, Austin.

WUNDERLIN, R.P. 1998. Guide to the vascular plants of Florida. Univ. Press of Florida, Gainesville.

INDEX